A DICTIONARY OF CLIMATE CHANGE AND THE ENVIRONMENT

T0329690

For Ariana and Brecon, Alexa and Hannah

A Dictionary of Climate Change and the Environment

Economics, Science, and Policy

R. Quentin Grafton

Crawford School of Economics and Government,
The Australian National University, Australia

Harry W. Nelson

Department of Forest Resources Management,
University of British Columbia, Canada

N. Ross Lambie

Crawford School of Economics and Government,
The Australian National University, Australia

Paul R. Wyrwoll

Crawford School of Economics and Government,
The Australian National University, Australia

Edward Elgar

Cheltenham, UK • Northampton, MA, USA

Published by
Edward Elgar Publishing Limited
The Lypiatts
15 Lansdown Road
Cheltenham
Glos GL50 2JA
UK

Edward Elgar Publishing, Inc.
William Pratt House
9 Dewey Court
Northampton
Massachusetts 01060
USA

A catalogue record for this book
is available from the British Library

Library of Congress Control Number: 2011932893

ISBN 978 1 84980 387 8 (cased)

Typeset by Servis Filmsetting Ltd, Stockport, Cheshire
Printed and bound by MPG Books Group, UK

Contents

Acknowledgements

We are very grateful to the many authors whose papers and books have helped us to understand the economics, science and policy concerning climate change and the environment, so that we could help others.

Introduction

This book is a much updated version of an earlier edition authored by R. Quentin Grafton, Linwood Pendleton and Harry Nelson, published by Edward Elgar a decade ago. Although the primer on economics for the environment written by Linwood Pendleton in the first edition has been retained in the current version, we have updated and revised all the words in the dictionary, the other primers, and appendices. Our biggest change has been to include an additional 700 words or so on climate change policy and science, as well as a primer on climate change policy.

Given the existence of Wikipedia and other internet sources, you may ask why do we need a dictionary? Our view is that many technical words are not available on Wikipedia. Further, many internet sources on climate change have been developed by people or organizations with a particular view or perspective. By contrast, our definitions are clear, unbiased, and definitive.

We strongly believe that without a holistic view of the environment that transcends disciplines, we can never hope to have a comprehensive understanding of our world. Unfortunately, the language of each discipline includes its own set of jargon and words that are commonly used by people within a discipline but, frequently, cannot be understood by people from the outside. Too often, these barriers prevent us from understanding and developing a systematic view of our environment.

A Dictionary of Climate Change and the Environment: Economics, Science, and Policy offers a reference that bridges the gap between the many disciplines encompassing climate change, environmental economics, environmental sciences, and environmental studies. It provides a comprehensive set of words used in environmental, ecological, and resource economics. In addition, the dictionary includes a selection of some of the most important and frequently used words in the environmental sciences and environmental studies. The book includes over 3700 words from *abatement* to *zooplankton* and, where additional explanation is required, further reading and references are provided. The dictionary is not only a ready-reference for students, but should prove useful to policy makers and professionals who need to understand the many different terms and concepts concerning the environment.

The book has a number of desirable features. Included are four primers that provide introductions to the topics of Environmental Economics; Climate Change Policy; International Environmental Problems; and Environmental Systems, Dynamics, and Modeling. These primers can be used as reviews or introductions in many different courses and programs on the environment. In addition, the dictionary has a list of annotated references that provides a useful introduction to the topics covered in the dictionary, listing a comprehensive list of references referred to in the definitions, and seven appendices that include: the Greek alphabet, Roman numerals, the système internationale (SI) and other units, prefixes for the SI, common abbreviations for measures and units, the geological time-scale, and a listing and description of selected environmental treaties. The completely revised appendix on environmental treaties is intended for readers interested in the legal measures being undertaken to sustain the global environment.

We trust that our book, its definitions, primers, and appendices, will prove a ready-reference that will help to bridge the knowledge and understanding gaps between different academic disciplines related to the environment, and will help you gain a better understanding of the world we share and the environmental challenges that we face.

R.Q.G., Canberra, Australia
H.W.N., Vancouver, Canada
N.R.L., Canberra, Australia
P.R.W., Canberra, Australia

Economics for the Environment: A Primer

Linwood Pendleton

1. WELFARE ECONOMICS AND MARGINAL ANALYSIS

Much of what we know as environmental economics can best be thought of as applied economics. The positive side of environmental economics uses micro and macroeconomic theory to learn more about the ways in which economic factors influence the consumption of environmental goods and services. Positive environmental economics is largely descriptive and predictive. The normative aspect of environmental economics is largely prescriptive and applies principles of welfare economics to determine the allocation of environmental goods and services that generates the greatest economic good for society.

Normative welfare economics provides a framework to determine how society "should" allocate resources, assuming that economic welfare is a good measure of society's well-being. **Pareto** pointed out that an optimal allocation of goods exists when any reallocation would only make one person better off at the expense of another; we call this allocation a **Pareto efficient** or **economically efficient** allocation. At sub-optimal allocations, we could potentially reallocate resources to make everyone better off as whole. Reallocations away from the sub-optimal towards the optimal are known as **Pareto improvements**. Using the ideas of Pareto, normative environmental economics provides a framework for finding the economically optimal allocation of environmental resources.

Talking about the greatest good to society and measuring this good are quite different things. To make discussions of well-being more tangible, economists need to find concrete ways of capturing the values people give to goods, resources, and even services. Generally, we say that someone values a good if they are willing to give up something they already have or are entitled to have to obtain the good in question. In subsistence economies, people may trade their time (say, time spent hunting) in order to get things they value (such as game). In barter systems, people trade among each other by offering one good for another. In market systems, we trade currency (whether be it Euros or Treasury Notes) for goods. The reasons people are willing to make sacrifices to get goods or services may be complex. Nevertheless, the fact that people are willing to make tangible trades gives us clues about people's valuations and choices.

a.   Measuring the Value of Goods

The market price of a good is an important indication of the value a consumer places on that good. Since people presumably do not value currency in and of itself, economists accept that the amount paid in currency is a simple reflection of the **current value** a consumer places on a good. When consumers hand over 50 cents to buy an apple, they reveal that they place a value on that apple of at least 50 cents. Even when goods are sold individually it is common for consumers to buy more than one unit of a good, either at once or over the course of some period of time. The total number of units bought depends on the price; when prices are high consumers buy less of a good and when prices are low consumers buy more. As before, the consumer only buys apples as long as the value he or she places on each apple exceeds the price per apple in the market. Of course, the consumer never buys an infinite number of apples, so we know immediately that the value a consumer places on any individual apple must fall as that consumer acquires more and more apples. We call the value that consumers place on each newly purchased apple their **marginal willingness to pay** for that last apple. The consumer will keep buying a good as long as his/her **marginal willingness to pay** for the good is at least equal to the price of the good. Therefore, if a consumer buys 10 apples at 5 cents/apple we know that the value of the last apple is 5 cents/apple. Does that mean the value of all 10 apples is 5 cents times 10 or 50 cents? Not necessarily. To understand why, we need to think about why a consumer buys more when the price is low and less when the price is high. We call the study of this relationship of prices and quantities "demand analysis", and the relationship a **demand function**.

b.   Demand

In the market, we observe regular patterns in which the quantity of a good purchased by a consumer varies with the price of the good. We call the quantity of a good that a consumer would like to buy for a given price, the demand for that good. When prices are high, demand is small, and vice versa. While prices usually determine demand, we have already shown that when a consumer makes a purchase, the purchase price reveals the value that the consumer places on a good. A consumer places a higher value on a single apple than on one of 10 apples, especially if the 10th apple is consumed after the first 9 have been eaten.

We call the schedule of quantities demanded at different prices the **Marshallian demand** function or simply the demand curve for a good.

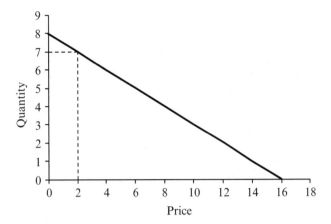

Figure 1a   Demand

The **demand function** reveals how much of a good a consumer would buy at a given price (e.g., when the price is 2, the demand is 7). Figure 1a shows that at a price (drawn on the horizontal axis) of 0, the consumer would demand (drawn on the vertical axis) 8 units and at a price of 16 the demand falls to 0 (we call this the **choke price**). The actual schedule of prices and quantities demanded depends on the consumer's personal characteristics, including income. For instance, rich people may buy more of a good at a given price than poor people. The slope and the shape of the demand curve (a demand curve can bend in towards the origin) depends on such things as how essential the good is or how many substitutes for the good are available to the consumer.

If we switch the axes of our graph, we now have a schedule that reveals the value a consumer places on each unit of a good, depending on how many units already have been consumed. As before, we call the value that is placed on one more unit of a good, the **marginal willingness to pay** for that unit, and the new schedule of prices as a function of quantity the **inverse demand function** or **curve**.

Once again, when 7 units are purchased, the 7th unit is valued at 2. Note that the 8th unit has zero value, that is, the consumer would consume up to 8 units if they were free, but any more than that would not make the consumer any better off.

The inverse demand curve also reveals both the value of the last marginal unit **(the marginal willingness to pay)** and also the cumulative value that the consumer would place on the consumption of all the units consumed **(the total willingness to pay)**. The marginal willingness to pay is simply read off the inverse demand curve. The total willingness to pay, however, must

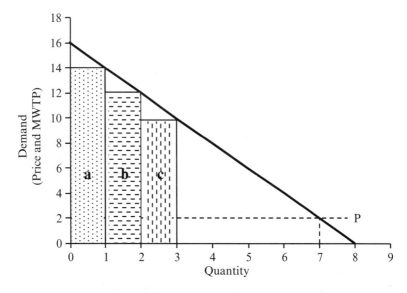

Figure 1b   Inverse Demand

be found by summing together the **marginal willingness to pay** for every individual unit. For example, the total willingness to pay for 3 units would be equal to the marginal willingness to pay for each unit separately (Figure 1b, areas a + b + c). A continuous inverse demand function, however, shows that the **marginal willingness to pay** varies for even small fractions of a unit.

The stair steps of areas a, b and c in Figure 1b only approximate the **total willingness to pay** for 3 units of the good. The marginal willingness to pay is an exact measure only of the infinitely small fraction of the last unit consumed. If we make the units smaller and smaller, the exact **total willingness to pay** is found by summing up increasingly tiny slivers under the demand curve. The summation of these slivers equals the area under the inverse demand curve up to the quantity consumed.[1] In Figure 1c, we draw the **total willingness to pay** for 3 units.

While total willingness to pay represents the gross value a consumer places on the consumption of a good, economists generally are interested in the benefit of a good or resource net of the cost of obtaining that good. From the consumers' perspective, the cost of a good is the total price paid for a good (price × quantity). If only a single, uniform price is charged for each

---

1.   Readers familiar with calculus may recognize that the total willingness to pay is simply the integral of the marginal willingness to pay evaluated from 0 to the quantity consumed. Conversely, marginal willingness to pay is the derivative of total willingness to pay.

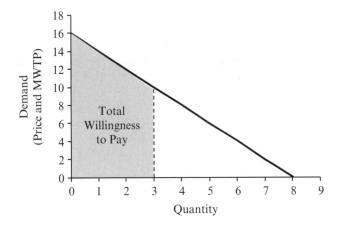

Figure 1c   Total Willingness to Pay

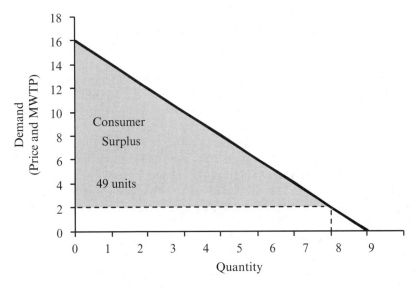

Figure 1d   Consumer Surplus

unit of a good, then the total willingness to pay for a good will exceed the cost to the consumer (price paid). The difference between total willingness to pay and the actual amount paid is called **consumer surplus** (Figure 1d).

This is the most basic measure of the contribution of a good to a consumer's well-being. Other more precise measures of welfare include the compensated welfare measures (e.g., **compensating variation** and **equivalent variation**).

c.   Aggregate Demand and Society's Welfare

The leap from measuring the welfare of a consumer to measuring the welfare of society requires some sort of aggregation of values across people. If we examine the problem strictly from the perspective of market demand, we can simply add together the demand that each consumer expresses for a good at the given price. For instance, if at a price of 2 cents, José demands 5 apples and Sue wants 3 apples, then their combined demand is simply 8 apples when the price is 2 cents. By adding up across individual consumers, we can derive a market demand function that has all of the properties we described above (e.g., inverse demand still reflects marginal willingness to pay, but now for all consumers). In a similar manner, we calculate **consumer surplus** for all consumers that participate in the market.

d.   Supply

An economy consists of more than consumers; we also need to look at the supply of goods and the costs of providing goods. Supply analysis can be graphically demonstrated using the same marginal framework we developed for demand. The cost of providing one more good is the marginal cost of that good. As before, we draw the quantity of a good provided on the horizontal axis. The cost of providing one more unit of a good, its marginal cost, is drawn on the vertical axis. The relationship between quantity and marginal cost is called the supply function or the marginal cost function. Marginal cost may increase, remain constant, decrease, or even develop an irregular shape known as a **backward bending supply curve**. The total cost of supplying a good is found by determining the area under the supply curve between zero and the quantity supplied.

Producers will supply goods as long as the price on the market is greater than the marginal cost of production. If the supply curve is upward sloping, then the price will exceed the marginal cost for every unit but the last and the producer will earn a benefit known as **rent** or **producer surplus**. Figure 2 demonstrates an upward sloping supply function, price, and the associated producer surplus that comes from supplying goods up to the point where marginal cost equals price. In the aggregate, there exists a market supply function that reflects the marginal costs of supplying goods when all producers are considered. Some producers may be able to produce goods at low marginal costs and others may produce at higher marginal costs. Those that can produce goods more cheaply will earn more benefits and thus more **rent** or **producer surplus**.

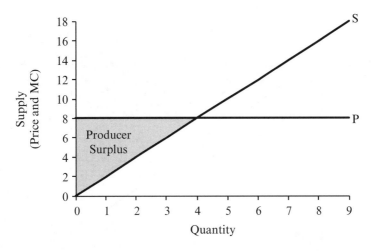

Figure 2   Producer Surplus

2.   MARKET EQUILIBRIUM AND MARKET FAILURE

In market economies, prices are not set arbitrarily, but are the outcome of the forces of supply and demand. Producers will supply goods as long as consumers are willing to pay at least the marginal costs of supply. Consumers will buy goods as long as their marginal willingness to pay (marginal benefit) exceeds the costs charged by producers. Through market processes of bids and offers, consumers and producers make trades. If consumers demand more goods than a given set of producers can provide, new producers will enter the market as long as they can supply goods at a cost less than the consumers' willingness to pay. As more and more goods are produced and consumed, two things may happen: (1) because of the property of declining marginal benefits, the consumers' willingness to pay will fall; and (2) the marginal costs of production may remain constant or increase, perhaps due to technological reasons. Eventually market trades reach a point where the marginal cost of production equals the marginal benefit to consumers. We call this an equilibrium point (demonstrated in Figure 3a) where q^* is the equilibrium allocation and p^* is the equilibrium price.

   If the amount supplied is less than q^*, a producer can earn rent by increasing the supply. If more than the equilibrium quantity is supplied, consumers will not pay a price sufficient to cover marginal costs and some producers will lose money and reduce their supply.

   As long as market costs reflect the actual cost of goods, consumers and producers will trade at an allocation that maximizes welfare to society,

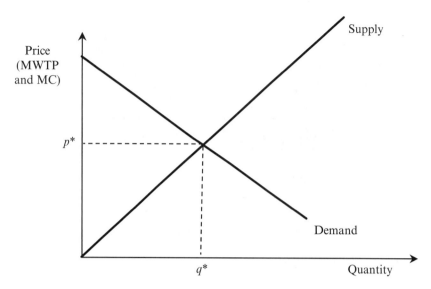

Figure 3a   Supply and Demand

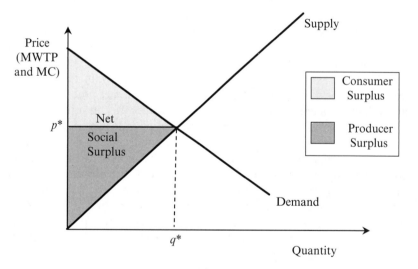

Figure 3b   Net Social Surplus

where welfare is **consumer surplus** plus **producer surplus** or **net social surplus** (Figure 3b).

(When net social surplus is maximized, we say we have an **economically efficient** allocation.) This welfare maximizing property of free markets has led many to support market-based mechanisms for the allocation of resources (**market environmentalists**), and has given rise to a movement known as Free Market Environmentalism. The conditions under which a market must operate to achieve the maximum welfare for society are, however, difficult to find. At the minimum, a well functioning market needs to be one in which **property rights** for resources are well defined, there are many buyers and sellers, and prices most accurately represent the costs of providing goods. Frequently these conditions are not met and we say the economy experiences **market failure**.

The efficient allocation of natural resources and environmental goods is routinely impeded by **market failures**. Property rights do not exist for many resources (e.g., deep sea minerals or open ocean fisheries). If resources are open to all (**open-access** resources), the market will not allocate these resources in a way that maximizes the benefits to society. Sometimes the lack of property rights stems from a difficulty in laying claim to **fugitive**, mobile and remote resources. When this is the case, steps can be taken to assign property rights, thus improving the degree to which the market functions properly. In other cases, resources lack "owners" because their very nature defies ownership. When consumers cannot be excluded from enjoying a resource and when the enjoyment of that resource by one consumer does not diminish the enjoyment of that resource by another, we call the good in question a **public good**. Examples of **public goods** include national defense, air quality, and whale conservation (public bads include things like global warming and coastal pollution). Because property rights cannot be assigned to public goods, the market is unlikely to allocate public good resources in a way that is **economically efficient** (i.e., the welfare maximizing allocation is not chosen by the market).

Often environmental impacts are the byproduct of other activities. Recycling trucks create noise when they drive through your neighborhood to pick up bottles; power plants emit smoke and steam that can obstruct views; surfers can create entertaining viewing for beachgoers. In all three cases, we find that the people who bear the costs or enjoy the benefits of the "byproducts" of these activities may be different than the people whose activities create these "byproducts". When costs and benefits of an activity are not borne by the immediate consumer or producer, we call these costs **negative externalities** and the benefits **positive externalities**. As in the examples of open access and public goods, the problem is that property rights to the externalities have not been properly defined. When property rights for an externality are well defined (e.g., people have property rights to be free from noise or recycling trucks have the right to make noise), we expect a well functioning market to allocate resources efficiently (see **Coase Theorem**). Property rights can sometimes be assigned directly to externalities. In other cases, government can internalize externalities by determining the economic value of an impact and directly charging those responsible for negative externalities (see, for example, **Pigouvian taxes**) or by subsidizing positive externalities.

3.   NON-MARKET VALUES AND TOTAL ECONOMIC VALUE

Many environmental and natural resources are not traded in markets, either because they are public goods or because property rights to these goods have not been properly assigned. Unlike most standard market goods (say a candy bar), certain environmental goods may possess values that can be considered as **public goods** and other values that are purely **private** in nature. It would be hard to imagine how a candy bar could be anything other than a **private good**. On the other hand, consider a tree. When a tree is cut for timber, it is consumed as a **private good**. That same tree, however, could also have been viewed by hundreds of hikers had it remained alive in the forest. Viewing that tree would represent a **public good** value associated with the tree. The fact that natural and environmental resources tend to have properties that are both public and private, utilitarian and spiritual, practical and intangible means these goods are far more complex than standard market goods. To understand when and why markets allocate and misallocate environmental and natural resources, we need to start by understanding the types of values these environmental goods can possess.

To start, we break down the **total economic value** of an environmental good or service into those values that people derive from direct use (called **use values**) and those values that do not require direct use by the consumer (**non-use** or **passive use values**). **Use values** include all of those transitive, active uses of a good or resource. Some use values have properties of private goods. These values typically involve some kind of extraction or consumption that allows the consumer to exclude others from using the good or resource. Timbering a tree is one example, but many less dramatic examples also exist, including picking up seashells or diverting water from a lake to irrigate your garden. When a good is removed from its place in nature and consumed in a way that makes exclusion possible, we call the value associated with that good an **extractive use** value. If environmental goods had only extractive use values, we would expect that the market might be able to allocate these goods in an **economically efficient** way. However, the very same goods may also have **non-extractive use values**. **Non-extractive use values** are those values that emanate from the use of a good that do not require removal from nature and subsequently are difficult to exclude from use by others. Viewing a tree is a **non-extractive use value**, wading in the ocean and sailing on a lake are both activities that generate non-extractive use values.

Environmental goods differ from other common market goods in that many environmental goods possess **non-use** or **passive use values**. Non-use values are values that a consumer derives from a good that they never use or do not plan to use directly. For instance, some people may derive utility just from knowing that sea otters exist somewhere in California (we call this **existence value**). These people may even be willing to pay to protect these animals (direct evidence of the minimum value they place on this passive use). Other **passive use values** include **bequest value** and **option value**. Whether **non-use values** are true values, distinct from some sort of expected use value, is a matter of some debate. Furthermore, the universal acceptance of **non-use** values by many policy makers is hindered by the difficulty economists have had in finding reliable techniques to measure these values. Nevertheless, a "blue ribbon" panel convened by the National Oceanic and Atmospheric Agency of the United States (**NOAA panel**) found that **non-use values** were at least sufficiently legitimate to be used as a starting point for assessing economic losses in a judicial setting.

4.   MEASURING NON-MARKET VALUES

Measuring the non-market values of environmental goods has been one of the greatest challenges for environmental economics during the last 30 years. To capture the very real value of these goods, environmental economists have undertaken two avenues of empirical elicitation that can loosely be divided into "revealed preference, or non-hypothetical approaches" and "hypothetical, or contingent market" approaches. (As the science advances, these two approaches continue to blend together, but we treat them separately here to get several key points across.)

By definition, prices for "non-market" goods simply do not exist. Nevertheless, environmental values do influence market activities by influencing prices for certain goods (as **complements** or **substitutes** for market goods) and by influencing spending on other market goods. For instance, all things being equal, houses where the air is clean have higher market prices than houses in smoggy neighborhoods. When the quantity or quality of environmental goods and services affects consumer behavior we often can tease out the economic "value" of the environmental good through **econometrics**. We say that people "reveal their preferences" for environmental goods by their consumption behavior. Statistical methods can then be applied to econometric models of consumption that account for the ways in which environmental goods influence consumers' preferences for other goods. The class of methods that employs empirical statistical analysis of consumption behavior with models of consumption is known as the class of **revealed preference** methods. In the field of environmental economics, standard **travel cost methods**, **random utility models** and **hedonic methods** are three important revealed preference methods used to elicit the value of non-market environmental goods.

In many cases, it is impossible to observe the ways in which environmental goods influence market goods. Sometimes, the environmental good of interest is independent of other market goods. Other times, the provision or loss of the good is rare or even entirely novel to the consumer – consider the loss of wildlife in Prince William Sound following the grounding of the *Exxon Valdez* or the hypothetical extinction of the blue whale. When the provision or loss of an environmental good is beyond our experience or falls outside of what we would normally consider complements or substitutes to market goods, we must turn to more hypothetical, or contingent, approaches to valuing environmental goods. The **contingent methods** use a variety of techniques to determine how consumers would value hypothetical changes in the quantity or quality of goods. The most basic of the contingent models use survey techniques to estimate the respondents' **willingness to pay** for the provision of an environmental good or the removal of an environmental bad, or their **willingness to accept** compensation for the opposite. Other contingent methods elicit values for environmental goods and services by having respondents rank environmental non-market goods with other market goods (**contingent ranking**). Increasingly, the contingent methods are designed to look more and more like market transactions. In the **stated preference method**, respondents are given a choice between hypothetical "bundles" of goods with each bundle containing at least one environmental good, and with hypothetical prices for every bundle. The consumer then chooses the bundle he or she prefers. In the stated preference approach, consumers' hypothetical choices can be treated much like the actual choices analyzed in the revealed preference methods. Finally, environmental economists can combine both revealed preference and stated preference approaches to fully utilize the full suite of available empirical and hypothetical data to provide clues about the values people give to non-market environmental goods.

5.   SUSTAINABILITY

To address issues of sustainability, we must expand our discussion of resource allocation to include the idea that there may be an optimal way of distributing resources across time as well as among people. Under the umbrella of sustainability, we can consider two principal questions: (1) how long will our resources last, given current consumption patterns? and (2) how should we manage our resources so that future generations have access to the same quality of life as present generations?

The first question of resource longevity is really one of prediction and accounting; how do technology, taste, population, and natural regeneration influence the stock of environmental goods that will be available from one year to the next? Questions about how long resources will last have long concerned planners and economists who have worried that the world is running out of agricultural land, fish, and other important environmental goods. While **scarcity** may simply be defined as a quantity below which a good has a price in the market, the topic of **sustainability** attempts to frame the discussion as the degree to which a good is becoming more scarce. **Malthus** wrote that land for agriculture would become scarce, not because the quantity was disappearing but because the demand for land, driven by a growing population, was increasing dramatically. David Ricardo, on the other hand, pointed out that land may indeed become scarce, but not because it was in short supply. On the contrary, Ricardo believed land itself was abundant, but good, high quality land for agriculture was scarce and becoming scarcer for the same reasons proposed by Malthus. In either case, the concern was that the supply of a resource was not growing to keep pace with increasing demand.

A variety of indices have been developed to measure the degree to which resources are being consumed sustainably. In their most basic form, these indices simply divide stocks by consumption rates. More elaborate indices take into account the fact that changes in taste or technology may reduce future demand while natural regeneration, especially for biological resources, may increase stocks to keep pace with increasing demand. Generally, poor foresight has left predictions about resource sustainability wanting in accuracy. In a now famous bet, Professors Paul Ehrlich and Julian Simon wagered respectively that five important resource commodities were growing more scarce or less scarce. Ehrlich chose to bet that $1000 invested equally in five metal commodities would increase in real value over a 10-year period due to increasing scarcity. In fact, the real value of these commodities declined by more than 50 percent to $423.93. Clearly, a better understanding of trends in demand and technology are necessary before we can say with confidence how environmental quality and goods will fare in a consumption-based world.

The falling commodity prices shown in the wager between Ehrlich and Simon may have been misleading. Ten years is a short time frame. What might have happened after 50 or 100 years? Also, market prices may signal market scarcity, but fail to indicate local scarcity. As markets grow and become global in scope, market signals for scarcity will also become global, revealing global conditions but masking local conditions of supply and demand.

The second question of sustainability "how should we manage our resources?" requires that we have some management target for the way in which resources are distributed across generations. The concept of **Pareto optimality**, useful in the analysis of welfare within a generation, is more difficult to apply when we do not know the tastes, preferences or technologies of future generations. Consumers tend to be impatient, preferring consumption this year over next (this is why we borrow). Are we to ignore this impatience when thinking of sustainability or should we include society's impatience by **discounting** the value of future consumption when we decide how to manage for sustainability? What about expected improvements in technology? Could saving too much of our natural and environmental resources make future generations better off at the expense of present generations?

The "we" in the second question of sustainability "how should *we* manage. . ." implies that there is some sort of shared vision about the goals of sustainability. While oft attempted, efforts to define a single target for sustainability have failed to achieve global consensus. Neoclassical economists often see the maximization of **net present value** as a metric against which to measure the sustainability of proposed resource use over time. Others have set out to define more broadly appealing, but less easily operable, definitions of sustainability.

The United Nations, under the mantle of the United Nations Commission on the Environment and Development (**UNCED**), turned to Gro Harlem Brundtland and a task force of experts (also known as the **Brundtland Commission**) to define sustainability. The **Brundtland Commission**, in turn, sought to frame the debate on sustainability in the context of economic development. While still vague in its prescriptions, the Brundtland Commission's positions on **sustainable development** made the trade-offs clear: economic growth should not come at the expense of future generations. Moreover, since the Commission's report sustainable development has been expanded to encompass three pillars: social, environmental and economic development Of course, determining its definition and what paths of resource use constitute a cost to the future still remain topics of debate. However, whether sustainability and sustainable development ought to refer to the maintenance of the status quo, economic growth, redistribution of wealth, the protection of capital stocks or the preservation of **natural capital**, will vary by individual. Related to these concepts is the notion of **weak sustainability**. It supposes that we can substitute between human-made capital and natural capital in production and consumption, such that economic growth can be associated with improvements in environmental quality. By contrast, **strong sustainability** posits that natural and human-made capital are complements and cannot be substituted for each other in either production or consumption. Consequently, economic growth that uses natural resources and generates wastes must increase environmental degradation.

Regardless of the targeted goal of sustainability, the debate surrounding the topic has led to broad changes in the ways in which scholars think about resource use and the ways planners move forward with development projects. Today, a growing field of study known as **Ecological Economics** acts as a forum for exploring new ways to combine ecological knowledge with economic policy and analysis. In 1992, the United Nations hosted the Conference on Environment and Development in which many concepts of sustainability were formally institutionalized through an international declaration and a unified program of action.

Outside of the policy sphere, industry is weaving sustainability canon into its business practices. In fact, through networks, such as the Global Reporting Initiative (GRI) and the World Business Council for Sustainable Development (WBCSD) among others, many firms are contributing to the definition of **sustainability**. Corporate social responsibility (CSR) is now underscored by the triple bottom line, an accounting framework encompassing the three pillars: economic, environmental and social development. Other corporate conscience initiatives, including carbon footprinting and fair trade, have gained traction among progressive enterprises. The effectiveness of such practices is still debatable; however, industry is tepidly moving towards a more holistic operating vision. Shareholders' interests are no longer the sole concern of business; stakeholders are slowly gaining influence.The debate on **sustainability** and **sustainable development** is far from over. Few agree on the ability of technology to slow resource use or the ability of human-made capital to substitute for natural capital. For others, the debate centers on the fact that most programs designed to achieve a more sustainable world require substantial sacrifice that may not be shared equally among all. Nevertheless, the renewed focus on resource limits, first highlighted by Malthus and Ricardo, has led to the careful reconsideration of resource limits in models of development, growth and economic thought. In doing so, the scholarly and pragmatic pursuit of environmental understanding is increasingly realistic, more thorough and more in tune with the ways in which economics, human society and the environment are linked.

Climate Change Policy: A Primer

1. THE SCIENCE OF CLIMATE CHANGE

The term **climate change** refers to a global change in the Earth's climate patterns that can manifest as cooling or warming trends, as well as other meteorological changes. **Climate change** is an inherent feature of the Earth's history, as shown by evidence of past significant warming and cooling episodes. Since the Industrial Revolution, human activities such as fossil fuel combustion and deforestation have begun to influence the global climate; the concern being that this influence will significantly increase in the future. The problem with these activities is that they contribute to higher concentrations of **greenhouse gases (GHGs)** in the **atmosphere**. A very likely consequence of increased atmospheric concentrations of GHGs is an increase in the Earth's average temperature, which may in turn produce more frequent and extreme climatic events and **sea level rise**. The behavior and response of the Earth system defines the links between human activities influencing the climate system, and climatic influences on society. An understanding of human induced (or **anthropogenic**) **climate change** and the policy response to it therefore begins with the **greenhouse effect**.

The Earth's climate system is driven by the Earth's energy balance, which depends on the difference between the amount of incoming and outgoing radiation (the **radiation budget** or radiative balance). Incoming radiation is in the form of short-wave **solar radiation**. Approximately 46 percent of the incoming radiation reaching the **atmosphere** is either absorbed (20 percent) or reflected back into space by clouds, air molecules, and airborne particles (26 percent). The remaining solar radiation that reaches the Earth's surface (this energy is called solar insolation) is either reflected (4 percent) or absorbed by surface objects, and, in the latter case, as these objects warm the radiation is reemitted to the **atmosphere** in the form of long-wave **infra-red radiation (terrestrial radiation)**.

About 90 percent of the **terrestrial radiation** in the **atmosphere** is partially absorbed and reemitted by natural and **synthetic chemicals** termed **greenhouse gases (GHGs)**. The most abundant **GHGs** in the **atmosphere** are: the natural gases comprising **water vapor (H_2O)**, **carbon dioxide (CO_2)**, **methane (CH_4)**, **nitrous oxide (N_2O)**, and **ozone (O_3)**; and the synthetic gases of **chlorofluorocarbons (CFCs)**, **hydrofluorocarbons (HFCs)**, **perfluorocarbons (PFCs)** and **sulfur hexafluoride (SF_6)**. The first three synthetic gases are collectively termed **halocarbons**. The energy reemitted from GHGs warms both the Earth's **atmosphere** and surface until the outgoing and incoming energy flows are balanced. This process is called the **greenhouse effect** and produces a mean global surface temperature that is 33°C higher than if most of the **infrared radiation** was lost into space. Temperature is the controlling factor that ensures the **radiation budget** is balanced.

Radiative forcing measures the imbalance in the **radiation budget** at the boundary of the **troposphere** and the **stratosphere**, and is expressed in units of Watts per square meter (Wm^{-2}) of the Earth's surface. The contribution of each GHG to overall **radiative forcing** is measured in relation to a baseline year (often 1750, the start of the Industrial Revolution). For example, the change in the **radiative forcing** of **carbon dioxide** and **methane** from the year 1750 to 2005 is 1.66 [+/− 0.17] Wm^{-2} and 0.48 [+/− 0.05] Wm^{-2}, respectively. In addition to GHGs, **radiative forcing** may also be affected by: the release of aerosol particles from natural causes, such as volcanic eruptions, or human activities; alteration of surface albedo from changes in land use; linear contrails (persistent trails of condensation) from aircraft; and changes in solar output (solar irradiance).

Each GHG has a different atmospheric lifetime, which is measured as the length of time taken for chemical reactions to remove a kilogram of the gas from the **atmosphere**. The long-lived GHGs comprise CO_2, CH_4, N_2O, SF_6 and **halocarbons**. Both the **radiative forcing** and the atmospheric lifetime of each gas are significant factors in determining its **global warming potential** (GWP). GWP measures the radiative effect of a given mass of a **GHG** over a specific time period relative to that of the same mass of **carbon dioxide**. It therefore provides an index of the amount of energy trapped over specific time periods for each gas. For example, CO_2 always has a GWP of 1, whereas the GWP for methane is 72 times larger over 25 years and 25 times larger over 100 years.

Climate change metrics are most commonly expressed as **carbon dioxide equivalent** (CO_2-e) or CO_2 concentrations in units of parts per million (ppm), or total forcing in Wm^{-2}. GHG concentrations expressed in terms of CO_2-e represent the concentration of CO_2 that would produce the same effect on the Earth's **radiative budget** as the combined concentration of all the GHGs being accounted for. For any given effect on the Earth's **radiative budget**, CO_2 concentrations are always relatively lower than CO_2-e concentrations because the latter is composed of more constituent gases. Total forcing is the net effect on the Earth's **radiative budget** from GHGs and other direct and indirect off-setting factors that include surface and cloud **albedo** change, and radiatively active **aerosols** such as sulfate, nitrate, black carbon, organic carbon and dust.

Carbon dioxide accounts for just over 75 percent of annual **anthropogenic** GHGs. Fossil fuel use is the largest contributor of CO_2 at approximately 56 percent of annual GHG emissions, followed by land use change at just under 31 percent. Recent estimates of atmospheric concentrations of GHGs show that CO_2 increased to approximately 384 ppm in 2008, from a 'pre-industrial' level of approximately 280 ppm. Recommendations for reducing the risk of dangerous **climate change** aim to stabilize global CO_2 emissions to between 350 to 400 ppm (450 to 500 ppm CO_2-e).

2.   HUMAN-INDUCED CLIMATE CHANGE AS A PROBLEM

Our awareness that human activities could influence the climate originated in the nineteenth and early twentieth centuries out of a growing scientific interest in **climate change**, particularly the causes of the ice ages. During this time research on **climate change** was divided into theories involving three main causes: astronomical, terrestrial, and molecular. The molecular theories focused on changes in the amounts of water vapor (H_2O) or **carbon dioxide (CO_2)** in the **atmosphere** and the associated **greenhouse effect**. It was not until the late nineteenth century that the effect of industrial emissions of CO_2 on the **carbon cycle** was first considered by scientists. This work gave rise to the **hypothesis** that changes to the atmospheric concentration of CO_2 from industrial emissions could significantly impact on surface temperatures if they persisted over a very long time. The **hypothesis** fell out of favor in the early twentieth century due to: the overly simplistic model of the climate system that it was based on; contrary experimental findings that were later found to be flawed; and the widely held belief of a self-stabilizing climate system, in which plant fertilization and the ocean played key roles in the absorption of CO_2. As a result, the dominant scientific view until the late 1950s was that **anthropogenic emissions** of CO_2 did not present a problem to the global climate.

Post-war advancements in theoretical physics and technology (particularly digital computers and measuring equipment) provided improved insights into the carbon cycle and the absorption of **infrared radiation** in the **atmosphere**. Evidence started accumulating that CO_2 in the upper **atmosphere** could significantly alter the Earth's surface temperature, and that the ocean's absorption of CO_2 was slower than previously thought. These findings led a few scientists in both America and Russia to publicly warn that **global warming** may be experienced much sooner than expected and that the consequences may be severe. Doubts, however, continued being expressed by some prominent scientists due to the ongoing problem of obtaining reliable measurements of atmospheric concentrations of CO_2. From the late 1950s, the research focus moved to resolving the measurement problem, and gaining a better understanding of the carbon cycle through the use of computer modeling.

The increasing use of computer modeling to understand the climate system facilitated research on **climate change** and its possible effects in a wide range of interrelated fields. Adverse impacts experienced during major climatic events in the 1970s such as droughts, desertification and fisheries collapses saw an increased interest in the **short-run** fluctuations in climate. Scientific interest also focused on the role of **synthetic chemicals** such as **chlorofluorocarbons (CFCs)** as **GHGs**, as well as aerosols in mitigating any warming effect in the short term. By the early 1980s, the **GHGs** associated with **anthropogenic emissions** had expanded from **carbon dioxide (CO_2)** to include: **nitrous oxide (N_2O)**, **methane (CH_4)**, **sulfur hexafluoride (SF_6)**, **hydrofluorocarbons (HFCs)**, **perfluorocarbons (PFCs)**, and **chlorofluorocarbons (CFCs)**.

The accumulating evidence that humans were influencing the climate, and that the associated impacts could be severe, led to increasing demands by scientists for governments to take action to avoid the possibility of dangerous climate change. Experts from 50 countries met in 1979 at the World Climate Conference held in Geneva where it was agreed that urgent action was needed to avoid human-induced **climate change**. This was followed by the Villach Conference in 1985 where scientists from 26 countries concluded that there was evidence of a causal link between increasing concentrations of **GHGs** in the **atmosphere** and **global warming**. In 1986 the first cooperative framework to identify key policy issues informed by science was established in the form of the Advisory Group on Greenhouse Gases.

Science continues to play a key role in the political recognition of **climate change** as a serious problem, and informing the policy process on the level and timing of the **GHG** emissions reductions required. The **Intergovernmental Panel on Climate change (IPCC)**, a cooperative scientific institution formed by the **United Nations Environment Programme (UNEP)** and the World Meteorological Organization (WMO), released four assessment reports (1990, 1995, 2001 and 2007) summarizing the scientific knowledge, likely impacts, and policy aspects of climate change.

3.   THE EVOLUTION OF CLIMATE CHANGE POLICY

Human-induced **climate change** has been described as a "wicked policy problem". The form in which it will manifest, and its extent and timing are all uncertain. Furthermore, an effective policy response needs comprehensive international cooperation that imposes different relative costs among countries, which are incurred in the near future to obtain benefits that are not realized until the more distant future. International cooperation is problematic as there is an incentive for each country to be a **free-rider** and benefit from the actions taken by other countries while avoiding their share of the cost burden. This gives rise to the difficult task of resolving a **prisoner's dilemma** style problem, whereby countries have an incentive to engage in uncooperative behavior even though cooperation may result in a better outcome. The global nature of the problem has meant that international meetings and conferences are critical for the development and agreement of policy settings and measures, and these have allowed the articulation of fundamental ideas on key issues such as **mitigation**, **adaptability**, and cooperation.

The 1979 World Climate Conference was the first in a series of conferences to determine the state of scientific knowledge on **climate change**, its implications for humanity and, eventually, the appropriate policy response. In 1983 the General Assembly of the United Nations established the World Commission on Environment and Development. The 1987 report by the Commission drew attention to the potential threat to humans from **climate change**. In response, the Canadian Government organized the World Conference on the Changing Atmosphere (the Toronto Conference) in June 1988. It produced two significant outcomes: **climate change** was recognized as a serious problem, requiring developed countries to lower their CO_2 emissions by 20 percent of 1988 levels by 2005; and the establishment of the **Intergovernmental Panel on Climate Change (IPCC)**.

By 1990 the **precautionary principle** was used to guide policy actions on climate change. Most **OECD** countries proposed to either stabilize CO_2 emissions at 1990 levels by 2000, or to set targets for reducing them from 1990 levels by 2005. The policies to achieve these outcomes focused mostly on adaptation to **climate change** (primarily with respect to impacts on coastal management and resource use and management), international action and cooperation (in the areas of information, development and transfer of technology, economic and financial mechanisms, and legal and institutional mechanisms) and **mitigation** (the options available in the energy, industry, agriculture, and forestry sectors).

An international environmental treaty – the **United Nations Framework Convention on Climate change (UNFCCC)** – was adopted at the **United Nations Conference on Environment and Development (UNCED)** held in Rio de Janerio in 1992, which later came into force in March 1994. The objective of the UNFCCC (stated in Article 2) is: to stabilize GHG concentrations in the **atmosphere** at a level, and in a time frame, that would prevent dangerous **anthropogenic** interference with the climate system; and allow ecosystems to adapt, food systems to be maintained, and

sustainable development to proceed. Parties to the UNFCCC are divided into developed countries (**Annex I countries**) and **developing countries** (**non-Annex I countries**). The UNFCCC is a non-binding treaty that calls on **Annex I countries** to aspire to stabilizing their 2000 emissions at 1990 levels. No obligation is placed on **non-Annex I countries** to reduce their GHG emissions. A third classification of Parties is the **Annex II countries**, which are a sub-group of the **Annex I countries**, and comprise members of the **OECD** that were not economies in transition (EIT) in 1992. The **Annex II countries** may assist **non-Annex I countries** to reduce their emissions and adapt to **climate change** by providing financial and technological assistance.

Five principles are used to guide the actions of Parties to the UNFCCC: (1) while all countries share responsibilities for addressing climate change, they should be differentiated with respect to their contribution to the problem and capabilities to address the problem (often referred to as common but differentiated responsibilities); (2) the needs of developing countries that are particularly vulnerable should be considered; (3) the **precautionary principle** should be adopted subject to measures taken being **cost-effective**; (4) the right of all countries to pursue **sustainable development** should be recognized; and, (5) an open international economic system should be supported.

The "supreme body" of the UNFCCC is the **Conference of the Parties** (**COP**). It is the highest decision-making authority and meets annually to assess progress on addressing climate change. A move to legally binding commitments for GHG reductions was adopted at COP-1 in Berlin. Key policy debates during this period involved: defining **equity**, the appropriateness of applying **cost–benefit analysis** to policy measures, and determining the **social costs** of **climate change**. In 1997 the United States adopted the Byrd–Hagel Resolution, which prevented it from agreeing to any future binding GHG emissions targets until the large emitting **developing countries** also committed to participating meaningfully in GHG reductions.

The **Kyoto Protocol**, the most recent international agreement linked to the UNFCCC, was adopted by the third Conference of the Parties (COP-3) in December 1997 and entered into force in February 2005. The Protocol presently commits 37 industrialized countries together with the European Union to binding targets for GHG emissions reductions for the period 2008 to 2012. To date, 194 Parties have ratified the UNFCCC and of these 193 have ratified the **Kyoto Protocol**. The United States withdrew from the Protocol in 2001. The Protocol does not include any new long-term objectives or principles, but rather develops those of the UNFCCC. Key developments include an increasing emphasis on: the interaction between **climate change** and **economic development**, net reductions of

GHG emissions, use of **market-based instruments** to reduce emissions, and opportunities for adaptation.

Where the UNFCCC encourages developed countries to stabilize GHG emissions, the **Kyoto Protocol** sets binding targets for countries that have committed to them. Developed countries that become signatories (**Annex B countries**) are committed to jointly reducing aggregate emissions of GHGs by 5.2 percent of 1990 levels over the 2008–2012 commitment period. It should be noted that while some countries with high per capita emissions did not immediately ratify the Protocol, such as Australia and Canada, or refused to as with the United States, many **climate change** initiatives were undertaken in these countries at a sub-national level. Bargaining between **Annex B countries** during negotiations on the Kyoto Protocol established different targets for each country on the level of allowed GHG emissions over the commitment period (**Quantified Emission Limitation and Reduction Commitments**). This level is expressed as a quantity of **assigned amount units** (AAUs) for each country, where each unit represents one metric tonne of CO_2-e emissions.

The Protocol departs from previous approaches on determining a country's target by allowing for CO_2 sequestered by **afforestation, reforestation** and **deforestation** activities since 1990 to be offset against its GHG emissions. The term CO_2-e emissions accounts for GHGs regulated under the **Kyoto Protocol**, which comprise four specific gases (**carbon dioxide (CO_2), nitrous oxide (N_2O), methane (CH_4) and sulfur hexafluoride (SF_6)**) and two groups of gases (**hydrofluorocarbons (HFCs)** and **perfluorocarbons (PFCs)**). These gases are often referred to as the Kyoto gases and do not include the long-lived GHGs regulated under the **Montreal Protocol** (primarily **chlorofluorocarbons (CFCs)** and **hydrochlorofluorocarbons (HCFCs)** that are due to be mostly phased out by 2010 and 2040, respectively) or **aerosol** emissions. The effect of the Kyoto gases on the Earth's **radiative budget** is called Kyoto forcing, which is the concentration of CO_2 that would produce the same effect on the **radiative budget** as the combined concentrations of the Kyoto gases.

The Protocol specifies a range of domestic policies and measures that **Annex B countries** can use to achieve their emissions target. These policies and measures are aimed at: enhancing **energy efficiency**; bringing about changes to land use, transportation, waste management, and other GHG-intensive sectors such as stationary energy; promoting **technological change**; and, removing market imperfections and economic distortions. There are also five Protocol mechanisms in addition to the domestic initiatives. They comprise: joint fulfillment (Article 4), which allows individual countries to have different targets while adopting a joint overall target for emissions reductions; a financial mechanism to assist with implementation

(Article 11); and three **flexibility mechanisms** (Kyoto mechanisms), namely, **joint implementation** (JI) (Article 6), the **clean development mechanism** (CDM) (Article 12), and **emissions trading** (ET) (Article 17). The Kyoto mechanisms are also termed market-based mechanisms because they facilitate the creation of an international **carbon market**.

Joint implementation allows an **Annex B country** to earn **emissions reduction units** (ERUs) from undertaking emissions reduction projects in other **Annex B countries**. The **clean development mechanism** allows an **Annex B country** to earn tradable **certified emissions reduction** (CERs) credits from undertaking emissions reduction projects in **developing countries**. Both JI and the CDM create tradable rights to emissions that count towards a country's Kyoto target. The Protocol also allows for the creation of tradable **removal units** (RMUs) from projects that reduce emissions through **land use, land-use change and forestry** (LULUCF) activities. **Emissions trading (ET)** allows countries that have a surplus of emissions units (AAUs in excess of the amount needed to satisfy both the target and a commitment period reserve) to sell them to countries that are unable to reach their target with domestic action alone. Each **emissions reduction credit (ERC)** created from ERUs, CERs or RMUs (collectively termed other trading units) is also specified as a tonne of CO_2-e emissions and is additional to the AAUs. Details regarding the actual working of the Protocol's mechanisms and compliance processes were left to subsequent negotiations.

A Conference of the Parties to the Kyoto Protocol (known as the CMP) meets parallel to the UNFCCC COP. The CMP serves a similar function as the COP, however decisions relate specifically to the Protocol. In 2001 at COP-7, the framework for implementing the Protocol – the Marrakesh accords – were adopted. Other important COP/CMP meetings include the following. The COP-11 at Montreal in 2005 decided that Annex B Parties would begin negotiations on a second period of commitments. At COP-13 in Bali in 2007, the Bali Action Plan was adopted. It called for a decision on the amount of GHG emissions that would need to be reduced to avoid unsafe **climate change**, and began a two year process for a post-Kyoto agreement that was to be adopted at Copenhagen in 2009. A key element was that **developing countries** were to adopt **nationally appropriate mitigation actions** (NAMAs). Many of these countries, however, continued to apply the principle of common but differentiated responsibilities in articulating these actions.

The COP-15 meeting in Copenhagen failed to adopt, but rather noted, the Copenhagen Accord, which accepted that average global temperatures should not be allowed to exceed a 2°C rise above preindustrial levels, and that deep cuts in global emissions were required to achieve this. Instead of setting medium or short-term targets for emissions reductions, countries were asked to make voluntary offers on targets and actions for reducing emissions by 2020. This bottom-up approach, which allows countries to use different metrics and baselines to define their Copenhagen targets, and different ways to account for emissions, departs from the top-down approach previously applied in negotiations under the Kyoto Protocol.

A country's ambition for emissions reduction is often expressed as a change in a metric such as **absolute emissions**, **emissions intensity**, **business-as-usual emissions**, or **per capita emissions**. All except the last of these metrics are used in country submissions on Copenhagen targets. For example, the United States' target is an absolute reduction in emissions of 17 percent from 2005 levels by 2020, the European Union propose an absolute reduction of 20 to 30 percent from 1990 levels, China's target is a reduction in emissions intensity of 40 to 45 percent from 2005 levels, and Brazil's target is a 36 to 39 percent reduction in emissions from **business-as-usual** levels by 2020.

The COP-16 held in Cancun, Mexico, in late 2010 agreed to both the aspirational emissions reduction targets submitted by countries under the Copenhagen Accord, and not to exceed a 2°C rise in global average temperature above preindustrial levels (equivalent to a 450 ppm CO_2-e concentration). For the first time, the world's largest GHG emitters in both developed and developing countries (including United States, China, India, European Union, and Brazil) have pledged reduction targets by 2020. When differences between each country's metric and baseline are adjusted for, comparative results show that while reductions in absolute emissions between 2005 and 2020 are substantially different between countries, these differences are significantly reduced when expressed in terms of per capita emissions, and demonstrate broadly similar levels of ambition when expressed as reductions in emissions intensity or business-as-usual emissions. A major criticism of the Copenhagen targets is that they are not stringent enough to prevent the global average temperature rising above 2°C. Subsequent analysis shows the targets are consistent with a long-term CO_2-e concentration of 550 ppm (or a 3°C warming scenario).

The Cancun Agreements did, however, deliver progress towards international action on **climate change**. These Agreements include: a move to better mechanisms for monitoring, verification, and reporting that also apply to **developing countries**; a Green Climate Fund for the delivery of US$100 billion per year by 2020 to help with **mitigation** and adaptation efforts in developing countries (which includes pledges by developed countries of close to US$30 billion of "fast-start" finance up to 2012); the development of a program to reduce **deforestation and forest degradation (REDD)**; and, an initiative to identify how to transfer both clean energy and adaptation technologies to **developing countries**.

While the United States withdrawal from Kyoto negotiations in 2001 undermined the notion that developed countries should take the lead in reducing emissions, the principle of common but differentiated responsibilities remained a strong influence on policy development. The long-held distinction between developed and **developing countries**, based on a country's historical contribution to GHG emissions and per capita wealth, is now becoming less relevant. The Cancun Agreements recognize that the common but differentiated responsibilities relate to not only historic emissions but also future emissions. The Agreements also recognize that GHG reduction initiatives outside the UNFCCC will make a necessary contribution to effective **climate change** policy. Many countries have been, and continue to be involved in bilateral and multilateral activities on addressing climate change. For example, although the United States has remained outside the Kyoto Protocol, it has been engaged in agreements on Methane-to-Markets, the International Partnership on the Hydrogen Economy, and the Asia-Pacific Partnership on Clean Development and Climate.

4.   THE ROLE OF MARKET-BASED INSTRUMENTS

The policy architecture provided by the **UNFCCC** requires that actions taken by governments to reduce GHG emissions must be based on **cost-effective policies**. Policies implemented to date include an array of approaches consisting of regulations and standards, taxes and charges, **tradable permits**, voluntary agreements, subsidies and incentives, and research and development. Although it is acknowledged by the IPCC that national policies to reduce emissions will depend on each country's circumstance, those providing an effective price signal for **carbon dioxide** are seen to have greater potential to significantly reduce emissions in all sectors of the economy.

Economics has contributed significantly to the development of pollution control policy in general by showing that pollution taxes and tradable **pollution permits**, which use price signals to encourage changes in behavior, in principle provide governments with methods to achieve their objectives at least cost. In the case of pollutants such as GHGs that are uniformly mixed in the **atmosphere**, emissions reductions may be achieved by implementing **cost-effective policies**, which equalize the marginal cost of reducing emissions across emitters. The burden of reducing emissions then falls on emitters for which it is relatively less costly, thereby minimizing the total cost of achieving the desired reduction.

A **carbon tax** was initially proposed by economists as the instrument of choice for countries to reduce their GHGs. However, countries seeking **cost-effective policies** have tended to show preference for GHG **emissions trading scheme**s (ETSs). An advantage of **emissions trading** is that it provides greater certainty about the quantity of GHG reductions achieved compared to an emissions tax; whereas a **tax** provides greater certainty about the cost of reducing emissions, but not the quantity of emissions abated. The **Kyoto Protocol** reflects the least-cost approach by using a market-based mechanism, the international trade in **emission reduction credits**, to facilitate international cooperation on reducing emissions, and requiring the reduction or phasing out of national policies that may impede a county's adoption of **market-based instruments** as part of its **climate change** policy. The Cancun Agreements continue to endorse this approach.

There are two basic approaches underpinning the design of a national ETS. The first is a **cap-and-trade** scheme where a target (cap) is set on the total amount of GHG emissions from specific sources (usually firms in emissions-intensive industries) allowed over a particular period, and a quantity of permits is supplied to the market that in total ensures this target is strictly met. **Emission permits** may be allocated to participants in the scheme through a pricing mechanism (such as an auction or at a fixed charge), for free, or a combination of both. **Compliance** with the target is achieved by requiring specified emitters to surrender (acquit) to a prescribed authority an **emission permit** for each unit of GHG they produce at the end of each acquittal period.

The second approach is a **baseline-and-credit** scheme, which is theoretically equivalent to a cap-and-trade scheme. It allocates free permits on the basis of a participant's historical emissions (**grandfathering**). A baseline-and-credit scheme assigns each participant a **baseline** for their allowable GHG emissions in each of the scheme's acquittal periods. The sum of all participants' baselines for each period equals the target set for the total amount of GHG emissions allowed for the period. Rather than requiring participants to acquit an amount of permits equal to their total emissions in a period, a baseline-and-credit scheme uses the difference between actual emissions and the baseline emissions to determine the amount of permits allocated and acquitted. If a participant's actual emissions are under the baseline it earns free credits (tradable permits) equal to the emissions saving, which may then be sold in the market. If a participant's emissions are over the baseline they are required to purchase an amount of credits from the market equal to the excess and surrender these to the prescribed authority.

Under both approaches it is the operation of the permit market and the method used to allocate permits that provide the necessary price signals for participants to efficiently make their **short-run** operating and **long-run** investment decisions. These decisions affect levels of both GHG emissions and economic output, and, therefore, the effectiveness and **efficiency** of the scheme, respectively.

A comprehensive international trading scheme that links national emissions trading schemes is still a long way off. The European Union and New Zealand are the only countries presently with schemes capable of linking. While international emissions trading is yet to play the significant role in reducing global emissions envisaged earlier, there are encouraging signs that emissions trading may still be a key policy measure in GHG mitigation at the national level. Several major emitters in developed and **developing countries** (including India and China) are planning or proposing the implementation of national or regional schemes.

Climate change policy is now at a very important juncture with respect to whether an international approach to GHG mitigation can be achieved, and the role of **market-based instruments** in domestic mitigation efforts. The limited use of **market-based instruments** may be due to a general unwillingness among countries to accept that stringent action on climate change is required. Instead, **no-regrets policy** options are being considered or pursued such as: reducing **deforestation** and forest degradation, increasing energy efficiency, increasing the use of **biofuels**, developing carbon capture and storage (CCS) technologies, preparing adaptation strategies, and researching opportunities for **geo-engineering**. Other approaches that involve potentially more costly measures focus on sector-specific technological solutions to mitigate GHGs, including: mandating levels of renewable energy use or emissions-intensity in the electricity sector; investing in or funding the development and production of low emissions technologies, such as electric and hybrid vehicles and solar generating facilities; and, providing subsidies to incentivize emission reducing practices in industry, agriculture, and forestry.

Experience gained from using **market-based instruments** to address environmental problems has shown that they are often a necessary but not sufficient measure when more than one market failure exists. This is especially applicable to climate change policy, and a point that is often lost in debates over which particular measures should be implemented. For example, while carbon pricing cost-effectively addresses the externality arising from emissions, government support for private research and development (R&D) that may lead to innovation in lower emission technologies addresses the market failure arising from the public good nature of the information acquired. In this case, carbon pricing alone may not provide adequate incentives for firms to undertake an efficient level of R&D. The lesson for climate change policy is that while **market-based instruments** are necessary for achieving efficient mitigation, depending on the market failures that exist other measures may also be required to achieve an efficient and effective reduction in GHG emissions.

International Environmental Problems: A Primer

One of the main virtues of the market system is that it allows people and firms to exchange goods at a relatively low cost by relying on a series of decentralized markets. The presence of these markets enables people to improve their welfare by exchanging the goods that they produce for goods that they can use and value more highly. The theory of international trade shows that the same principle is at work when countries engage in free trade: by specializing in the production of various goods in which they may enjoy either a **comparative advantage** or **absolute advantage** they are able to realize greater gains from trading goods with one another than simply consuming the goods themselves.

At the same time, many of the environmental problems we observe arise from some form of **market failure**. While markets can be efficient ways of distributing goods, they may not necessarily lead to optimal outcomes if the price of goods does not include all of the environmental costs associated with their production or consumption. One of the main ways in which all of the costs may not be included and markets might fail is when the rights to a resource are not well defined. Rights might not exist in the case of **open-access resources**, where no one party owns the resource and the resource is free to be exploited by all, such as fisheries on the open sea. In this case, the price of the fish might not take into account the over-exploitation of fish stocks and the risk of the stock falling below some critical level. Problems can also arise around **common-pool resources** where different parties each have access rights but do not manage the resource collectively, either because a common governance framework has not been established or is not strong enough to overcome individual incentives that contribute to overexploitation and degradation of the resource. Examples include shared aquifers across international boundaries and regional fisheries. Alternatively, the resource may be a **public good**, where the nature of the resource is such that it is impossible to exclude people from using or benefiting from it. Examples of such goods are the protective ozone layer in the Earth's atmosphere or the genetic capital embodied in biodiversity. There may also be **public bads**, as in the case of greenhouse gases, when everybody potentially suffers the consequences of climate change. One consequence of public goods is that countries have incentives to let others bear the cost of any actions, such as maintaining biodiversity, reducing greenhouse gas emissions, or eliminating ozone-depleting substances, since they can enjoy the benefits from actions taken by other countries without incurring any cost themselves. This type of behavior is called **free-riding**, and can either lead to overexploitation of the resource (in the case of a public bad) or under-provision of the resource (in the case of a public good). Collectively, those common-pool resources and public goods that are truly international in nature are considered the **global commons**: resources shared equally by all countries that can be potentially degraded or destroyed.

It may also be the case that the pollution or environmental degradation associated with the production or consumption of a good is not included in the price (**negative externalities**). Problems also arise when the scale of activity becomes great enough to have a measurable and negative impact on local ecosystems: for example, higher harvesting levels may reduce fish or wildlife populations below some critical level; or the cumulative effect from effluent flows discharged into a river from different countries may exceed the assimilative ability of a river to break down waste. Finally, problems can also occur from **policy failures**: government programs or regulations that have negative effects upon the environment. An example of such a failure might be subsidies to encourage fishers to switch from inshore fisheries to ocean fisheries, which increases pressure upon marine fish stocks.

Rapid population growth and subsequent development in countries around the world and the concomitant increase in trade have led to impacts outside the local environment and, increasingly, upon the global environment. At the same time, an increasing awareness of environmental issues and the interaction and interdependence of ecosystems, many of which cross national boundaries, has focused attention on the global or international nature of many of these problems. Solutions to environmental problems can be difficult in and of themselves: the nature of the environmental processes at work might not be well understood; it may be costly to identify the parties involved (either those who benefit or bear the costs); and it may be costly to create some kind of mechanism through which a solution can be implemented (whether it is through regulation, the allocation of rights, the establishment and collection of fees, or some other mechanism). These problems are only compounded when the problems cross political boundaries. The issues become particularly troublesome when the benefits and environmental costs are distributed such that some countries enjoy much of the benefit while others bear much of the environmental cost. Those bearing the damage may be countries downstream, or downwind, that receive the pollution from industrial activity in polluting countries that have little incentive to change their behavior if they do not suffer the consequences from their actions.

While there is no difference between the nature of the issues that arise in terms of global environmental problems and those that might arise in a community or country, there is one fundamental difference in terms of the environment in which they occur. International environmental problems, by definition, are those that involve more than one country; consequently, there is no one overarching set of international laws or institutions that has the ability to address problems, enact solutions, and enforce them. To respond to those problems, countries have to pursue cooperative arrangements such as **Multilateral Environmental Agreements (MEAs)**, which are agreements between different countries over how to handle various common environmental problems. As of today, there are thousands of MEAs. A selection of these agreements is summarized in Appendix 7.

1. INTERNATIONAL AGREEMENTS

International agreements may take various forms. The most common are **treaties**, **conventions**, and **protocols**; all of which are agreements between member countries. Treaties are legal agreements under international law in which the countries that sign the documents spell out their obligations or responsibilities. Conventions are agreements that establish a general framework or objectives, for which subsequent protocols will embody the details of how the objectives are to be achieved. Agreements do not become effective (that is, binding upon the countries that signed the agreement) until they come into force, and agreements do not come into force unless a sufficient number of countries have ratified the agreement. A treaty or convention only applies to those countries that are signatories, although provisions in the agreement may govern trade with non-parties, and countries may be able to make reservations at the time they sign an agreement (making a unilateral statement that modifies or excludes a certain portion of the agreement) as well as renounce their participation in an agreement at a later date.

The first set of international agreements between countries are primarily concerned with conflicts over shared resources, such as how to address the problems created by resources (e.g., fish stocks and migratory wildlife) that move between different countries or waterways that pass across national boundaries. In these cases, the **common-pool** nature of the resource means that the first to capture the resource can lay claim to it, which can lead to overexploitation as countries "race for the resource". Examples of such agreements include the 1923 Convention for the Preservation of the Halibut Fishery of the Northern Pacific Ocean and the Bering Sea, and the treaties signed between the USA and Canada over shared boundary waters in 1909, and the USA and Mexico over the shared waters of the Rio Grande in 1906. In recent years, a number of new agreements have been struck between countries sharing transboundary water resources, reflecting the growing importance of water as pressure on water resources grows.

Efforts to address high seas fisheries have been less successful. The approach here has been for member countries that exploit the same fishery to establish regional fisheries management organizations; examples of fisheries targeted include tuna, salmon, and other migratory fish stocks with organizations covering, among other areas, the South and Central Pacific, Northwest Atlantic, and Southeast Atlantic (**SEAFO**). Growing public concern over unsustainable exploitation of marine resources has led to efforts to strike broader agreements; a notable example is the **Agreement On Port State Measures To Prevent, Deter and Eliminate Illegal, Unreported and Unregulated Fishing**, established in 2009, although it has yet to enter into force.

In the past four decades, MEAs have increasingly been concerned with issues of pollution, environmental degradation, and the global commons. The UN Conference on the Human Environment, held in Stockholm in 1972, helped establish the **United Nations Environmental Program** (**UNEP**) and created a public awareness of the global nature of many of these problems. In 1983, the UN established the World Commission on Environment and Development (subsequently renamed the Bruntland Commission after its head) that issued the report "Our Common Future" in 1987 promoting the idea of sustainable development. That was followed by the **Earth Summit** in 1992 in Rio de Janiero, where countries committed to addressing three major global environmental issues: climate change (the **United Nations Framework Convention on Climate Change**); biodiversity (the **Convention for the Protection of Biodiversity**); and desertification (the **United Nations Convention to Combat Desertification**).

Countries have used MEAs to address shared environmental problems such as the **long-range transport of pollutants**. These are pollutants that can travel significant distances before deposition and can cause air pollution, marine pollution, or water pollution. One of the best known examples of such a problem is acid rain in both Eastern North America and in Scandinavia, which is caused by industrial activity elsewhere. In the case of Eastern Canada and the United States, the **acid rain** is caused by the emission of coal-fired power plants in the US Midwest that create sulfur oxides by burning coal. Prevailing weather patterns carry the emissions several hundred miles to the east and north, where the sulfur compounds react and eventually become deposited along with precipitation, making the rain highly acidic in places where it can have a significant effect on both freshwater lakes and forest ecosystems. More broadly, because these harmful activities often take place outside of the countries or regional jurisdiction where the costs are incurred, various governance authorities need to reach an agreement to help solve the problem. This problem has been addressed on a regional basis in both North America and in Europe, where countries have undertaken steps to identify and reduce the amount of emissions through the installation of scrubbers and other equipment, and utilizing alternative fuels.

Another example of international action to address pollutants is the **Basel Convention** (1989). It was convened to address the issue of hazardous wastes and the perceived threat from illegal shipments and improper disposal of these wastes. One concern that has been raised is that some countries could become **pollution havens**, such that political entities would set lower environmental standards and thereby become locations where toxic industries would locate, as well as receiving toxic waste from other countries. The objectives of the agreement are to limit the transboundary shipment between countries and to create a means of defining and tracking the movement of such wastes, as well as allowing countries to refuse to accept these wastes.

Perhaps the most successful example of international action to address a global environmental problem is the **Montreal Protocol** (1987), which was designed to address the issue of ozone-depleting substances. Chlorofluorocarbon compounds (**CFCs**) were widely used cooling substances because they are inert, non-toxic, and inexpensive to produce. However, it has been shown that the breakdown of CFCs in the atmosphere is causing a thinning in the **ozone layer** of the Earth's atmosphere. This received widespread publicity with the discovery of a "hole" in the **ozone layer** over the Antarctic in 1985. The ozone layer reduces the amount of **ultraviolet radiation** that reaches the Earth's surface, and thinning of this layer increases the amount of ultraviolet radiation reaching the surface, with negative effects for all life on Earth. The ozone layer is a classic example of a public good; all countries benefit from its presence. The Montreal Protocol specified a series of reductions in the use of CFC compounds with an eventual ban upon the use of such substances. The Protocol is widely acknowledged as leading to a sharp reduction in the production and use of CFC compounds in recent years, and a corresponding increase in the ozone layer.

It has been more difficult to establish agreements around the global commons, notably biodiversity and climate change. People have placed increasing value upon preserving biodiversity because of the possible richness of the genetic material embodied in wild populations, as well as the inherent value given to maintaining wildlife. One of the best known agreements is the **Convention on International Trade in Endangered Species**

(**CITES**) (1973). This agreement is designed, through the use of permits, to regulate trade in both animals and plants that are considered endangered and, if needed, to impose export bans based upon the status of the particular species (which are listed in various appendices depending upon the degree to which they are considered threatened). While CITES has been relatively successful, especially in protecting certain CITES listed species, such as crocodilians, there is increasing concern that biodiversity continues to be lost. This point has been underscored by the most recent **Millennium Ecosystem Assessment**, which issued a report examining the state and trends in ecosystems around the world in 2005 and found biodiversity in many regions facing severe threats. The **Convention for the Protection of Biodiversity (CPB)** (1992), initiated at the Earth Summit, has tried to develop a framework for providing a mechanism by which biodiversity can be maintained in member countries. The convention envisages a mechanism by which wealthier countries can fund efforts to preserve biodiversity in less developed countries, and establishes the rights to the ownership of genetic material found within countries to create an economic incentive for sustaining biodiversity. However, to date, these efforts have failed to reverse the downward trend in biodiversity, especially within the developing world.

As outlined in the preceding section, the **United Nations Framework Convention on Climate Change (UNFCCC)**, signed by over 160 countries at the 1992 Earth Summit in Rio de Janeiro and coming into force in 1994 following ratification by over 50 nations, is perhaps the most high-profile international agreement today. Under the agreement, Annex I countries (more **developed countries**) committed themselves to return their emissions of **greenhouse gases** to 1990 levels by the year 2000. This commitment was modified by the **Kyoto Protocol** in 1997. A new round of negotiations is underway to determine the successor agreement to the Kyoto Protocol, which expires in 2012. It was hoped that the **Copenhagen Conference** of 2009 would result in a binding global climate change agreement; however, negotiations were unsuccessful. The ultimate objective of the UNFCCC is to stabilize greenhouse gas concentrations at a level such that there is no detrimental impact upon the global climate. As can be expected, this is the classic **public goods** problem – countries would benefit from the actions taken to reduce emissions regardless of where they take place. Many **developing countries** have argued that the more developed countries should take the lead in reducing emissions and that it is unfair to expect them to forgo development opportunities (through the use of inexpensive fossil fuels like coal) that were available to developed countries in their earlier stages of development. Developed countries argue among themselves how best to measure credits and debits for **greenhouse gas** emissions because different methods of calculation can make a significant difference to countries in terms of the greenhouse gas reductions they need to make. Even so, progress vis-à-vis carbon reduction policies and technology have recently been made.

The **Copenhagen Accord** stipulated the creation of a $100 billion per year fund by 2020 to help developing countries mitigate climate change. There have also been pushes for further research and adoption of **carbon capture and storage (CCS)** and **reducing emissions from deforestation and forest degradation (REDD)**; however, countries are still reluctant to commit to more meaningful targets as CO_2 levels continue to rise, let alone to funding any significant international efforts to address climate change.

2. TRADE AND THE ENVIRONMENT

At the same time that an increasing number of **MEA**s are being pursued, either internationally or on a regional basis, countries are also engaged in similar efforts to boost trade among themselves. As in the case of MEAs, such efforts have taken one of two forms: the first has been through either bilateral or regional trade pacts such as the **North American Free Trade Agreement (NAFTA)** – a trade agreement between Canada, the United States and Mexico – and other agreements such as the **European Union**; while the second has been a sustained effort to establish an agreement governing international trade more generally. The first outcome of this effort was **GATT (General Agreement on Tariffs and Trade)**. **GATT** was formed after the Second World War, and established a framework and series of rules for trade between signatory countries through various rounds (meetings held between member countries over the past 50 years designed to further rules within a particular area). The **Uruguay Round**, which ended with the **Marrakesh Agreement** in 1994, provided the basis for the **World Trade Organization (WTO)**, an organization designed to enforce the trade rules agreed to by member countries through adjudicating trade disputes between member countries and enforcing any judgments.

These efforts to create rules to improve trade between countries have become increasingly contentious because of a perceived conflict between the wish to liberalize trade by reducing the amount of rules and government discretion over trade, and the need to take into account environmental considerations. While **Article XX** of GATT does allow for countries to make exceptions in terms of trade for goods that pose a risk to human, animal, and plant life, or if conservation measures are required for the protection of exhaustible resources, it forbids such exceptions where they are used as a pretext to create trade barriers.

The best known example of a conflict between environmental and trade issues is the **Tuna–Dolphin dispute** between the United States and Mexico. The United States had passed a law prohibiting the importation of tuna from countries where the incidental kill of dolphins exceeded a certain amount (dolphins follow tuna, and certain catching methods, such as the purse seine net, can trap the dolphins along with the tuna). A US environmental group sued to enforce the law, arguing that the nets used by Mexican fishers were killing too many dolphins, and a US federal judge agreed, banning tuna imports from Mexico. Mexico brought the dispute to the GATT panel (a special body struck to hear disputes) and the GATT panel in 1991 found that the ban contravened GATT rules (although it allowed a labeling scheme that identified dolphin-friendly catching systems to stand). While the decision was never formally adopted by GATT, because the United States and Mexico reached an agreement, the case has been used as an example where trade rules have supplanted a government's ability to make environmental policy. This concern has carried over into the WTO, which has created the **Committee on Trade and the Environment (CTE)**, which is trying to recognize environmental concerns without creating a series of exceptions that make any trade rules meaningless.

At the same time, due to concerns raised about the ability of countries to make environmental policy given international trade rules, an increasing number of **MEA**s are also incorporating trade measures into their agreements. Examples of such MEAs include **CITES**, the **Montreal Protocol**, and the **Basel Convention** (all described above). In addition, agreements such as the **United Nations Framework Convention on Climate Change** envisage market-based mechanisms as one possible way to address environmental problems.

A variety of trade measures can be used to further environmental objectives. These measures include information-gathering measures such as labeling, reporting, and the requirement for notification and prior informed consent (for receiving toxic or potentially hazardous goods). More stringent trade measures may include the requirement of either export and import permits or licenses, as well as selective import or export bans which may be between parties to an agreement, as well as trade bans with countries that are not party to the agreement. These trade measures serve various purposes, but the most important objective is to make sure that all of the environmental factors are known and presumably incorporated into the exchange when trade takes place. Other objectives may include: collecting and disseminating environmental information; monitoring certain environmental factors; and the prevention of trade diversion and industrial relocation (where stricter environmental measures simply serve to redirect production or pollution to different locations rather than leading to a change in behavior and reduction in pollution).

The effectiveness of **MEA**s depends both upon the full participation of countries that are affected by, or contribute to, the environmental problem, as well as compliance with the agreement. The absence of a critical country from, or significant non-compliance with, an agreement may mean that the environmental problem only worsens. Given the interest in expanding trade found in most countries today, the ability to enter into agreements that facilitate trade can act as a strong inducement for countries that might have reservations about the costs associated with addressing environmental issues, as well as any possible diminution in government sovereignty. In addition, by setting environmental standards and policies within these different agreements, there is evidence that other countries, even if unwilling to sign formal agreements, may still adopt similar standards and policies. Finally, to the extent that concerns about environmental issues are associated with higher income levels, increased international trade may help address environmental issues through improving people's welfare.

In recent years, these efforts to resolve international environmental problems have broadened beyond states working exclusively in formal bilateral processes, and now include **non-governmental organizations (NGOs)** and businesses in efforts to promote and identify more sustainable business practices and products. These activities focus on creating codes of conduct, developing eco-labeling, and promulgating principles that take into account not only the environmental, but also social, aspects of producing and selling goods and services, both domestically and internationally. These efforts reflect some of the frustration among NGOs with the slow pace of international processes in addressing some of the more pressing environmental issues. There are a number of such initiatives under way, ranging from agreements identifying lending criteria (the **Equator Principles**), the emission of greenhouse gases (the **Carbon Disclosure Project**), certifying sustainable forest management (such as **Sustainable Forestry Initiative (SFI)**, **Forestry Stewardship Council (FSC)**, and **Programme for the Endorsement of Forest Certification (PEFC)**, sustainable fisheries and other traded goods. In some cases, these efforts may involve national governments in developing national schemes or have even been sanctioned by the UN. Examples of the latter include the UN working with multinational corporations and leading NGOs to develop the **United Nations Global Compact**, elaborating standards of socially responsible corporate behavior, as well as the **United Nations Principles of Responsible Investment (UNPRI)**. Overall, these efforts are likely to continue and expand, providing not only an alternative system of governance (otherwise known as NSMD or non-state market driven approach) while states continue to work to overcome the challenges of fashioning international agreements, but also helping provide an understanding of how to effectively approach these complex issues and providing the basis on which countries may ultimately be able to address problems such as global warming.

Environmental Systems, Dynamics, and Modeling: A Primer

1. ENVIRONMENTAL SYSTEMS

Our **environment** is a product of many different plants and animals, shaped by the past, the climate, the physical landscape and, of course, human activity. As **species adapt** to circumstances, they also influence their **habitats** and reshape the **environments** in which they live. **Ecosystems** (and our planet) are also subject to **random shocks** that can radically change the course of life. Indeed, the universal constant of life is change. The mass deaths that led to the extinction of the dinosaurs almost certainly contributed to **mammal speciation** and an increase in **mammal** size. Thus, if the **Alvarez hypothesis** is correct, and if a meteor had not hit the Earth 65 million years ago at the end of the Cretaceous period, it is extremely unlikely that modern humans would exist and that you would be reading this book!

One of the most important processes by which life changes over time is **natural selection**. This simply means that those members of a **species** whom are fortunate enough to **reproduce** will bequeath their genetic material to the next generation. Members of a **species** that by chance, or physical characteristics, or behavior are less successful at reproducing, will be less favored in the **gene pool** in the subsequent generation. In this manner, life evolves over time. The process of **evolution** involves many different paths and interactions with **predators** and **prey**, and with **parasites** and **hosts** co-evolving. For example, slower deer are more likely to be caught by, say, wolves. As a result, slower deer are less likely to be successful in reproducing and thus, over time, they are eliminated and the **species** becomes faster at running. Similarly, wolves that are less able to cooperate in their hunting may be less successful at catching game and, thus, over time wolves that cooperate (assuming that cooperation is an inherited characteristic) reproduce more and dominate their **species gene pool**. **Co-evolution** can also be in the form of **symbiosis** or **mutualism**, such as with flowering plants that provide food (in the form of nectar) for **insects** that, in turn, increase the reproductive success of the plants that attract such **insects**.

The greater the period of time, all other things equal, the greater will be the effect of **evolution** on life. Nevertheless, even in very short time periods (on a geological timescale), **species** can evolve rapidly. For example, the widespread use (or rather misuse) of **antibiotics**, particularly in the livestock industry where animals are given low doses of antibiotics to increase weight gain, has, in a short period of time, led to antibiotic-resistant **bacteria**. Rapid changes may also occur in multicellular **species**. A classic example is the peppered moth found in the United Kingdom that, prior to the **Industrial Revolution**, had light-colored wings and rested on **lichen** which helped camouflage them from predators, such as birds. Increased air **pollution** killed off much of the lichen on which they rested and the lighter-colored moths became conspicuous to birds, reducing their breeding success. In turn, in a few decades, the **species**, as a whole, evolved to have much darker colored wings that were less noticeable to predators in the black background of industrial Britain.

If the golden rule of ecosystems is change then evolution's maxim is **biological diversity** – the genetic variation that exists both across **species** and among individuals within **species**. Life on Earth started at least 3.5 billion years ago and began with **bacteria** that still remain, in terms of their total biomass weight, the most successful life forms on our planet. By contrast, modern humans are the latest gatecrashers to a party that has been going on for almost 4 billion years! Beginning in the **Cambrian period**, some 570 million years ago, the first multicellular organisms appeared, ultimately leading to the plants and animals we see today. Almost all the life on Earth (**bacteria** deep in the Earth's surface may be an exception) directly or indirectly survive from the energy that reaches us from the sun. Indeed, organisms that can **photosynthesize** (or take energy from sunlight) that first evolved some 2 billion years ago, represent a fundamental turning point of life on Earth. As important is the sun to the evolution of life, so has been the spin of the Earth and its tilt (that gives us seasons) and the interaction of our moon that collectively help generate the energy, mixing and turbulence in the **atmosphere** and **oceans** that help sustain life.

Along the many paths of evolution, **species** evolved into other **species** such that their archaic forms passed away. In addition, many **species** became extinct in the sense that they left no descendants. Indeed, the ratio of extinct to living **species** is likely to be at least 1000 to 1. In other words, if the Earth has between 10~100 million **species** alive today, between 10 thousand million and 100 thousand million **species** that previously existed are now extinct. Given the genetic variation that exists within individual members of a **species**, the diversity of life on our planet since it first began is truly staggering.

If the species alive today are just 0.1 percent of all the **species** that have ever lived it begs the question, what strategies can individuals and **species** use to ensure their survival? The most obvious mechanism for survival is to produce many offspring. Indeed, the larger the number of individuals, all other things equal, the greater is the chance a **species** has of surviving into the future. Animals that use this approach are called **r-strategists** and include most **species**. An alternative strategy is to focus less on the number of offspring and instead to increase the **probability** of offspring successfully reproducing. **Species** that follow this approach, like humans, are called **k-strategists**. These strategies, however necessary, are insufficient to ensure the long-term survival of **species**. This is because any **environment** is subject to perturbations and shocks that can lead to mass deaths. For example, life on an island could be destroyed by a volcanic eruption, or many of the **species** on an entire continent could be obliterated by a meteor, or the Earth's climate could suddenly become much colder leading to mass deaths (events that have all happened more than once on our planet). Unless members of a **species** can **acclimatize** and **adapt** quickly, or the species can evolve to adjust to the changed circumstances, the **species** will become **extinct**. Thus, to ensure **resilience** to local or regional shocks or events, **species** need to have **meta-populations** or to be widely distributed spatially. In other words, **species** (no matter how numerous) that are only found in geographically limited locales are at particular risk of **extinction**.

Resilience is a term that can be applied equally well to **ecosystems** as to **species**. Interestingly, some studies suggest that increased diversity of **species** (at least to a certain point) can result in a greater **biomass** from ecosystems and may foster **stability** in the sense that plant and animal communities can "bounce back" from shocks and perturbations. **Resilience** is, in part, determined by **population dynamics**. To a greater or lesser extent, species are regulated by their **population density**. Ultimately, since no population can increase indefinitely, the total population growth rate must eventually decline with **population density**. However, at low densities, increases in density may actually increase the rate of population growth. For example, at low densities some species are highly vulnerable to predation because to survive they need to aggregate, or school or herd together for protection (such as with schools of sardines or herds of zebra). Moreover, the rate of growth of populations is likely to depend on the density of other **species**, such as zebras and lions or deer and wolves, in what are called **predator–prey** relationships. These linkages across species, and with **abiotic** or non-biological factors, can lead to **non-linearities** in both growth and population numbers. For example, below a **critical point** in terms of numbers, a population may simply be unsustainable. Similarly, a change in the environment beyond a **threshold** or **critical point** (such as a decline in the **pH** level of a lake) can lead to mass deaths. The **system dynamics** of populations and their **environments** is extremely complicated and requires an understanding at the cellular level (such as **genetics**) and systems level (such as **ecology**). It also requires a comprehension of the many cycles and **feedbacks** on Earth that help maintain our ecosystems, such as the **water cycle** that refreshes and recirculates water on our planet. The feedbacks in cycles are illustrated with the **carbon cycle** (shown below) that represents the interaction between **organisms** that helps determine the concentration of **carbon dioxide** in the **atmosphere**.

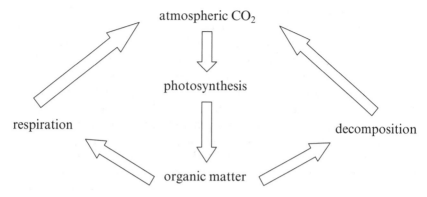

Figure 4   Simplified Model of the Carbon Cycle

At the ecosystem level, an understanding of the **food web** is required to appreciate the **feedbacks** and links across **species** and to understand how perturbations or shocks can travel throughout an **ecosystem** or **environment**. These links can be highly complex and may even be **chaotic** such that there exists neither an **equilibrium** nor **periodicity** in how populations change over time, and the future population numbers may be highly sensitive to past values.

How life interconnects is a subject of much debate. Some writers have gone so far as to view the world as one giant interconnected organism, as in the so-called **Gaia hypothesis**. By contrast, others view **ecosystems** as merely overlapping **habitats** of different **species**. Whatever the interpretation, there is no question that how **species** interconnect plays an important role in their survival and their ability to adjust to shocks. An important step to understanding these interactions is to model the relationships and linkages among the many factors that constitute our **environment**.

2.   MODELING AND DYNAMICS

Modeling is a process by which an abstract or simplification of the world around us, or events, is developed. **Models** can simply be "mental maps" or rules we develop about how the world operates. Unfortunately, the insights from models that exist in someone's head can be very difficult to impart to others. Moreover, although we are very good at expressing mental models in terms of cause and effect (I jump out of a window, I fall to the ground), we are often unable to develop mental models that reflect the **feedbacks** and interconnectedness that exist in even the simplest **systems**. Thus, to model an **ecosystem** or economic system, we need tools to help us visualize relationships and to quantify how the parts of the system affect each other. These tools include, but are not restricted to, mathematics, to help us be as explicit and as clear as possible about the nature of the **system**, and computers to help us make the many calculations required to quantify how a **system** changes over time.

Models are used for two principal purposes: **optimization** and **simulation**. **Optimization** models include an **objective function** that needs to be optimized (such as **social welfare** or species diversity) subject to a set of constraints (such as the resources available) using a set of **control** or **choice** **variables** (factors that can be changed, such as tax rate or the area of land in conservation areas). **Optimization** models are prescriptive and are thus part of **normative analysis** and help answer the questions of what *should* happen. **Simulation** models are used to represent a **system** or systems and, to be useful, must adequately reflect the relationships among the variables within the **system(s)**. Such models are often used for predictive purposes and help answer *what if* questions. For example, **general circulation models** in **climate modeling** help us answer the question what will be the Earth's **climate** in 50 years should we continue to emit **greenhouse gases** in a **business-as-usual** scenario?

Models are often criticized as not being an adequate representation of reality. No matter how complex the model, it must (inevitably) be a simplification of the **system** being modeled. The art and science of modeling, however, is *not* to include every possible variable or relationship (which is impossible) but to include those **variables** and links that are important or significant within the system. In this sense, what is "important" or "significant" will depend on the purpose of the **model** and availability of data. For instance, a **general circulation model** used for predicting surface temperatures a century hence must include the atmospheric concentrations of **greenhouse gases** in the **atmosphere**. By contrast, a **model** used for predicting whether it will rain in a particular region in the next three days can ignore the interactions between atmospheric **greenhouse gas** concentrations and the local **weather**. Choosing what should and what should not be included in a **model** (**variables** and **feedbacks**), and whether **variables** should be treated as **exogenous** (determined outside the model) or **endogenous** (determined within the model), are questions that all modelers face. Good models are able to represent the **system** parsimoniously, or in as simple a way as possible, while still being able to effectively answer the questions for which the **model** was built.

Given that the **environment**, the **economy**, and human activity are always changing, models need to be **dynamic** or inter-temporal in order to adequately reflect these systems. In other words, **variables** within the **model** should change over time. All dynamic models include **stocks** and **flows**.

Stocks are **state variables** and may be called reservoirs, levels or inventories, depending on what **system** is being modeled. They characterize "the state of the world" such as the **biomass** of fish in a fishery model, the volume of standing trees in a forestry model, or the surface temperature in a climate model. The value or level of a **stock** depends on the **initial conditions** or past values, and can change over time depending upon additions and subtractions from **flow variables**. For instance, recruitment represents a **flow** into a fishery as it increases the **stock** of fish, while harvesting of fish is a **flow** out of the fishery as it reduces the **stock**, as shown below.

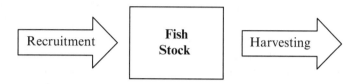

Figure 5a   Stock and Flows in a Fishery Model

Flows represent rates of change in a **system** and could be the **birth rate** or **death rate** in a population model, or the water used and discharged by a **pulp and paper mill** in a model of water use. Although stocks are changed by inflows and outflows, **stocks** may also influence the flow variables through feedbacks. For instance, the **death rate (flow)** of a **population** is likely to be influenced by the total size of the **population (stock)** such that there is a feedback in the model.

Stocks that are uniquely determined in the model by flows and feedbacks are **state variables** of a system. A **model** may also include auxiliary **variables** that are functions of both **stocks** and **flows**. For instance, the recruitment into a fishery is itself influenced by a **birth rate** that is a function of both the total population (**state-determined stock**) and environmental factors (such as ocean temperature) that may be treated as **exogenous** to the **model**, as shown below.

Modeling requires tests of the **robustness** and the effect of changes in a **model**, and its assumptions, in terms of results. These tests should include running or simulating the model under a variety of scenarios such as different **initial conditions, parameter** values or constants. Further, a model's predictions can sometimes (where the data is available) be compared to past values and used as a possible indicator of performance for predictions or forecasts. However, just as **correlation** between variables does not imply cause and effect, neither does past performance of a model ensure accurate predictions. In addition, a **model** should be tested to explore the

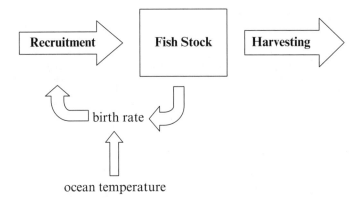

Figure 5b   Fishery Model with Exogenous Variables

effect of changes in the **feedbacks** and interrelationships of the **system**. By performing such tests and **sensitivity analysis**, researchers are better able to understand the limitations of their **models** and, thus, to comprehend better what questions the **models** can help answer and what questions they cannot resolve.

An understanding of our environment, and how to model it, is the basis of scientific research. Clearly, neither this primer nor this dictionary can do justice to the myriad of topics that encompass the many disciplines needed to understand the world around us. Readers interested in understanding more about the environment and environmental economics should, as a first step, use the references, words and appendices in this dictionary as tools in their journey of learning.

Annotated References: A Starting Point

Below is a list of annotated references that will provide a useful introduction to some of the topics covered within this dictionary. These books, in general, are not seminal works or classic texts within their disciplines, but provide a useful introduction without demanding a special expertise in the area.

The texts are grouped into nine headings, under three main areas. The first area covers the biological sciences; the second, the physical sciences; and the third, economics and modeling as it applies to environmental and natural resource issues, as well as international agreements. In addition to the texts listed below, a detailed list of references is supplied at the back of the dictionary.

1. BIOLOGY

a. General
Campbell, N.A., J.B. Reece, L.A. Urry, M.L. Cain, S.A. Wasserman, P.V. Minorsky and R.B. Jackson (2010), *Biology*, 9th edn, San Francisco: Benjamin Cummings.

An introduction to biology and its application in understanding how life exists on Earth.

b. Cellular/genetic
Griffiths, A.J.F., S.R. Wessler, R.C. Lewontin, W.D. Fixsen and S.B. Carroll (2008), *Introduction to Genetic Analysis*, 9th edn, New York: W.H. Freeman.

A text that introduces the reader to genetic theory and cellular development.

c. Evolutionary
Freeman, S. and J.C. Herron (2007), *Evolutionary Analysis*, 4th edn, Upper Saddle River, New Jersey: Prentice Hall.

An undergraduate text that examines the concepts and ideas underlying evolution.

d. Wildlife
Sinclair, A.R.E., J.M. Fryxell and G. Caughley (2006), *Wildlife Ecology, Conservation and Management*, Malden, Massachusetts: Blackwell.

Provides a holistic view of wildlife management, with a particular emphasis on the underlying ecological systems.

e.   Fisheries

King, M. (2007), *Fisheries Biology, Assessment and Management*, Malden, Massachusetts: Blackwell.

An introduction to fisheries conservation and fisheries management, including the biological systems involving fish and their habitats.

2.   BOTANY

a.   General

Taiz, L. and E. Zeigler (2010), *Plant Physiology*, 5th edn, Sunderland, Massachusetts: Sinauer Associates.

An introduction to understanding how plants function.

3.   FISHERIES, FORESTRY AND WATER

a.   Fisheries

Grafton, R.Q., R. Hilborn, D. Squires, M. Tait and M. Williams (2010), *Handbook of Marine Fisheries Conservation and Management*, New York: Oxford University Press.

The most comprehensive and interdisciplinary work on marine conservation and fisheries management ever compiled.

b.   Forestry

Kimmins, J.P. (2004), *Forest Ecology: A Foundation for Sustainable Management*, 3rd edn, Upper Saddle River, New Jersey: Prentice Hall.

A well written book examining the economic and biological issues involved in forest management.

Perry, D., R. Oren and S. Hart (2008), *Forest Ecosystems,* Baltimore, Maryland: Johns Hopkins Press.

A comprehensive overview of forest ecosystems and related issues.

Smith, D.M., B.C. Larson, M.J. Kelty, P. Mark and S. Ashton (1997), *The Practice of Silviculture: Applied Forest Ecology*, New York: Wiley.

The techniques and application of silvicultural practices in North America.

c. Water

Grafton, R.Q. and K. Hussey (eds) (2011), *Water Resources Planning and Management,* Cambridge: Cambridge University Press.

A comprehensive and up-to-date volume of 35 chapters, including foundational chapters (such as the water cycle), water management chapters (such as adaptive water management), and case studies.

4. ECOLOGY

a. General

Krebs, C.M. (2009), *Ecology*, San Francisco: Benjamin Cummings.

An introductory text to the framework used in ecological analysis.

Smith, R.L. and T.M. Smith (2008), *Elements of Ecology*, 7th edn, San Francisco: Benjamin Cummings.

A book that examines environmental issues today from a perspective of ecological principles.

5. GEOLOGY

a. General

Lutgens, F.K., E.J Tarbuck and D. Tasa (2011), *Essentials of Geology*, 11th edn, Upper Saddle River, New Jersey: Prentice Hall.

An introductory text to the basic principles underlying geological processes.

b. Hydrology

Fetter, C.W. (2001), *Applied Hydrogeology*, Upper Saddle River, New Jersey: Prentice Hall.

A more advanced text that outlines the basic models and science that are used in hydrology.

c.   Geomorphology

Bloom, A.L. (2004), *Geomorphology: A Systematic Analysis of Late Cenozoic Landforms*, 3rd edn, Long Grove, Illinois: Waveland Press.

An introductory text that covers geological processes and the resulting landforms we see around us.

6.   CLIMATE AND METEOROLOGY

a.   General

Ahrens, C.D. (2007), *Meteorology Today: An Introduction to Weather, Climate, and the Environment*, 8th edn, Belmont, California: Thomson.

An introductory text to the basic concepts of meteorology and cycles underlying weather patterns.

Barry, R.G. and R.J. Chorley (2003), *Atmosphere, Weather and Climate*, 8th edn, London: Routledge.

A highly informative text covering all principles of weather and climate.

b.   Climate change

Archer, D and S. Rahmstorf (2010), *The Climate Crisis: An Introductory Guide to Climate Change*, Cambridge: Cambridge University Press.

A highly accessible summary of climate change science, future impacts and mitigation.

Houghton, J. (2009), *Global Warming: The Complete Briefing*, 4th edn, Cambridge: Cambridge University Press.

Possibly the most readable and informative guide to understanding climate change science.

Ruddiman, W.F. (2007), *Earth's Climate: Past and Future*, 2nd edn, New York: W.H. Freeman.

The definitive scientific textbook introducing the dynamics and variation of climate across time.

c. Climate Change Policy

Garnaut, R. (2008), *The Garnaut Climate Change Review*, Melbourne, Australia: Cambridge University Press.

A comprehensive guide to climate change policy in Australia, with a strong emphasis on the international case for action on climate change.

Stern, N. (2007), *The Economics of Climate Change: The Stern Review*, Cambridge: Cambridge University Press.

The groundbreaking treatise on the economic case for climate change mitigation.

7. ECONOMICS

a. International Economics (Including Trade)

Krugman, P., M. Obstfeld and M. Melitz (2010), *International Economics: Theory and Policy*, 9th edn, Upper Saddle River, New Jersey: Prentice Hall.

An introductory text that uses examples to examine the issues involved in international economics, including trade and finance.

b. Introductory Economics

Mankiw, N.G. (2009), *Principles of Economics*, 5th edn, Mason, Ohio: Cengage Learning.

An introductory text illustrating core principles with practical examples.

Katz, M.L. and H.S. Rosen (1998), *Microeconomics*, 3rd edn, Boston, Massachusetts: Irwin McGraw-Hill.

An excellent introduction to the important topics in microeconomics. The text includes a chapter on game theory and a chapter on externalities and public goods.

c. Non-market Valuation

Garrod, G. and K.G. Willis (1999), *Economic Valuation of the Environment: Methods and Case Studies*, Northampton, Massachusetts: Edward Elgar.

A comprehensive overview of the main methods of pricing non-market goods and the application of the techniques in case studies that include amenity values, biodiversity, and water quality.

d. Econometrics
Kennedy, P. (2003), *A Guide to Econometrics*, 5th edn, Cambridge, Massachusetts: MIT Press.

An introduction to econometric techniques that is very readable, covers the main types of estimation techniques and addresses the most common problems encountered in empirical work.

Studenmund, A.H. (2010), *Using Econometrics: Practical Guide,* 6th edn, Upper Saddle River, New Jersey: Prentice Hall.

A good beginner's introduction to econometrics.

e. Natural Resource Economics
Hartwick, J.M. and N. Olewiler (1998), *The Economics of Natural Resource Use*, Reading, Massachusetts: Addison-Wesley Longman.

A comprehensive examination of the economics underlying the development and use of natural resources, including case studies.

Field, B.C. (2008), *Natural Resource Economics: An Introduction*, 2nd edn, Long Grove, Illinois: Waveland Press.

An excellent introduction to natural resource economics that includes many examples.

f. Environmental Economics
Tietenberg, T. and L. Lewis (2009), *Environmental Economics and Policy*, 6th edn, Upper Saddle River, New Jersey: Prentice Hall.

A very readable introduction to the economics of the environment, with many examples.

Grafton, R.Q., W. Adamowicz, D. Dupont, H. Nelson, R.J. Hill and S. Renzetti (2004), *The Economics of the Environment and Natural Resources*, Malden, Massachusetts: Blackwell Publishing.

A higher level text covering the essential topics in a detailed and concise manner.

g.   Game Theory
Bierman, H.S. and L. Fernandez (1993), *Game Theory with Economic Applications*, Reading, Massachusetts: Addison-Wesley Publishing.

A gentle introduction to game theory that includes many useful examples from economics.

h.   International Development
Perkins, D.H., S.C. Radelet and D.L. Lindauer (2006), *Economics of Development*, 6th edn, New York: W.W. Norton.

A useful book that examines the issues involved in economic development and includes many examples.

i.   Cost–Benefit Analysis
Hanley, N. and E.B. Barbier (2009), *Pricing Nature: Cost–Benefit Analysis and Environmental Policy*, Cheltenham, UK: Edward Elgar.

A thorough review of cost–benefit analysis and its application to the environment.

j.   Forest Economics
Amacher, G., M. Ollikainen and E. Koskela (2009), *Economics of Forest Resources,* Cambridge, MA: MIT Press.

The most current review of developments in the field as well as a comprehensive review of theory.

8.   SYSTEMS DYNAMICS

a.   Environmental Modeling
Ford, A. (2009), *Modeling the Environment: An Introduction to System Dynamics Modeling of Environmental Systems*, 2nd edn, Washington DC: Island Press.

A good introduction to system modeling of the environment, with examples.

9.   INTERNATIONAL AGREEMENTS AND PROCESSES

Bodansky, D. (2009), *The Art and Craft of International Law*, Cambridge, Massachusetts: Harvard University Press.

Goldstein, J., M. Kahler, R. Keohane and A.M. Salughter (2001), *Legalization and World Politics*, Cambridge, Massachusetts: MIT Press.

Victor, D., Raustiala, K. and E. Skolinoff (1998), *The Implementation and Effectiveness of International Environmental Commitments: Theory and Practice*, Cambridge, Massachusetts: MIT Press.

Internet Resources: Environmental News Sources and Blogs

The growth of the internet has greatly increased our capacity to share information concerning environmental issues. However, a great many websites are dedicated to this purpose and one can easily be overwhelmed with the sheer volume of information available. With the aim of assisting the reader to access the best information in the most efficient manner, we list below a number of sites that we find helpful. They are quality resources for keeping up to date with contemporary environmental news, issues, and opinions. We list them under three categories: general sites that deal with a broad range of issues, those focusing on climate change, and other issue specific sites. They appear in alphabetical order within either category to avoid the appearance of recommending or ranking individual sites. This catalogue is not intended to be comprehensive and, in many cases, the online resources we mention below are linked to other useful sites.

1. GENERAL ENVIRONMENT

Environmental News Network (ENN)　(www.enn.com)

One of the longest running online environmental news services. ENN publishes original articles and reproduces items from other sources. Also provides a daily email newsletter containing links to new stories.

Yale Environment 360　(e360.yale.edu)

An online publication of the Yale School of Forestry and Environmental Studies. Features opinion and analysis of environmental issues by academics, journalists, and policy makers.

2. CLIMATE CHANGE

Climate Progress　(climateprogress.org)

Widely acknowledged as one of the most influential climate change blogs. Renowned for a high level of activity, it hosts multiple daily posts concerning both policy and science.

RealClimate   (www.realclimate.org)

A blog site run by working climate scientists. Intended as a forum to explain the science and provide commentary on emerging developments.

SolveClimate News   (solveclimate.com)

High volume source of daily news and commentary on climate policy and science.

3.   ISSUE SPECIFC

China Dialogue   (www.chinadialogue.net)

A bilingual site that provides articles, links, and opinions concerning environment issues in China and the rest of the world.

Circle of Blue – WaterNews   (www.circleofblue.org/waternews/)

The news and analysis arm of an international network of scientists and journalists working on global freshwater issues.

Global Water Forum   (www.globalwaterforum.org)

A blog featuring articles authored by the world's leading water researchers and practitioners. This site acts as a multidisciplinary forum for sharing ideas about local, regional, and global water governance issues.

GreenBiz.com   (www.greenbiz.com)

News and opinion site looking at environmental issues from a business perspective.

Mongabay.com   (www.mongabay.com)

A highly subscribed news and information site focusing mainly on rainforest and wildlife conservation.

SciDev.Net   (www.scidev.net)

An online source of news and opinion on science and technology in developing countries, with a strong emphasis on the environment.

A

abatement. A method or process that controls or reduces discharge and emission of pollutants. Often used in reference to the abatement of greenhouse gas emissions. See *climate change mitigation*.

ability-to-pay principle. The notion that publicly provided goods and services should be financed on the basis of what people are able to pay. See *willingness to pay*.

abiotic factors. Non-living components of the natural environment.

ablation. The combination of processes that remove snow and ice from a glacier or ice sheet.

about as likely as not. See *likelihood*.

abrupt climate change. Sudden adjustments or major events originating from the nonlinear features of the climate system. See *climate feedback* and *climate threshold*.

absolute advantage. The notion that an individual, firm or country is able to produce more of or a better product using identical quantities of inputs than others. See *comparative advantage (theory of)*.

absolute emissions. Term for the total volume of greenhouse gas emissions by an entity, such as a state or business. A common metric for national emissions, including under the Kyoto Protocol.

absolute zero. The lowest possible temperature defined as $0°$ Kelvin or $-273°C$.

absorption. 1. The process whereby a substance is assimilated and retained by another. 2. The process whereby radiant energy is retained by a substance, such as the absorption of solar radiation by the Earth's landmasses and oceans.

absorptive capacity. The maximum amount that can be absorbed by a substance, such as the absorptive capacity of soil for water.

abundance.   A measure of the number of individuals, or groups of individuals, in a given population.

abyssal.   A zone under the ocean where there is no sunlight and that is commonly defined as being deeper than 2000 meters.

acceleration principle.   A hypothesis that investment varies with the rate of change in output in an economy.

access right.   A property right that entitles the holder of the right access or entry to a natural resource or environmental asset, such as an entry permit to a National Park.

acclimatization.   The physiological and behavioral, but not genetic, adaptations of flora and fauna due to changes in the environment. See *adaptation.*

accuracy.   Precision of a measurement, computation, estimate, or a prediction to actual quantities and observed facts.

acidic.   A substance which has a low pH value. See *pH.*

acidification.   The process by which a substance, such as water is rendered acidic.

acid precipitation.   Precipitation in the form of snow and rain that has a pH of less than 5.0. See *acid rain.*

acid rain.   Rainfall that is acidic and which has a pH level less than 5.0. The pH level of acid rain falling in eastern North America is 4.0 to 4.5. Acid rain can occur naturally but the principal cause of acid rain is the anthropogenic emissions of nitrogen oxides (NO_x) and sulfur dioxide (SO_2). Acid rain is a major problem in eastern Canada, the north-east United States, Scotland and Scandinavia.

acid shock.   The biological shock that arises from the rapid acidification of water systems, such as might arise from spring runoff.

Acquired Immune Deficiency Syndrome (AIDS).   An inevitable consequence of contracting the human immunodeficiency virus (HIV), in which the immune system is weakened and eventually fails due to opportunistic infections. There is currently no cure; however, antiret-

roviral drugs are used to prolong life. The disease has its most devastating effects in Africa, where infection rates are high and control
mechanisms are unavailable or poorly understood. See *human immunodeficiency virus*.

acre. A British measure of surface area equal to 4840 square yards or
0.407 hectare (247.11 acres equals 1 square kilometer).

acre-foot. Amount of water needed to cover one acre of land to the depth
of 1 foot or 0.3048 meter.

activated carbon. Carbon treated to increase its adsorption and which is
used to remove impurities from water, air and solids. See *adsorption*.

activated sludge. Sludge or sewage effluent that contains a range of
micro-organisms that are used to help break down solids in the secondary
treatment of sewage.

active gear. Fishing or harvesting gear that must be moved through the
water to be able to catch fish. An example of active gear is a trawl which is
towed behind or alongside fishing vessels to catch those animals which are
too large to swim through the net. See *passive gear*.

active layer. The layer of soil above permafrost that is subject to freezing
and thawing on a seasonal basis.

active value. See *use value*.

activity analysis. A method of analysis used to find a maximum or
minimum subject to a set of linear constraints. See *linear programming*.

acute toxicity. A condition in which an organism's health is adversely
affected following short-term exposure to a toxic substance.

adaptability. The extent to which a population or species can adjust to
changes or shocks in the environment. See *resilience* and *adaptation*.

adaptation. The process of genetic change of flora and fauna due to
changes in the environment. This term has also become widely used in
the context of reducing the vulnerability of human societies and natural
systems to future climate change. See *acclimatization* and *climate change
adaptation*.

adaptation benefits. Avoided damages from climate change obtained through implementing adaptation projects. See *climate change adaptation*.

adaptation costs. The costs associated with planning and employing climate change adaptation measures.

Adaptation Fund. A central finance facility established under the Kyoto Protocol to fund adaptation projects and programs in the developing countries most vulnerable to climate change.

adaptive expectations. The hypothesis that people base their expectations of the future on the basis of past outcomes. Depending upon the nature of the adaptive expectations, it can imply that people make systematic errors in their predictions. See *rational expectations*.

adaptive management. A term used to describe approaches to the management of natural resources which explicitly recognize uncertainty. Actively adaptive management requires:
1. that managers develop alternative models about how natural populations change over time;
2. that experiments (where appropriate) are used to provide more information about the relative value of the models; and
3. that these models are used to compare different policies and simulate the effects of shocks to the environment.

FURTHER READING
Walters and Hilborn (1976).

adaptive radiation. Term for the diversification and creation of different species from a common ancestor over time as each species adapts to a particular ecological niche.

ad hoc. Latin term meaning "for this purpose".

ad infinitum. Latin term meaning "without end".

ad libitum (ad lib). Latin term commonly meaning "undertaken without preparation".

ad nauseam. Latin term meaning "to the point of being sick".

ad valorem tax. A tax calculated as a proportion of a defined value, such as a 5 percent charge on the value of timber sold.

additionality. A term defining whether a project or policy directed at improving environment outcomes would not have been undertaken except for that specific purpose. Additionality is an important concept with regards to climate change mitigation because the availability of financial resources for these purposes can create moral hazard, whereby nations or firms have an incentive to claim external funding for activities that, for other reasons, they would be prepared to finance themselves. See *Clean Development Mechanism*, *moral hazard*, and *Reducing Emissions from Deforestation and Degradation*.

adiabatic process. A process that occurs without the transfer of heat across the system's boundaries.

adjustment time. A unit of measurement describing the timescale across which an external increase in a trace gas is removed from a reservoir, such as carbon dioxide in the atmosphere. See *lifetime* and *turnover time*.

adsorption. Formation of a layer of a solid, liquid or gas on the surface of a solid without penetrating the solid. Absorbents, such as activated carbon, are highly effective in removing undesirable pollutants from the water and air. See *activated carbon*.

advanced countries. So-called advanced countries are characterized by high levels of per capita income and standard of living indicators. Advanced countries would include Japan, United States, members of the European Union and countries in the Organization for Economic Co-operation and Development. See *developing countries*.

Advanced Metering Infrastructure (AMI). Metering technology that provides electricity consumers and suppliers with detailed information on electricity usage, thus facilitating environmentally and economically efficient decision making. See *smart grid*.

Advanced Spaceborne Thermal Emission and Reflection Radiometer (ASTER). An imaging instrument aboard a NASA satellite that provides scientific information pertaining to various branches of climate science.

advection. The transportation of an atmospheric property or seawater through mass motion. In meteorology, advection refers to mostly horizontal motions of the atmosphere, while in oceanography it may refer to horizontal or vertical flows in the form of currents.

adventure tourism.   Tourism in which visitors participate in activities that involve at least some degree of risk or danger that is perceived to be out of the ordinary. Adventure tourism usually involves participation in outdoor activities.

adverse selection.   A problem which may arise in a transaction when one party has imperfect information but the other party has full information. For example, in the second-hand car market sellers know whether their car is of high or low quality but this is difficult for prospective buyers to determine. As a result, buyers will pay the same price for high or low quality cars but this discourages sellers of high quality cars from selling. See *asymmetric information*.

FURTHER READING:
Akerlof (1970).

aerobic.   Organisms or processes that require oxygen to function.

aerosol.   The suspension of tiny solid or liquid particles dispersed in the atmosphere. Mostly of natural origin (e.g., sea salt, dust and volcanic ash), although may also result from human activity (e.g., fossil fuel and biomass combustion). Some aerosols have a cooling effect on the climate system through reflection of incoming solar radiation and acting as a catalyst for cloud formation. Others retain heat in the system through adsorption of radiation. Although the dynamics are very complicated, scientists believe that aerosols have an overall cooling effect upon the climate system. See *Asian Brown Cloud, black carbon, geoengineering,* and *global dimming*.

afforestation.   Term used to describe the conversion of land to forest, where the land has not contained forest before. See *deforestation* and *reforestation*.

aflatoxins.   Toxic substances that occur naturally in some molds and which are produced in conditions of relatively high humidity.

African Development Bank (AfDB).   Bank established in 1964 by member countries that is mandated to further the economic and social development of nations in Africa.

age distribution. The distribution of individuals in a population according to their age.

Agenda 21. A document that outlines policies to help achieve sustainable development and which was released at the 1992 Rio de Janeiro Earth Summit. See *United Nations Conference on Environment and Development.*

agent-based modeling. Computational models that simulate the interaction of autonomous decision makers and assesses system outcomes.

Agent Orange. A herbicide which has the toxic chemical dioxin as one of its active ingredients, a carcinogen that is suspected of causing birth mutations, skin diseases (such as chloracne) and problems with the immune system. Agent Orange was used by the American forces in the Vietnam war as a defoliant to reduce the tree and vegetative cover available to the Viet Cong. See *2,4,5-T.*

age structure models. Models of population dynamics whereby individuals are grouped according to their age.

agglomeration economies. Economic benefits and synergies that arise from the proximity of firms of the same type to each other. For example, high technology firms choose to locate in the "Silicon Valley", despite high land costs, because of site-specific agglomeration economies.

aggregate consumption. Total expenditures on consumption in an economy in a given period of time.

aggregate demand. The total expenditures within an economy in a defined period of time.

aggregate production function. A relationship that relates aggregate output to the factors of production including land, labor and capital.

aggregate supply. The total value of all goods and services produced in an economy over a defined period of time.

Agreement on Technical Barriers to Trade (TBT). A World Trade Organization (WTO) agreement that determines which non-tariff barriers to trade are acceptable and which are not.

Agreement on the Application of Sanitary and Phytosanitary Measures (SPS). An agreement that is part of the World Trade Organization (WTO) that allows countries to restrict imports to provide protection against pest and diseases, and risks associated with chemicals, fertilizers, animal feeds, pesticides and herbicides.

agricultural byproduct. A secondary material or consequence of agricultural activities, such as damages to river ecosystems from the use of agrochemicals.

agricultural drought. A deficiency of water that reduces crop production and is caused by insufficient rainfall or poor water and land management practices.

agricultural yield. The volume of goods produced by agricultural activities per unit area.

agriculture. The production of food and materials from crops and animals.

agricultural sector. A part of the primary sector of an economy that is concerned with the production, packaging and processing of agricultural goods.

Agriculture, Forestry and Other Land Uses (AFOLU). A term used in the context of climate change policy making to categorize net carbon emissions from agriculture, forestry and changing land use. See *Land Use, Land Use Change and Forestry* and *Reducing Emissions from Deforestation and Forest Degradation*.

agrochemicals. Chemicals used in the agricultural industry such as herbicides, pesticides and artificial fertilizers.

agroforestry. The cultivation of trees as an agricultural crop, such as the use of leaves as litter and animal feed, and the use of branches for fuel.

AIDS. See *Acquired Immune Deficiency Syndrome*.

air. A common name for the constituent gases in the Earth's atmosphere that consists of nitrogen (78.08 percent), oxygen (20.95 percent), argon (0.93 percent) and trace gases, including greenhouse gases.

air quality control region (AQCR).   Regions in the United States used to delineate communities with similar air pollution problems, and to implement federal pollution regulations.

air quality standard.   Maximum permissible concentration of a pollutant or pollutants in the atmosphere in a defined airshed at a given period of time.

air mass.   A large body of air with broadly homogenous characteristics at a horizontal level, particularly with regards to temperature and moisture.

air pollution.   A substance or chemical present in the air to a concentration such that harm may be caused to human health and the natural environment.

air pollution abatement equipment.   Technology such as filters that are fitted to sources of pollution in order to reduce harmful emissions. See *carbon capture and storage* and *scrubbers*.

air pressure.   See *atmospheric pressure*.

airborne fraction.   The proportion of carbon dioxide that remains in the atmosphere as a gas relative to a given exogenous increase from anthropogenic sources, presently around 45 percent.

airshed.   A geographic area defined on the basis of local atmospheric conditions and used in the control of air pollution.

albedo.   A measure of the reflective ability of a body defined by the ratio of reflected radiation to incoming radiation. Freshly fallen snow may have an albedo as high as 90 percent while black asphalt may have an albedo as low as 10 percent.

albedo feedback.   A process whereby changes in the Earth's cryosphere accentuate global warming or cooling. In a warming climate, the melting of ice and snow reduces the amount of solar radiation reflected by the Earth's surface, which further increases global temperature. Alternatively, an increase in ice and snow cover amplifies a cooling period by strengthening reflectivity. See *climate feedback*, *cryosphere*, *glacier retreat*, and *Milankovitch Theory*.

alcohol fuels.   Fuels, such as ethanol or methanol, that are produced by the fermentation of sugars, starches or cellulose and that can be used to replace fossil fuels.

alcohols.   Organic compounds that are produced during fermentation and that have a wide variety of uses.

aldehydes.   Reactive compounds used in many different applications, such as in resins and glues, and which are formed by the oxidation of alcohols.

algae.   A range of organisms, usually found in aquatic environments, that are capable of photosynthesis.

algal bloom.   A massive proliferation of algae in water bodies due to pollution (such as phosphates), nutrification, eutrophication, and environmental factors.

algorithm.   A program or set of rules that can be used to solve mathematical or computational problems.

ALH84001.   Identification for a meteorite that originated on the planet Mars and landed in Antarctica some 13 000 years ago. The meteorite includes polycyclic aromatic hydrocarbons that can be produced by decaying organisms, as well as inorganic processes. Their presence, and other evidence, has been used to support the hypothesis that microbial life once existed on Mars.

FURTHER READING
Davis (2000).

alienate.   The ability to transfer a good or asset to another person.

alkaline.   See *base*.

allele.   A specific form of a gene, that may differ across members of the same species.

allelopathy.   The process by which plants emit chemicals to prevent or retard the growth of competing plants.

Allen's rule.   The so-called rule that animals in cold climates tend to have shorter and less exposed extremities than animals of related species in warmer climes. See *Bergman's rule*.

Alliance of Small Island States (AOSIS).   A coalition of small island and low-lying coastal countries that are especially vulnerable to climate change.

This grouping of 42 states and observers acts as a negotiating bloc within the UNFCCC framework. A notable position of this coalition is that global warming should be restricted to below 1.5°C, beyond which many of its member countries may become uninhabitable due to rising sea levels. See *United Nations Framework Convention on Climate Change*.

allocation.   The procedure for distributing permits to participants in an emissions trading scheme, such as through an auction. May also refer to the actual number of permits allocated. See *grandfathering*.

allocative efficiency.   An enterprise is allocatively efficient if the ratio of the marginal products of its inputs equals the input price ratio, e.g., marginal product labor/marginal product capital = price of labor/price of capital. Or, in other words, when inputs or factors of production are combined in proportions that minimize the cost of producing a given level of output. In a fishery, for example, if the per unit cost of crew rises, a vessel that is allocatively efficient will substitute (to the extent possible) toward capital and technology that is labor-saving. See *economic efficiency*, *marginal physical product*, *scale efficiency*, and *technical efficiency*.

allopatric speciation.   The evolution of new species following the geographical separation of populations.

allowable annual cut (AAC).   The allowable annual harvest of timber permitted on an area of land. An AAC is often applied when the owner of the land is the state and timber companies have cutting rights on the land.

alpha diversity.   The number of species in a defined area, commonly used as a measure of biodiversity.

alpha radiation.   The least dangerous form of the three main types of nuclear radiation. It consists of a stream of alpha particles released from radioactive isotopes. See *beta radiation* and *gamma radiation*.

alternative development path.   A sustainable, non-carbon emissions intensive process of economic development. It is widely acknowledged that developing countries must adopt such a path to prevent catastrophic global climate change. See *development pathway*.

alternative energy.   Energy obtained from sources other than fossil fuels (e.g., wind, solar, geothermal). Contemporary definitions sometimes include nuclear energy.

alternative fuels.   Energy derived from sources other than hydro-carbons (e.g., fuel cells, methanol, biodiesel).

alternative fuel vehicle (AFV).   A vehicle that is powered by a fuel not derived from crude oil. This term also applies to hybrid electric vehicles (HEVs) that supplement petroleum fuel with electrical energy from batteries and solar power.

alternative hypothesis.   The contrary hypothesis to the null hypothesis in a statistical test and often denoted by H_1. For example, if the null hypothesis to be tested is that the means of two populations are equal, the alternative hypothesis is that the means are not equal. If we reject the null hypothesis in a statistical test we (conditionally) accept the alternative hypothesis. See *hypothesis* and *null hypothesis*.

altimetry.   The measurement of altitude or height.

altruism.   A term used to describe deliberate act of individuals that benefits others without the expectation of it being reciprocated.

Alvarez hypothesis.   The hypothesis that the Earth was hit by at least one body from outer space some 65 million years ago and that this event led to mass extinctions.

Amazonia.   An area defined by the watershed of the Amazon River and including the largest area of tropical rainforest in the world.

ambient.   The background level of an environmental measure, such as the air temperature or the level of particulate matter in the air.

ambient permits.   Pollution permits denominated in the amount of a unit of pollutant that is permitted at defined receptor sites in the environment.

ambient standard.   A standard that regulates the maximum permitted pollution at defined receptor sites in the environment.

amenity value.   The non-market value attached to an environmental asset, such as the value of walking in a forest.

amensalism.   An interaction where one organism's growth is inhibited by another. Such interaction is common among plant species.

ammonia.   A gas used in the manufacture of plastics and fertilizers that when liquified at a low temperature can be used as a refrigerant. The production and application of ammonia-based fertilizer is a source of greenhouse gas emissions. Ammonia naturally occurs in the environment as a result of microbial breakdown of proteins. See *nitrous oxide.*

Amoco Cadiz.   Name of an oil tanker that ran aground in France in 1987 and released over 220 000 tons of crude oil, which devastated the Brittany coast.

amoeba.   A one-cell organism.

amortization.   The provision of the repayment of a debt (such as a mortgage) over time. Amortization may also refer to the depreciation of an asset or good.

amphibian.   A common term for toads, frogs and other members of the class *Amphibia.*

anabatic wind.   A flow of warmer air in a valley directed upslope.

anadromous fish.   Fish that live primarily in salt water but migrate to freshwater in order to spawn. Salmon is a commercially important anadromous fish.

anaerobic.   An environment that contains no oxygen. Organisms that can live in anaerobic conditions are called anaerobes.

analysis of variance (ANOVA).   A statistical method for analyzing the sources of variance in studies or experiments.

analyte.   A material, substance, or chemical that is the subject of analysis.

ancient forest.   Forests characterized both by species longevity and large intervals between catastrophic disturbance. Some trees in ancient forests can attain lifespans in excess of 600 years.

ancillary impacts.   A secondary consequence of an action or policy. With regards to climate change mitigation, ancillary benefits may include: energy security, improved efficiency of resource use and reduced air pollution.

anemometer.   General term used for instruments that measure wind speed.

angiosperm.   Any flowering plant, including grasses and deciduous trees.

animal feces.   Animal excrement that is dried and used as fuel in many developing countries, particularly Asia. Despite being practical and cheap, this practice causes significant pollution haze and, hence, respiratory problems. Also used globally as fertilizer in agriculture. See *Asian Brown Cloud*.

animal unit (AU).   A measure of the amount of feed used by grazing animals set at 1000 pounds (454 kilograms), the combined weight of one cow and one calf.

Annex B countries.   Signatories to the 1997 Kyoto Protocol that collectively agreed to reduce their GHG emissions to 5 percent less than 1990 levels over the 2008 to 2012 commitment period.

Annex I countries.   Signatories to the United Nations Framework Convention on Climate Change (UNFCCC) which committed to stabilize their emissions to 1990 levels by the year 2000. Although it is not strictly correct to do so, this term is also often used in reference to Annex B countries. See *United Nations Framework Convention on Climate Change*.

Annex II countries.   Countries from the Organization of Economic Co-operation and Development (OECD) that are listed in the United Nations Framework Convention on Climate Change. Annex II countries are expected to help developing countries develop national reports on their GHG emissions and promote the use of more energy efficient technologies.

annual.   A plant that lives only one season.

annuity.   A constant annual payment. Annuities can be obtained by paying a lump sum to a financial institution in return for a constant and certain stream of annual payments.

annular modes.   Hemispheric-scale weather circulation patterns determined by variability in westerly winds at the mid-latitudes. These winds are caused by the temperature difference between tropical and polar regions. The two major types are the Northern Annular Mode (also known as the Arctic Oscillation or the North Atlantic Oscillation) and the Southern Annular Mode (also known as the Antarctic Oscillation or the High Latitude Mode). See *modes of climate variability*.

FURTHER READING
Ruddiman (2007).

anomaly.   Deviation of a measurement or situation from normal or expected conditions.

anoxic event.   Term used to describe past events within the geological record when the oceans were completely devoid of oxygen, causing mass extinctions of marine life. Such incidents occurred rapidly in periods of higher temperatures and greater atmospheric carbon dioxide concentration.

Antarctic Cold Reversal.   A significant episode of cooling that took place during the deglaciation at the end of the last glacial.

Antarctic Oscillation.   See *annular modes*.

Antarctic ozone hole.   See *ozone hole*.

Antarctic Treaty.   A treaty first signed in 1959 that encourages scientific research in the Antarctic in the absence of territorial and sovereignty claims by nations.

anthracite.   A hard, valuable type of coal that has the least impurities.

Anthropocene.   The current geological age. It is defined as commencing with the industrial revolution, when humans began to exert growing influence on the ecosphere.

anthropocentric ethic.   Term used to describe a set of human beliefs and associated behaviors that are focused upon the well-being of human beings, to the possible exclusion of other species.

anthropogenic.   A term that indicates that an object is of human origin. For example, anthropogenic greenhouse gas emissions arise from human activity, such as the burning of fossil fuels.

anthropogenic emissions.   Term used to refer to emissions of greenhouse gases that are generated by human activity. See *greenhouse gases*.

anthropogenic fires.   Fires started by humans, often to maintain ground cover of a desired vegetation type, such as savannas.

anthropogeomorphology.   The study of the human role in the alteration of physical processes and landforms.

anthropomorphic.   Term used to describe the attribution of human emotions and responses to other animals. Anthropomorphism has long been criticized by biologists despite the fact that recent animal studies suggest that some life forms (such as elephants) may, indeed, experience similar emotions to human beings. For example, elephants appear to exhibit a form of grief following the death of a member of the family that has similarities to grief exhibited by human beings.

antibiotic.   Chemicals used to control diseases caused by bacteria and fungi. The widespread use of antibiotics began after World War II. The over-use of antibiotics has hastened the development of antibiotic-resistant strains of bacteria.

antibody.   Proteins produced by cells that can neutralize and destroy viruses and bacteria. A heightened level of antibodies due to environmental sensitivities can be deleterious to health and is associated with allergic reactions.

anticipatory adaptation.   Adaptation that occurs prior to the impacts of climate change being observed. See *climate change adaptation*.

anticyclone.   A large-scale system of wind circulation around a central region of high atmospheric pressure.

anti-dumping.   Legislation and regulations designed to prevent foreign firms from selling goods at a price below cost.

antigen.   Any substance that results in an immune response in multicelled organisms.

anti-trust.   US legislation designed to restrict the use of monopoly power and control anti-competitive behavior.

AOSIS.   See *Alliance of Small Island States*.

APEC.   See *Asia-Pacific Economic Cooperation*.

aphelion.   The point during orbit when a planet is the greatest distance from the Sun. See *Milankovitch Theory*.

applied science. Research that seeks to use basic theories to solve practical problems.

appreciation. The increase in price or value of an asset or good over time.

appropriability. The ability of individuals to benefit from the flow of benefits from an asset or resource.

aquaculture. The managed raising and harvesting of fish and shellfish.

aquatic angiosperms. Flowering plants found in aquatic environments.

acquiclude. A by and large non-porous rock that can adsorb water very slowly over time, but does not transmit it.

aquifer. A porous body of rock or sediments capable of storing water. Aquifers are the source of water for springs and wells. Over-use of aquifers by withdrawing more groundwater than is replenished naturally is a major concern in many countries.

aquitard. An underground geological formation containing water, but which is unable to transmit substantial quantities.

arable land. Land suitable for the cultivation of crops.

aragonite. A form of calcium carbonate used to form shells or skeletons by marine organisms such as coral and squid. See *ocean acidification*.

Aral Sea. An inland sea formally located on the borders of Uzbekistan and Kazakhstan. The diversion of the waters entering the sea for irrigation purposes caused it to be divided it into four lakes that comprise less than 10 percent of the sea's original volume. The exposed sea bed is now a desolate salt plain contaminated with chemicals and poses a major health risk to the few communities remaining in the area. Both agriculture in surrounding areas and a formerly thriving fishing industry in the sea itself have now collapsed.

arbitrage. The purchase of goods or assets at one location to sell at another so as to generate a profit. See *speculation*.

arbovirus. A virus transmitted by blood-sucking insects such as mosquitoes, e.g., dengue fever. See *dengue fever*.

archaea.   A domain of micro-organisms that may be some of the earliest life forms on Earth. Archaea include organisms capable of living at extreme temperatures.

Archaen era.   A geological era from 3800 to 2500 million years ago that began with the first life on Earth.

Arctic haze.   Term given to an apparent haze in the winter months in the northern polar regions that is associated with air pollution in the mid latitudes.

Arctic Oscillation (AO).   See *annular modes.*

Arctic shrinkage.   A term for the decline in area of the Arctic ice sheet currently taking place. Scientific forecasts project that summer ice cover will become absent in the Arctic at some stage between the mid to late twenty-first century as a result of global warming. Declining ice coverage will reinforce climate change by reducing the Earth's albedo. See *albedo feedback*, *climate feedback*, and *permafrost.*

arid region.   A land region of low precipitation, generally defined as receiving less than 250 mm of rainfall. In agriculture, this term is used to describe extremely dry areas where land cultivation is impracticable without irrigation. See *drought.*

arithmetic growth.   Growth which increases by a constant amount per unit of time. For example the series 2, 4, 6, 8, . . . exhibits arithmetic growth, as the next number in the series is always the current number plus 2. See *geometric growth.*

arithmetic mean.   The sum of the values of all observations in a sample or population divided by the number of observations. See *geometric mean.*

aromatic hydrocarbons.   Group of hydrocarbons, including benzene, so named for their strong odor. See *hydrocarbons.*

Arrow–Debreu equilibrium.   A general equilibrium named after two Nobel Laureates in Economics, Kenneth Arrow and Gérard Debreu, who developed the mathematical proofs for the existence of a general market equilibrium.

artesian.   An aquifer where water, because of pressure, can rise to the surface.

artificial (forest) regeneration. The establishment of new forests by artificial means, such as the planting of seedlings.

artisanal fisheries. Fisheries, often located in developing countries, where the methods of fishing are small scale and limited to coastal or inshore areas.

asbestos. A general name given to a group of materials derived from silicate minerals which are resistant to heat, and which were extensively used to provide thermal insulation and the primary material in brake pads. Its use in thermal insulation is now prohibited in a number of countries because of the risk associated with chronic lung diseases and lung cancer from repeated exposure to asbestos fibers.

ASEAN. See *Association of Southeast Asian Nations.*

asexual reproduction. Reproduction without fertilization.

Asian Brown Cloud. A large pollution cloud that gathers during the winter monsoon over much of South and South-East Asia. The lack of rainfall at this time allows particles from burning of biomass fuel, car emissions and forest fires to gather in the atmosphere. It is estimated that this large pollution cloud causes up to 2 million deaths per year, mainly from respiratory illness. The net effect of this phenomenon upon global warming is uncertain. Although the deposition of black soot particles accelerates Himalayan glacier melt, atmospheric pollutants reflect solar radiation away from the Earth's surface. See *aerosol* and *black carbon.*

Asian crisis. A financial crisis that began in Thailand in 1997 and subsequently affected other Southeast Asian countries and, in particular, Malaysia, South Korea, and Indonesia over the period 1997–99. The crisis resulted in large falls in the value of the currencies and stock markets of the affected countries and led to tens of billions of dollars being granted in financial aid from the International Monetary Fund (IMF), Japan, United States, and other countries.

Asian Development Bank (ADB). Bank headquartered in the Philippines that has as its mandate to foster economic growth in the Asia and Pacific region.

Asia-Pacific Economic Cooperation (APEC). A forum formed in 1989 consisting of Pacific-rim countries including Australia, Brunei, Canada,

Chile, China, Indonesia, Japan, Malaysia, Mexico, New Zealand, Papua New Guinea, the Philippines, Singapore, South Korea, Taiwan, Thailand and the United States.

aspect.   The compass direction in which an area faces. Aspect is an important consideration in the planting of crops in temperate regions.

asphalt.   A dark material comprised predominantly of bitumen obtained through petroleum processing and mineral aggregates. Commonly referred to as bitumen, and also used to described the primary material found in locations such as the Canadian tar sands, or to the refined substance that is used in road construction.

assay.   A scientific test for: a specific chemical or compound, an organism, biological effect, or a natural or artificial process.

Assessment Report.   Individual reports in a series published by the IPCC that provide an updated summary of scientific information concerning climate change, likely impacts, and potential options for mitigation and adaptation. These reports are authored by leading experts selected by the IPCC. The First Assessment Report in 1990 provided the basis for establishing the UNFCC, successive reports have been released every 5 to 7 years. See *Fourth Assessment Report, Intergovernmental Panel on Climate Change*, and *United Nations Framework on Climate Change*.

Assigned Amount Unit (AAU).   A unit of allowed greenhouse gas emissions for Annex I countries under the Kyoto Protocol. Corresponds to a metric ton of carbon dioxide equivalents.

assimilation.   Ability of the environment to absorb pollution or wastes.

assimilative capacity.   The maximum amount of a pollutant that can be absorbed, neutralized or made inert by the environment in a given period of time.

assimilation energy.   Measure of the energy assimilated by animals and defined as the sum of net production and respiration. See *net production* and *respiration*.

Association of Southeast Asian Nations (ASEAN).   An association formed in 1967 that now includes Brunei, Burma, Indonesia,

Laos, Malaysia, the Philippines, Singapore, Thailand and Vietnam. ASEAN members are committed to reducing tariffs among member countries.

asteroid. A rock much smaller in size than a planet, found in the solar system.

asymmetrical games. Games or strategies undertaken by individuals where the information available among the players is not the same. See *asymmetric information.*

asymmetric information. A situation where one of the parties to a transaction does not have the same information about the transaction as another party. For example, persons seeking health insurance are likely to know more about the state of their health than an insurance company. See *moral hazard.*

FURTHER READING
Akerlof (1970).

asymptotic efficiency. Term used to describe an estimated parameter in a statistical model that is normally distributed and with a mean value equal to the true value of the parameter.

Atlantic Multidecadal Oscillation (AMO). A cyclical and naturally occurring pattern of variability in the sea surface temperatures of the Atlantic Ocean by approximately 0.4°C. Scientific measurements have recorded cool phases from 1905 to 1925 and 1970 to 1990, and warm phases from 1860 to 1880 and 1930 to 1960.

atmosphere. The body of gas and liquid that surrounds the Earth, consisting of 78.08 percent nitrogen, 21.95 percent oxygen, 0.93 percent argon and small amount of trace gases including carbon dioxide. The atmosphere is divided into four zones: the troposphere, stratosphere, mesosphere and thermosphere.

atmospheric boundary layer. The bottom layer of the troposphere that has contact with the Earth's surface, often an area of significant wind turbulence.

atmospheric deposition. The deposition of nutrients and pollutants from the atmosphere, usually in the form of precipitation.

atmospheric lifetime.   See *lifetime*.

atmospheric pressure.   The total force per unit area directed by the motion of atmospheric gas molecules. Low pressure areas often lead to unstable and stormy weather while high pressure areas often provide stable weather.

atmospheric science.   A broad term used to describe the group of scientific disciplines relating to study of the physics, chemistry and dynamics of the Earth's atmosphere. See *climatology* and *meteorology*.

atmospheric stabilization.   The process or realization of halting increases in global greenhouse gas emissions at a specific concentration to avoid the dangers of anthropogenic climate change. See *350 ppm*, *450 ppm*, and *550 ppm*.

atoll.   A coral island with a lagoon at the centre, commonly found in the Pacific Ocean. Atolls are highly susceptible to submergence by rising sea levels resulting from global warming. See *sea level rise*.

attractor.   A set of points to which a dynamic system converges.

attribution.   The act of stating or identifying the cause of an event or process. Climate scientists attribute with high probability that proximate climate change is caused by human activity.

autarky.   National economic policy that precludes international trade.

autocorrelation.   Also known as serial correlation. Autocorrelation exists when errors in a model are correlated. A common measure of first-order autocorrelation (correlation between the error term and its first lag) is the Durbin–Watson statistic where a value of 2 indicates no autocorrelation.

FURTHER READING
Kennedy (2003).

autonomous.   A dynamic system or variable which is not subject to outside forces.

autonomous adaptation.   Adaptation that emerges spontaneously from changes in natural and human systems, rather than as a planned response to climate change. See *climate change adaptation*.

autonomous mitigation.   Mitigation actions that are undertaken by a state, community or individual voluntarily without regard for the actions of other groups. See *climate change mitigation*.

autonomous technological change.   Innovation that emerges spontaneously from existing technologies and processes, rather than changing market conditions or government incentives. See *endogenous technological change*, *induced technological change*, and *technological change*.

autoregressive process (AR).   A process by which the current value of a random variable depends on the past values of the random variable and a disturbance term. An nth order autoregressive process is defined by

$$x_t = a_1 x_{t-1} + a_2 x_{t-2} + \ldots + a_n x_{t-n} + e_t$$

where x_t is the random variable in time t and e_t is white noise. See *autocorrelation* and *white noise series*.

autotroph.   An organism capable of obtaining its nutrients directly from inorganic materials or substances, such as green plants and algae.

autotrophic respiration.   The process whereby carbon dioxide is produced during plant metabolism. See *metabolism* and *respiration***.**

average.   See *arithmetic mean*.

average cost.   The total costs of production divided by the total output. Average costs can either rise or fall with output, or in standard cases, initially fall as output rises, reaching a minimum, then rise as output increases. Economies of scale result in a downward sloping average cost curve.

average-cost pricing.   The pricing of goods sold by firms that just covers the average cost per unit of production.

average fixed cost.   The fixed costs of production divided by the total output.

average product.   The total output or production divided by the amount of an input required to produce the output.

average propensity to consume.   The proportion of total income spent on consumption.

average propensity to save.   The proportion of total income that is saved.

average revenue.   The total revenue of a firm divided by its total output.

average variable cost.   The costs of production that vary with output, divided by total output.

averting behavior costs.   Economic costs associated with actions or behaviors undertaken to avoid negative consequences of environmental phenomena, especially if caused by environmental change. Often cited as a proxy for the willingness to pay to avoid environmental degradation.

averting expenditures.   Market costs incurred in averting the effect of a deterioration in the environment. For example, the purchase of bottled water, because of concern over the quality of drinking water, represents an averting expenditure.

AVHRR.   Advanced Very High Resolution Radiometer. A sensor carried aboard NOAA meteorological satellites that provides images commonly used for vegetation mapping.

avoidance costs.   The costs involved in preventing degradation of environmental quality.

avoided deforestation.   Deforestation that would have occurred without the intervention of targeted policies or measures to prevent it. This term may cover unplanned yet projected deforestation from urban growth and other factors, in addition to planned activity. The provision of finance for avoided deforestation is one of the main focuses of the REDD mechanism. See *Reducing Emissions from Deforestation and Forest Degradation*.

B

back-casting. The process whereby a specific future goal is defined and measures are identified for transition from the current state. See *forecast*.

back-end charge. A charge that must be paid for the disposal of waste.

backscatter. The reflection or scattering of radiation. For example, the reflection of solar radiation away from the Earth's surface by clouds.

backstop resource. A natural resource that is currently too expensive to be used but, should the price of its substitutes rise sufficiently, would become economically efficient to exploit. See *backstop technology*.

backstop technology. A technology that is available but at current prices is not cost-effective to use as a substitute for existing technologies on a very large scale. For example, a range of renewable technologies are available for generating electricity. However, the relatively low long-run marginal cost of generating electricity from coal and natural gas currently renders many renewable technologies uneconomic.

backward bending supply. Used to describe circumstances under which the expected supply of a good or service falls as the price paid for the good or service increases beyond a certain point. Distinguished from normal supply curves in which the quantity supplied increases (or at least does not decrease) as the price increases.

bacteria. A group of one-celled organisms capable of multiplying by forming spores or by fission and which lack a distinct nucleus surrounded by a membrane.

bacteriaphage. A type of virus that feeds on bacteria.

bag limits. Limits imposed on hunters in terms of how many animals they are allowed to catch or kill. Bag limits are used as a control on the total level of exploitation of a population and are frequently used in recreational fisheries.

bag-tag. A system for the disposal of household waste in which only bags that are appropriately tagged are disposed. In some municipalities, households receive a fixed number of tags per week, month, or year, covered in their property taxes, and additional tags can only be acquired at a much higher marginal cost.

balanced budget.   A budget or forecast where a government's revenues equals its expenditures.

balanced growth.   Economic growth where all sectors of the economy grow at the same rate.

balance of payments.   A method of accounting for all debit and credit transactions of a domestic country with foreign countries. Transactions include the current account, or the trade in goods and services and profits and interest earned overseas net of what was paid overseas, and the capital account, which includes the inflows and outflows of loans and investments.

balance sheet.   A financial statement which lists the assets and liabilities of an enterprise.

baleen whales.   Sub-set of the order of Cetacea called mysticetes that includes the largest animal on Earth, the blue whale. Baleen whales feed through bonelike plates called baleen which trap small fish and shrimp. Baleen whales are divided into three families: Balaenidae that include right whales, which swim slowly and feed by swimming back and forth through schools of their prey, Balaenopteridae that include the blue whale and Eschrichtiidae that includes only the Pacific gray whale, which feeds along the ocean bottom. See *cetaceans*.

Bali Road Map.   A plan emanating from the COP-13 meeting of the UNFCCC in 2007 for securing a binding agreement upon international climate policy at the 2009 COP-15 in Copenhagen. The Bali Road Map outlined issues for negotiation including: a global adaptation fund, technology transfer, and reducing emissions from deforestation. The plan also acknowledged the responsibility of developing countries to pursue assisted mitigation measures that do not compromise development. See *Conference of the Parties*, *Copenhagen Accord*, *nationally appropriate mitigation actions*, and *United Nations Framework on Climate Change*.

BANANA.   An acronym for Build Absolutely Nothing Anywhere Near Anybody, used to describe a more extreme example of public unwillingness to accept any form of development within an area because of perceived deleterious effects (which could range from increased traffic congestion due to the construction of new homes to perceived environmental degradation from a new landfill site). See *NIMBY* and *locally unwanted land use (LULU)*.

bang-bang control. Term which refers to how the control variable suddenly changes at a certain point in time in an optimal control problem. See *optimal control theory*.

banking. A term for a system where polluting firms are credited for the difference between their actual and permitted emissions. The credited emissions may then be used at a later date or possibly traded, depending upon the rules of the emissions trading program.

bargaining. The interaction between parties to a transaction.

barrel (bbl). Standard unit of volume for oil and petroleum products that equals 159.1 liters. A ton of crude oil approximately equals 7 barrels.

barriers to entry. Factors which prevent firms from entering or competing in a market. For example, a barrier to entry for firms providing electric vehicle recharging services is the large capital cost required to provide an adequate number of battery recharging stations.

base. Base substances have a high pH value and can neutralize acids. See *pH*.

Basel Convention on the Control of Transboundary Movement of Hazardous Wastes and their Disposal. A 1992 convention, often known as the Basel Convention, which ultimately led to an agreement to phase out the dumping or inappropriate disposal of hazardous wastes in developing countries.

baseline. A reference point for measuring the effect of policies and changing conditions. Often used to describe the prevailing circumstances if the policy or event of interest did not occur. Alternatively, this term may simply describe a base level or year against which to quantify future changes. For example, baseline carbon emissions may refer to projected emissions without mitigation, or emissions in an arbitrary base year. See *base year* and *business-as-usual*.

baseline-and-credit. A form of emissions trading scheme whereby a baseline of expected or desired emissions is established and credits are generated by participants achieving emissions reductions relative to the baseline.

Critics argue that this system is less desirable than a cap-and-trade scheme because it is complicated to administer, involves greater uncertainty, and may generate less incentives for overall emissions reductions. See *cap-and-trade, emissions intensity,* and *emissions trading scheme.*

baseline emissions.   See *baseline.*

baseline scenario.   See *business-as-usual.*

basement complex.   A geological formation composed of basement rocks from the Precambrian period. Also known as Archean Complex.

Base-load power plants.   Electricity generation plants that have a relatively low operating cost and a high capacity factor to satisfy most of the electricity demand for a given period. Due to their high capital cost, these plants are typically operated continuously except for periods of maintenance. They have traditionally used nonrenewable fuels such as coal and nuclear energy, although natural gas base-load plants are now becoming common. See *peaking power plants.*

base period.   The period in an index to which all other index values are comparable. Base periods for indexes may vary with changes in the composition of the items that make up the index, such as a change in the composition of goods used to determine a consumer price index.

base year.   The year against which benchmarks are set for carbon emissions reductions. For example, under the Kyoto Protocol participating nations pledged emissions reductions with respect to 1990 levels.

basin.   A region drained by a single river system such as the Mekong River Basin.

basin-level efficiency.   The ratio of crop water evapotranspiration to total water depletion for irrigation in a basin, which accounts for the potential reuse of return flows and the potential decline in the water quality of return flows.

BASIC.   A collective term for Brazil, South Africa, India and China often used in the context of international climate policy.

basic science.   Scientific research that focuses primarily upon understanding the basic principles or theories of a discipline. See *applied science.*

Batesian mimicry.   An evolutionary phenomenon where a harmless organism closely resembles or mimics another species that is harmful to potential predators or is unpleasant to eat. See *Mullerian mimicry*.

Baumol effect.   Idea developed by William Baumol which states that as an economy matures productivity gains will be greatest in capital-intensive sectors thus reducing the price of goods produced in these sectors relative to services which are labor intensive. As a result, public spending which includes mostly expenditures on services will, over time, account for an increasing share of the gross domestic product.

Bayesian analysis.   Bayesian analysis is an approach to statistical analysis where the researcher forms a prior distribution which is then updated with additional data and Bayes theorem to generate a posterior distribution. A risk or loss function is often used to determine a parameter value from the posterior distribution. See *Bayes theorem*.

FURTHER READING
Kennedy (2003).

Bayes theorem.   Used in generating posterior distributions based on a prior distribution and available data. The theorem provides a method for updating the probability of an outcome following the acquisition of additional data or information.

FURTHER READING
Kennedy (2003).

Beaufort scale.   A system for measuring wind force developed by Sir Francis Beaufort, ranging from dead calm (0) to hurricane force (12).

behavioral change.   Term commonly used in the context of climate change referring to changes in individual, business, or community behavior necessary to mitigate carbon emissions. For example, consumers may turn off electrical appliances at a power source, or use public transport instead of driving. Behavioral change may occur as a result of government policies or through voluntary action.

Bellman's equation.   Bellman's functional recurrence equation is used in dynamic programming and defines the maximum value of a function. Its usefulness is that it provides an iterative method of solution to complex dynamic optimization problems in discrete time. See *dynamic programming*.

FURTHER READING
Chiang (2000).

bell-shaped distribution.   See *normal distribution.*

benchmark.   See *baseline.*

benefit-based standard.   A standard or regulation designed to lead to a desired environmental benefit, but developed with little or no consideration of the costs associated in achieving the standard. See *standard.*

benefit–cost analysis.   See *cost–benefit analysis.*

benefit–cost ratio.   The present value of benefits from an investment or project divided by the present value of the costs associated with the project. See *benefit analysis* and *present value.*

benefit stream.   The benefits that accrue over time from an investment or project. See *cost–benefit analysis.*

benefit transfer method.   An environmental valuation technique whereby ecosystem services benefits at one site are estimated by transferring estimated benefits from a similar site elsewhere. See *ecosystem services* and *environmental valuation.*

Benthamite social welfare function.   A social welfare function named after Jeremy Bentham, an eighteenth and nineteenth century political economist, that maximizes the aggregate utility over all individuals. See *social welfare function* and *utilitarianism.*

benthic community.   A term for all marine life living on or near the bottom of a lake or ocean.

benzene.   A flammable liquid derived from hydrocarbons that is both toxic and a known carcinogen.

bequest value.   Term used to describe a value that individuals may place on the environment that represents their consideration for its use by future generations.

Bergen declaration.   A declaration signed by 84 countries in 1990. Its goal is the promotion of sustainable development.

Bergman's rule. The so-called rule that animals in colder climates tend to be larger than similar species found in warmer climates. See *Allen's rule.*

beriberi. A disease common to poor countries caused by a deficiency of thiamine. If untreated, beriberi can cause nerve damage and death.

Bernoulli distribution. Named after Jacques Bernoulli, this is the distribution of a random variable that takes on only one of two possible values per observation (with probability p for one of the values and $1-p$ for the other) such as tossing a coin and whether it turns up heads ($p = 0.5$) or tails ($1-p = 0.5$).

Bertrand competition. Competition between oligopolists where firms independently select the price at which they sell their product. See *Cournot competition.*

best available demonstrated control technology (BADCT). A method of determining effluent guidelines from new point sources within a source category that is based on an analysis of control technologies currently utilized in the category.

best available technology (BAT). Term used in the regulation of pollutants. In some countries, sources of pollution are required to use the "best available" technology for controlling their emissions.

best conventional technology (BCT). Term used in the regulation of pollutants. In some jurisdictions, sources of pollution may be required to install equipment that meets best conventional technology to achieve pollution standards rather than use the often more expensive best available technology.

best linear unbiased estimator (BLUE). The estimator of a parameter in a linear statistical model that has the smallest variance.

best management practices (BMPs). The best practices in an industry which are often held up as an example for others to follow.

best practicable technology (BPT). Term used in the regulation of pollutants. In some jurisdictions, sources of pollution may be required to install equipment that meets BPT to achieve pollution standards. BPT is more strict than using the best conventional technology, but is often less costly than using the best available technology.

beta. A risk measure based on the covariance of returns of an asset and the returns from the market as a whole as a ratio of the overall variance of the market. A beta of 1 indicates that the asset, on average, moves in the same way as the market while a negative beta suggests that the asset moves in a different direction to the market.

beta diversity. An index of the degree of change in species composition where the greater the mix the greater the index.

beta radiation. One the three main types of nuclear radiation. It consists of a stream of beta particles released from radioactive isotopes. See *alpha radiation* and *gamma radiation*.

Bhopal. An Indian city that was the site of a Union Carbide plant. In 1984 a leak of methyl isocyanate gas caused the death of thousands of individuals and affected the health of many others in one of the world's worst industrial disasters.

biased estimator. A biased estimator is one where the expected value does not equal its true value.

bid rent models. Also known as random bidding models, discrete choice models in which the probability that an individual of a specified income level is the highest bidder for a specific property depends on the characteristics of the property. A different bid rent estimation is required for each income class.

biennial. A plant that germinates in one season and gives fruit in the following season.

bifurcation. A point where the qualitative behavior of a dynamic system changes.

FURTHER READING
Shone (2002).

Big Bang theory. The hypothesis that the universe began 12 billion years ago in a single explosion that created time and space.

bilaterism. An approach to international relations characterized by formal agreements between two countries. In the 1930s bilateralism focused on the granting of mutual trade privileges between two countries, which were denied to other countries. See *multilateralism*.

bilharziasis. See *schistosomiasis.*

binary choice models. See *dichotomous choice.*

binding constraint. A constraint in an optimization problem that holds as an equality at the optimum such that the value of the objective function would be greater if the constraint were relaxed. For most individuals, annual income represents a binding constraint on consumption.

binding emissions target. Compulsory quantitative reductions of carbon emissions by countries party to an international treaty. Reductions are made with reference to a specific base year. Within the contemporary UNFCCC framework, only developed countries are expected to formulate binding emissions targets, although developing countries are committed to pursuing non-binding mitigation measures. See *base year*, *Kyoto Protocol*, and *nationally appropriate mitigation actions.*

BINGOs. Acronym for business non-governmental organizations.

binomial distribution. The distribution of successes (such as the number of heads when tossing a coin) in a series of trials where the probability of success is p and the probability of failure is $1-p$.

bioabsorption. Term applied to the process of waste removal using biological or organic material. For example, the use of moss to remove metals in a solution is a form of bioabsorption.

bioaccumulation. The process by which persistent pollutants accumulate in the bodies of organisms. Some pollutants, such as PCBs and DDT, increase in concentration in animals higher up the food chain and can be millions of times greater than the concentrations found in animals lower in the food chain.

bioassay. A test for the existence or concentration of a substance by using the response of an organism as an indicator.

biocapacity. A land area's capacity to replenish biological resources and absorb excess waste.

biocentrism. A moral perspective that values all forms of life equally.

biochar. A variety of charcoal produced by heating organic materials in a low oxygen environment, known as pyrolysis. The principle objective of this activity is the sequestration of atmospheric carbon in the biochar, which can subsequently be used for enhancing soil fertility. See *carbon sequestration*.

biochemical oxygen demand (BOD).   A measure of the level of organic pollution in water where the greater the organic pollution, the larger will be the number of bacteria in the water and thus the higher the BOD. The measurement is made by calculating the oxygen utilized in a sample of water over a 5-day period at a constant temperature of 20°C.

FURTHER READING
Tomar (1999).

biochemistry.   The study of chemical reactions in organisms.

biocide.   A substance that inhibits the growth or causes the death of micro-organisms.

bioconcentration.   The process by which pollutants, such as organochlorines, concentrate in the fats of animals, including human beings.

biocover.   A cover containing gas or organic material that oxidizes methane gas emanating from landfill or agricultural waste and, hence, reduces greenhouse gas emissions. See *methane*.

biodegradable.   A substance is biodegradable if it can be degraded or broken down over time in the natural environment.

biodegradation.   The breakdown of materials in the presence of natural conditions.

biodiesel.   A biofuel used in diesel engines, typically made from soybean and sunflower seed oil. See *biofuel*.

biodiversity.   See *biological diversity*.

biodiversity hotspots.   Locations in the world with habitats that provide for a large number of different species. Most so-called hotspots exist in tropical rainforests such as in parts of Borneo, Indonesia and particular locales in Amazonia, Brazil.

biodiversity prospecting.   The cataloging, sampling, and collecting of plant and animal species, usually with the intent of finding economically valuable uses associated with rare or previously unknown species.

biodynamic farming.   A type of organic farming inspired by the writings of Rudolph Steiner. See *organic farming*.

bioeconomics.   A term used to describe economic modeling that includes biological models. Many of the pioneering developments in bioeconomics came from the subdiscipline of fisheries economics.

FURTHER READING
Clark (2010).

bioeconomic model.   A model combining economic and biological processes. In fisheries, refers to a combination of a biological stock assessment, a stock-recruitment relationship, and an economic profit equation, usually discounted over time, that determines the catch or effort level that maximizes resource rents in a fishery. Bioeconomic models range from relatively simple, deterministic, single species models to more complicated settings that allow for uncertainty and multiple species and fishing methods.

bioenergy.   Energy produced from biofuel.

bioethics.   A view of the world which emphasizes that all life on Earth is related and that human beings need to consider their partnerships with other organisms in their decision making.

biofuel.   A gas, liquid, or solid fuel created from biological materials. Due to being a renewable and, typically, low emissions energy source, biofuels have become a major focus of carbon mitigation efforts. However, the conversion of forests into agricultural land for biofuel production may negate, or even reverse, this mitigative capacity. Furthermore, the inflationary effect of biofuel production upon food prices may pose a risk to global welfare. See *biomass fuel*.

FURTHER READING
World Bank (2007).

biogas.   A combustible gas, consisting mostly of methane, that can be used as a fuel and is produced from the decomposition of organic matter.

biogenesis.   Hypothesis that only living creatures can create life.

biogenic.   Term used to refer to effects of biological origin.

biogeochemical cycle.   The passage through the various Earth systems of essential components of living organisms, such as oxygen and carbon.

biogeoclimatic ecosystem classification system.   A hierarchical system of ecosystem classification that includes information on local, regional, and chronological factors as well as data on climate, vegetation, and other site-specific factors.

biological capacity.   The ability of an ecosystem to sustainably produce resources.

biological control.   The use of biological agents to control pests.

biological diversity.   The total variation in all life on Earth or within a given area or ecosystem, typically expressed as the total number of species found within the area of interest or the genetic diversity within a species. See *extinction*.

biological growth function.   The mathematical description of how a population behaves over time. For example, the growth of a single species of fish may depend on the existing stock, its natural or intrinsic growth rate, and other environmental factors. See *population growth*.

biological legacies.   Material left behind following a natural disturbance, such as dead trees after a forest fire.

biological oxygen demand.   See *biochemical oxygen demand*.

biological pollution control.   Use of biological systems to break down waste and effluent into environmentally benign components through natural processes.

biological treatment.   Form of waste treatment which employs bacteria to break down organic materials.

biology.   The study of life and life processes.

biomagnification.   See *bioaccumulation*.

biomass.   The total weight of all individuals in a given population.

Biomass measures are common in species where there may be a large number of individuals, such as fisheries. Also used for referring to biomass fuel. See *biomass fuel*.

biomass dynamic models. Models that predict the total stock or biomass of natural populations. Such models have frequently been applied in fisheries and include the logistic growth model. See *logistic growth*.

biomass fuel. Organic material that can be used as a source of heat and energy, such as wood and soybean. Since the 1970s, Brazil has used sugar cane as a biomass fuel to produce ethanol, as a substitute to imported hydrocarbons. See *biofuel*.

biome. A terrestrial community characterized by consistency of animal and plant species, such as the boreal forest.

biomimicry. The imitation of the designs and processes of nature to help solve problems and promote sustainable development.

bionomic equilibrium. A sustainable equilibrium usually associated with an open access resource, in which economic profits are zero, or where total revenues just equal the total costs of harvest.

bioregion. A geographic area with distinctive attributes, such as climate, wildlife, vegetation, landforms, or other environmental characteristics.

bioremediation. The enhancement of the growth of micro-organisms in the environment so as to assist in the breakdown and clean-up of discharges, emissions and pollution. For example, in oil spills, the application of nitrogenous fertilizers can speed up the natural breakdown by bacteria of hydrocarbons.

Biosafety Protocol. Formally known as the Cartagena Protocol on Biosafety, this was signed by the Conference of the Parties to the Convention on Biological Diversity in May 2000. The Protocol seeks to regulate biodiversity via the regulation of the international transport of living genetically modified organisms and to assist in the protection of traditional knowledge.

biosequestration. The removal of atmospheric carbon dioxide through biological processes, such as photosynthesis in plants and the formation of biochar. Biosequestration via reversing deforestation has achieved great prominence within international climate policy, as evident in the formation of the REDD carbon mitigation mechanism. See *carbon sequestration* and *Reducing Emissions from Deforestation and Degradation.*

biosolids.   Dried solids from treated sewage that are used as a fertilizer.

biosorption.   See *bioabsorption.*

biosphere.   The area of land, sea and the air inhabited by living organisms.

biosynthesis.   Chemical reactions produced by living organisms.

biota.   Plants and animal life found in a given geographical area or time period.

biotechnology.   Term used to describe the processes and technologies used to develop or enhance specific genetic traits by genetic manipulation. See *genetic engineering.*

biotic.   The living community within an ecosystem.

biotic potential.   Maximum possible size of a population.

biotope/ CORINE biotope.   A scheme used by the European Commission to identify sites that are distinguished by their ecological diversity, richness of habitat, and relative ability to support threatened or endangered species.

birth rate.   The number of births per year divided by the estimated total population, usually expressed as a rate per 1000 individuals.

bitumen.   See *asphalt.*

bituminous coal.   Most common type of coal which contains between 18 and 35 percent volatiles or substances that are emitted when the coal is heated.

bivalve.   A mollusk that lives in a shell comprised of two hinged sections, such as an oyster. See *mollusk.*

bivariate data.   Data where observations are available on only two variables per subject.

black carbon. A variety of particulate air pollution formed from the incomplete combustion of fossil and biomass fuels, also commonly known as soot. Black carbon contributes to global warming by absorbing heat in the atmosphere, and causing an albedo feedback effect when deposited on snow or ice. As it remains in the atmosphere for a short period, reducing black carbon emissions presents a ready means of slowing proximate climate change. See *albedo feedback*.

black economy. Part of the economy where transactions are not officially recorded so as to avoid the payment of taxes or to engage in the trade of prohibited goods. Also known as the "underground economy" or "parallel economy".

black coal. A high quality grade of coal with a carbon content between 80 and 100 per cent. The term applies to anthracite, bituminous and sub-bituminous coals. Black coal has a higher energy and lower moisture content than brown coal and is primarily used to generate electricity in power stations. See *brown coal*.

black liquor. A biofuel which is produced as a byproduct from the conversion of wood into pulp in the paper making process. With a high energy content, it is often used as a fuel at paper making facilities and has been shown to have high greenhouse gas reduction potential. See *biofuel*.

black list. A European Union list of pollutants discharged into aquatic environments and which are considered a priority for control.

black swan event. A highly unexpected event, the exact timing of which is virtually impossible to predict.

blast furnace. A continuously operated furnace which produces pig iron, the primary ingredient of iron and steel. Operation of blast furnaces is a significant source of industrial carbon dioxide gas emissions. See *carbon capture and storage*.

blended value. A business model which extends the traditional return-on-investment approach to include social, environmental and cultural returns on investment. It can be distinguished from corporate social responsibility in that improving social impacts is a core strategy of the business model, rather than an additional component. See *corporate social responsibility*.

blizzard. A severe storm lasting for at least 3 hours which is characterized by: wind speeds in excess of 55 kilometres per hour; temperatures below −7°C; and heavy snowfall that reduces visibility to less than 1 kilometer.

blocking anticyclone.   A large high pressure system that remains nearly stationary for several days or weeks, effectively blocking the normal progression of migratory cyclones and causing affected areas to experience unchanging weather patterns for an extended period of time. See *anticyclone*.

blowdown.   Windthrow or uprooting of trees by the wind.

blue box.   A common term for curbside household waste recycling found in many towns and cities in North America.

blue water.   Fresh water directly accessible for extraction or diversion such as found in rivers, lakes and aquifers.

blue-green algae.   See *cyanobacteria*.

board feet.   A unit used in the lumber industry to measure lumber production. One board foot is the volume of wood contained in a board 12 inches long by 1 inch wide by 1 inch high. Board feet are typically expressed in thousands and written Mbf.

BOD5 test.   See *biochemical oxygen demand*.

bog.   A wetland or area of waterlogged soil with no significant inflows or outflows and an acidic composition, principally of moss or peat. Bogs are a natural source of the greenhouse gases methane and carbon dioxide. See *peatland*.

bond.   A financial instrument which can be issued by companies and various levels of government, that gives the buyer a fixed amount of interest per period.

bootstrapping.   A method by which random samples are generated from an existing data set to derive confidence intervals and estimates of the parameters of interest.

FURTHER READING
Greene (2003).

boreal forest.   A biome that includes mainly coniferous trees, character-ized by long, cold winters and relatively short summers. The boreal forests of Eurasia are commonly referred to as the *Taiga.*

borehole.   A hole that is drilled into the Earth's surface or ice in order to perform scientific tests or extract materials such as water and oil.

borehole technique.   A term for the procedure for measuring the char-acteristics of the ice or ground surrounding a drilled borehole. The sub-sequent data is used to make inferences pertaining to past climate. See *paleoclimatology.*

Boserup hypothesis.   A hypothesis developed by E. Boserup that sug-gests that increased population density leads to a reduction in the area of agricultural land in fallow, which in turn provides a stimulus for the development of agricultural innovations.

FURTHER READING
Boserup (2005).

botany.   The study of plants and plant life.

bottom-up.   A method of problem-solving that focuses on the constituent parts of the problem and builds up towards an overall solution. Commonly referred to in the context of "bottom-up" international climate policy, whereby nations identify individual strategies for climate change mitiga-tion that are then combined to form a non-predetermined level of collec-tive action. See *climate policy framework* and *top-down.*

boundary layer.   See *atmospheric boundary layer.*

bovine spongiform encephalopathy (BSE).   Commonly known as "mad cow" disease. BSE is thought to be transmitted via prions, a form of protein that has been linked to a fatal degenerative brain disease in humans called nvCJD. Transmission of BSE may have been assisted by the feeding of prion-contaminated meat and bonemeal to cattle. See *Creutzfeldt-Jakob disease.*

break-even point.   The point in production or output of a firm when its total revenue exactly equals its total cost.

breakwater.   A barrier constructed off coastlines to protect vessels and prevent erosion of coastal areas. The erection of such structures has been proposed as an important climate change adaptation activity for reducing damage from rising sea levels. See *erosion* and *sea level rise*.

breed.   An anthropogenic mating of chosen members of a species for a particular purpose. Most domesticated animals and plants are breeds.

breeder reactor.   A nuclear reactor capable of producing as much fuel (typically plutonium) as it uses.

Bretton Woods.   A place in New Hampshire that in 1944 was the meeting place of decision makers among the Allies who helped decide the post-World War II financial system. The system included fixed exchange rates and the establishment of the International Monetary Fund and the World Bank.

British Thermal Unit (BTU).   The amount of heat required to raise the temperature of one pound of water one degree Fahrenheit.

broad-leaved forest.   Forest dominated by non-coniferous species such as beech or oak.

bromofluorocarbons.   See *halons*.

brown coal.   A low grade form of coal, also called lignite, containing approximately 40 to 60 percent carbon. Brown coal has a higher moisture content, a lower calorific value and emits a greater amount of carbon dioxide emissions compared to other varieties of coal. See *black coal* and *coal-fired power generation*.

brownfield site.   Contaminated land in urban areas that potentially may be used for other purposes and redevelopment.

brown haze.   Atmospheric pollution distinguished by the presence of a fine brown cast to the surrounding air. The source may be microscopic particles or compounds created by photochemical reactions between various pollutants discharged into the atmosphere.

Brownian motion.   First identified by the botanist Robert Brown, this refers to a stochastic process that results in a seemingly random and erratic motion of small particles in various substances, such as liquids and gases.

FURTHER READING
Guttorp (1995).

Brundtland Commission.   The popular name for World Commission on Environment and Development and named after its Chair, Gro Harlem Brundtland. The Commission's final report called *Our Common Future* created the first commonly used definition for sustainability and became a catalyst behind the adoption of policies to promote sustainable development.

brushing.   The use of artificial (chemical or mechanical) means to reduce competing plants in a forest so as to promote growth in desired trees.

BTU.   See *British Thermal Unit*.

btu tax.   A tax or charge based on the number of British Thermal Units in a given volume of a fuel.

bubble policy.   A sub-program of emission credit trading set up in the United States to allow some flexibility in meeting air pollution control targets. It treats multiple plants owned by the same firm as one point source in the sense that the firm can adjust its emissions among its plants to minimize its pollution abatement costs, provided that the firm's total emissions meet the defined standard.

FURTHER READING
Tietenberg and Lewis (2009).

budget constraint.   A financial or revenue constraint which limits the total expenditure or spending of an individual, company, or organization.

budget deficit.   The amount by which government expenditures exceed government revenues.

budget surplus.   The amount by which government revenues exceed government expenditures.

buffer.   Any substance that is able to prevent rapid changes in base or acidity. The ability to buffer is extremely important for animals as they can often only tolerate relatively small changes in pH within their bodies.

buffer stocks.   Stocks of commodities that are held in reserve in anticipation of future events. For example, some countries maintain buffer stocks of commodities in the expectation they can sell at a higher price and some commodities, such as oil, are held in reserve for security purposes.

buffer zone.   A zone of vegetation or habitat that is conserved to protect environmentally important or sensitive areas. For example, logging may be prevented on land adjacent to a national park so as to provide a buffer from potentially harmful effects that might be detrimental to animals within the park.

business-as-usual.   A modeling scenario which predicts the future under the assumption that policies or actions do not fundamentally change from the present. In models of climate change mitigation, business-as-usual scenarios usually assume that no reductions in greenhouse gas emissions are undertaken. See *baseline* and *emissions projection*.

business cycle.   Fluctuations in the level of economic activity where the peak is characterized by high economic growth and the trough by low growth, or possibly a recession.

butterfly effect.   The notion that a small change in initial conditions, such as the flapping of the wings of a butterfly, can result in enormous effects in the future. This idea is associated with positive feedbacks and chaos theory. See *chaos* and *feedback*.

bycatch.   Species that are caught or harvested incidentally when targeting other species. See *targeted species*.

byproduct.   A secondary outcome of a process, or an additional consequence to an action. Often used in reference to physical processes such as manufacturing. For example, carbon dioxide gas is a byproduct of coal-fired power generation.

C

C₃ plants. Plants, such as wheat and beans, that produce three carbon molecules with photosynthesis. See *C₄ plants.*

C₄ plants. Plants, such as corn, that produce four carbon molecules with photosynthesis and are adapted to higher temperatures and greater sunlight. The productivity of C_4 plants can be significantly enhanced with higher levels of carbon dioxide. See *C₃ plants.*

$^{13}C/^{12}C$ ratio. A measurement of the relative abundance of two different carbon isotopes within a carbon dioxide molecule. The resultant value is used to provide inferences regarding the relative importance of different processes within the carbon cycle. See *carbon cycle.*

^{14}C. See *carbon-14.*

cadmium. A widely used but toxic metal that is a known carcinogen.

caesium 137. A radioactive isotope of caesium with a half-life of 30 years. Trace amounts of caesium 137 were deposited over most parts of northern Europe following the 1986 Chernobyl nuclear accident. See *Chernobyl.*

Cairns Group. A group of 18 countries, formed in Australia in 1986, organized to encourage agricultural exports.

calcite. A crystallized form of calcium carbonate that is the primary constituent of limestone. See *aragonite.*

calcrete. A layer or boundary in the soil formed by calcium carbonate.

calculus of variations. An approach to solving dynamic problems in continuous time. See *Euler's Equation.*

calorie. A measure of energy defined as the energy required to raise the temperature of one gram of water from $15.5°$ C to $16.5°$ C.

Cambrian. A period on the Earth 570 to 505 million years before the present.

Camelford incident. The accidental spillage of aluminum sulphate into the drinking water of the 20 000 residents in Camelford, England, in 1989. The residents suffered a number of serious health complaints including symptoms of joint and muscle pain and memory loss.

Campylobacter jejuni.   A widespread pathogen that is the most common cause of acute infectious diarrhea in developed countries. Infections from the bacteria can be fatal and are responsible for several hundred deaths a year in the United States. The bacterium is co-adapted with poultry and thus the principal source of infection is uncooked or undercooked poultry meat.

canopy.   The cover of branches and leaves in a forest. In equatorial forests the canopy may be very thick and can absorb as much as three quarters or more of the solar radiation, and over a third of the precipitation.

canopy depth.   The difference in height between the lowermost and uppermost branches of trees within a stand.

cap-and-trade.   A form of an emissions trading scheme whereby a limit (cap) is set on the total amount of emissions allowed from specific sources over a specified period. A quantity of permits equal to the capped amount of emissions is supplied to the market through an allocation method to ensure the target is strictly met. Each specified emitter must surrender to a prescribed authority at the end of each period an emissions permit for each unit of emissions they produced. Emissions permits are bought and sold in the market according to supply and demand from participants. Incentives are provided for firms to reduce their level of pollution if the cost of doing so is less than the revenue they receive from selling the permits. The capped level of overall pollution, and hence the number of credits, is steadily reduced over time. Although first introduced to combat acid rain in North America, cap-and-trade schemes are more commonly associated nowadays with climate change mitigation, for example, the European Union Emissions Trading Scheme. See *climate change mitigation* and *emissions trading scheme*.

capacity.   The ability of a firm, in the short run, to produce the maximum level of output without any restrictions to the amount of variable inputs that can be used. Overcapacity in harvesting is a frequent problem in the management of common-pool resources. See *commons, tragedy of*, and *common-pool resources*.

capacity building.   The development of the skills and resources required to undertake a specific activity. In the context of climate change, this term often refers to the process of assisting developing countries to undertake mitigation and adaptation activities.

capacity factor.   The ratio of the average capacity that an electricity generation unit operates at over a specified time interval (usually a year) to the total engineering capacity of the unit, which is expressed as a percentage. Alternative terms are utilization factor or capacity utilization.

capacity output.   The maximum output level that can be produced given fixed inputs, the existing technology, the availability of variable inputs and the prevailing technology.

capacity utilization.   See *capacity factor*.

Cape Domain.   A region of the Cape Province of South Africa known for the high number of endemic species.

capillary action.   Movement of liquids via molecular attraction, such as water through soils.

capital.   Economic assets, such as durable goods, machines, transport vehicles, and other plant and equipment that is used in combination with other inputs to produce goods and services.

capital asset pricing model (CAPM).   A model of asset pricing that suggests that the returns received from an asset compensate the investor for the risks in holding that asset in a perfectly diversified portfolio.

capital consumption.   The physical deterioration or reduction of capital stock through use, age or withdrawal and depreciation.

capital costs.   The costs associated with plant and equipment.

capital deepening.   An increase in the stock of produced capital, such as machines, relative to the stock of inputs in an economy.

capital good.   A good that can be used to produce other goods, such as machines in a car factory.

capital intensive.   A method of production which uses capital relatively more intensively than the other factors of production. See *factors of production*.

capital market.   The so-called market where companies and governments go to seek long-term financing or loans.

capital rationing.   Process whereby market imperfections or interventions restrict the amount of capital that would normally be available to borrowers.

capital services.   The flow of services or benefits provided by a capital stock (such as a fishing vessel) over a given period of time.

capital stuffing.   A consequence of imperfect markets and rivalry in use whereby rational firms competing for a limited share of a fixed harvest invest in gear and equipment, and other forms of capital, to achieve a competitive advantage. In aggregate, the investment in such capital reduces overall returns to the industry. This has commonly occurred in the past in the fishing industry as a result of inefficient regulation.

capital theory.   An aspect of economic theory that refers to the optimal investment in an economy over time. Capital theoretic problems often involve the use of optimal control theory.

capital utilization rate.   The ratio of the current or observed capital stock (such as vessel size) to the "ideal" level of capital. Overcapitalization arises whenever capital utilization exceeds one, such that too much capital is being used relative to the ideal level. Overcapitalization is common in "race to fish" fisheries and open access resources.

CAPM.   See *capital asset pricing model*.

capture.   The economic concept that governmental or regulatory agencies adopt the same goals and interests as the industry they regulate, and effectively lose their independence.

carbon.   One of the most common elements in the environment and one which can co-form a large variety of different molecules, such as hydrocarbons and carbohydrates. Carbon is present in all organisms and can exist in a pure crystalline form, such as diamonds and graphite, or in a non-crystalline form such as charcoal.

carbon-14.   A radioactive isotope of carbon used in dating organic material. See *isotope* and *radiocarbon dating*.

carbon accounting. The procedure for measuring, valuing and aggregating greenhouse gas emissions and reductions. Carbon accounting is a central compliance activity for states or businesses participating in mitigation regimes, such as an emissions trading or carbon offset scheme.

carbon budget. A mechanism for accounting for the net sources of carbon released into the atmosphere.

carbon capture and storage (CCS). A technique being developed to retrofit existing, and incorporate into new carbon emissions-intensive technologies, such as coal-fired power stations, by way of special filters that prevent carbon dioxide from entering the atmosphere. It is proposed that after capture these emissions would be stored underground or deep within the ocean.

carbon credit. See *emissions credit*.

carbon cycle. The flow of carbon through the atmosphere, ocean, biosphere and lithosphere. Carbon flows from the atmosphere (as carbon dioxide) through to plants in the form of carbohydrates via photosynthesis, which is then metabolized by animals and then released as carbon dioxide into the air. This dynamic circulation of carbon forms the basis of the conditions for life on Earth. Anthropogenic emissions from fossil fuel combustion and deforestation add to this natural flow raising atmospheric carbon dioxide levels..

FURTHER READING
Ruddiman (2007).

carbon dating. See *radiocarbon dating*.

carbon dioxide. A gas consisting of one carbon atom and two oxygen atoms (CO_2). Carbon dioxide accounts for around 0.035 percent (350 parts per million by volume) of the Earth's atmosphere, which represents about a 25 percent increase over its pre-industrial level. Carbon dioxide is one of the most abundant greenhouse gases. See *greenhouse gases*.

carbon dioxide equivalent. For comparative purposes, greenhouse gases are sometimes converted into a carbon dioxide equivalent that represents the equivalent amount of carbon dioxide required to give the same radiative forcing. For example, the carbon dioxide equivalent of 1 ton of methane is 21 tons of carbon dioxide. See *greenhouse gases* and *radiative forcing*.

carbon dioxide fertilization.   The effect whereby increased concentrations of carbon dioxide in the atmosphere increase the rate of growth of some plants.

Carbon Disclosure Project.   A not-for-profit organization publishing information arising from the measurement and disclosure by major businesses and organizations of their carbon emissions. See *corporate social responsibility*.

carbon emissions.   Anthropogenic emissions of carbon released into the atmosphere. The main sources of emissions include the use of fossil fuels and the clearing and burning of forests.

carbon equivalent of carbon dioxide.   A measure commonly used to convert tons of carbon dioxide into tons of carbon whereby 3.7 tons of carbon dioxide equals 1 ton of carbon.

carbon finance.   Term used to describe public and private financial resources that are made available for investments in activities directed at mitigating climate change.

carbon footprint.   The volume of greenhouse gas emissions attributable to the actions or behavior of an organization, individual, or state over a particular period.

carbon fund.   See *carbon finance*.

carbon intensity.   See *emissions intensity*.

carbon leakage.   The potential diversion of greenhouse gas producing activities by relocation from countries that undertake mitigation activities to those that do not. For example, prevention of deforestation in one country could simply prompt logging companies to switch their operations to countries without regulations. This topic is particularly pertinent to the issue of binding emissions targets in developed countries and potential leakage to developing countries. Lobby groups for emissions intensive industries regularly raise this issue as an argument against domestic mitigation, since, in their view, a carbon price would render local industry uncompetitive and emissions would simply shift offshore. Various measures have been proposed to mitigate potential economic losses from this situation, including border taxes, labeling and compensation.

FURTHER READING
Frankel (2009).

carbon market. The market for exchange of greenhouse gas emissions reductions, permits and offsets. See *emissions trading scheme*.

carbon monoxide. A colorless and odorless but highly poisonous gas formed by the incomplete combustion of carbon-based fuels that is found in exhaust fumes, cigarette smoke and other sources.

carbon neutral. The state of being whereby a single or collective group of actions and processes has a zero carbon footprint. In practice, this is usually achieved by offsetting emissions from carbon intensive activities with emissions reductions achieved elsewhere. For example, purchasing carbon emissions offsets from carbon sequestration activities undertaken in another region or country. See *carbon footprint* and *carbon offset scheme*.

carbon offset scheme. A formal or informal scheme wherein an individual, business, or state finances the offset of their greenhouse gas emissions by paying an intermediary to engage in emissions reductions activities on their behalf. Such activities may include: preventing deforestation or undertaking reforestation; replacing planned emissions intensive projects, such as a coal fired power plant, with renewable alternatives; or retrofitting emissions producing technologies. Formal schemes generally produce credits that can be used in emissions trading markets. See *Clean Development Mechanism, Joint Implementation,* and *Reducing Emissions from Deforestation and Forest Degradation.*

carbon pool. See *carbon reservoir*.

carbon price. The economic value associated with anthropogenic greenhouse gas emissions, usually expressed in terms of one metric tonne of carbon dioxide equivalent. A carbon price arises either explicitly through the imposition of a carbon tax, or implicitly through an emissions trading scheme and other climate change mitigation activities. Setting a value on greenhouse gas emissions attempts to incorporate the negative externalities associated with climate change into economic decisions, and thereby assists in moving an economy away from the consumption and production of carbon intensive to less carbon intensive goods and services. See *carbon tax, emissions trading scheme* and *externality*.

carbon reservoir.   Components of the carbon cycle that absorb and/or release carbon, such as the ocean and forests. See *carbon sinks*.

carbon sequestration.   The process by which the uptake of carbon dioxide in carbon sinks is increased, such as by planting trees, increasing the growth rate of trees, or by ocean fertilization. Sequestration may also involve the long-term storage of carbon or carbon dioxide in underground sites. This important component of climate change mitigation is the purpose of the REDD mechanism. See *Reducing Emissions from Deforestation and Forest Degradation.*

carbon sinks.   Components of the climate system other than the atmosphere that absorb more carbon dioxide than they release, such as growing forests and ocean sediments. See *carbon sequestration.*

carbon starvation.   Mechanism in which water-stressed trees shut down transpiration while continuing respiration, eventually killing the tree.

carbon stock.   The quantity of carbon contained in a carbon sink or reservoir. See *carbon sink.*

carbon tariff.   An import tax levied on greenhouse gas-intensive products by countries where local industries are subject to stricter climate change related regulations. See *carbon leakage.*

carbon tax.   A tax applied to emissions of carbon dioxide and other greenhouse gases by industrial producers that attempts to account for their external costs. The relatively higher prices for carbon intensive goods and services under a carbon tax provides an economic incentive for the consumption and production of less emissions intensive goods and services. See *carbon price* and *emissions trading scheme.*

carbon tetrachloride.   A toxic chemical formerly used in the dry-cleaning industry.

Carboniferous period.   Also known as the coal age, this occurred from 360 to 286 million years ago.

carcinogen.   A natural or synthetic substance that causes the growth of cancer cells.

carnivore.   Animals, such as wolves or the big cats, the diet of which consists primarily of the flesh of other animals.

carrying capacity.   Maximum number of individuals, or the biomass of a species, that can be supported by its natural environment. The term is sometimes applied to a collection of species or even the entire Earth.

Cartagena Protocol on Biosafety.   See *Biosafety Protocol*.

cartel.   A group of producers of a commodity or a good or service whose members collaborate to set prices and/or quantity produced by each member so as to increase their returns at the expense of their consumers. The organization of petroleum exporting countries (OPEC) was a highly effective cartel in the 1970s and was able to transfer a large amount of wealth from oil-consuming countries to oil-exporting countries.

cash flow.   The payments to and from a firm or organization over a given period of time. A positive (negative) cash flow exists when cash revenues exceed (are less than) cash payments.

casing.   A pipe cemented into a well that prevents the structure caving in, such as the case of wells used for hydrocarbon extraction.

catabolism.   The breakdown at the molecular level of complex substances into simpler forms as part of an organism's metabolism.

catadromous fish.   Fish that spend most of their adult life in freshwater but breed or spawn at sea, such as the American eel.

catalytic converter.   A device fitted to the exhausts of vehicles that reduces emissions by helping to burn hydrocarbons and converting carbon monoxide to carbon dioxide.

catastrophe theory.   The hypothesis that life on Earth is subject to sudden and massive evolutionary changes due to catastrophic events.

catastrophic climate change.   Sudden and highly damaging climate change arising from strong amplification of anthropogenic greenhouse gas emissions by feedback effects in the climate system. Scientists believe that potential consequences associated with this situation would include: widespread water and food insecurity; higher incidence of extreme events such as hurricanes, floods, fires and droughts; mass species extinction;

conflict arising from resource scarcity; flows of environmental refugees; and inundation of major urban centers by rising sea levels. The economic, environmental and social costs of this situation would pose an existential threat to human society. See *4° Celsius, climate feedback,* and *climate threshold.*

catch per unit of effort (CPUE).   A measure sometimes used in fisheries as an indicator of stock abundance, calculated by dividing the total catch by a standardized measure of fishing effort (such as days at sea). By itself, it is not a good measure of stock abundance because declines in some species are not reflected in concomitant declines in the CPUE. Further, if fishers become better at fishing over time, their ability to catch fish may increase per unit of fishing effort, which may, in turn, compensate for declines in the resource.

catchability coefficient.   A shift parameter that accounts for the largely unexplained harvest of fish, or catches that are attributable to things other than direct fishing effort and the stock of fish. A catchability coefficient is often used in harvest equations (as functions of effort and stock) to account for changes in the technology of catching fish. See *technical efficiency.*

catchment.   An area of land which drains into a water system, such as a stream or river.

CBD.   See *Convention on Biological Diversity.*

CCD.   See *Convention to Combat Desertification.*

CDC.   See *Center for Disease Control.*

cell fusion.   The practice of combining cells from plant species that would normally not interact.

cellulosic feedstock.   See *biomass fuel.*

Celsius.   Temperature scale named after Anders Celsius which defines the melting point of ice as 0 degrees Celsius and the boiling point of water as 100 degrees Celsius. A 1 degree change on the Celsius scale is equivalent to a 1.8 degree change in the Fahrenheit scale. See *Fahrenheit scale.*

cement.   Building material made from limestone and clay that is mixed with water to make concrete. Cement production has been estimated to contribute approximately 5 percent of global carbon dioxide emissions.

Cenozoic. Geological era covering the time period from 66.4 million years ago to the present.

censored regression model. See *Tobit model*.

census data. Data typically drawn or obtained by central governments or local government agencies from the general population on a regular basis. Variables typically surveyed are household size, income, and the level of education.

centimeter. Unit of length equal to 0.3987 inches.

Center for Disease Control (CDC). The US agency charged with promoting health by preventing and controlling disease.

center-periphery model. Model of economic development which suggests that the periphery (such as developing countries) are hindered in their development by the center (such as rich countries) through a variety of factors, including a deteriorating terms of trade. See *terms of trade*.

centralization. A process by which activities and decisions, over time, become located in a single place.

central limit theorem. The theorem states that for a random variable, with a given population mean and variance, the mean of a sample with sufficiently large enough observations will be approximately normal. Moreover, the sample mean will equal the population mean and the sample variance will equal the population variance divided by the number of observations in the sample, for a sufficiently large number of observations.

CERCLA. An acronym for the US Comprehensive Environmental Response, Compensation and Liability Act, and better known as the Superfund. The act, which came into effect in 1980, was designed to create funds to pay for the clean-up of hazardous waste sites.

CERCLIS. An acronym for the US national inventory of hazardous waste sites.

cerrado. Local term for a type of savanna found in northeastern Brazil.

certainty equivalent.   The smallest amount of money an individual would be prepared to accept for certain in an exchange for an uncertain return, such as from a lottery or gamble.

certified emissions reduction (CER).   An emissions credit issued for greenhouse gas emissions reductions that occur within the framework of the Clean Development Mechanism. See *Clean Development Mechanism* and *emissions credit*.

cetaceans.   Marine mammals that include whales, dolphins and porpoises and collectively represent 80 species. Toothed whales (*odontocetes*) include orca, sperm whales and all dolphins and porpoises. Baleen whales *(mysticetes)* have no teeth, instead filter feeding through plates called baleen which are lined with bristles. See *baleen whales.*

ceteris paribus.   Latin term meaning "all other things being equal". The term is frequently used in economics in partial equilibrium analysis when the effects of a change in one variable are analyzed holding other variables unchanged.

chain reaction.   A process by which a fission reaction begins in one nucleus of an atom but causes the release of at least two neutrons that, in turn, cause a fission reaction in two other atoms, and so the process continues. An uncontrolled chain reaction may result in an explosion due to a sudden and massive release of neutrons.

channelization.   The altering of stream or river flow by the construction of banks and levees and by the widening or dredging of river beds. Channelization is often done to improve navigation and transportation but can impose significant environmental costs.

chaos.   Refers to a particular behavior of dynamic systems and observed by Edward Lorenz in the 1960s. Chaos is characterized by a bounded steady state which is neither an equilibrium or periodic and where even smallest changes in the initial conditions can lead to completely different dynamic behavior.

FURTHER READING
Gleick (2008), Cohen and Stewart (2000).

chaotic system.   See *chaos*.

characteristic equation. An equation the roots of which are called the characteristic roots or eigenvalues. A characteristic equation is the determinant of $A - \lambda I = 0$ where A is the matrix of coefficients, λ corresponds to the eigenvalues and I is the identity matrix.

FURTHER READING
Simon and Blume (1994).

charge. See *Pigouvian tax*.

charismatic fauna. Term given to animals that have a strong appeal in soliciting support (such as baby seals) for protective measures.

chemical oxygen demand. A water pollution measure which calculates the dissolved oxygen taken up by organic matter in a sample of water.

Chernobyl. Site, in Ukraine, of the world's worst civilian nuclear disaster. The disaster arose in April 1986 after a series of explosions in one of the reactors at the Chernobyl nuclear plant scattered radioactive debris over a large area of land.

chicken game. A game which has its origins in the actions of two individuals who are racing their cars at each other. If one of them veers away, he or she is deemed to be a "chicken" and thus loses self-esteem. If both veer away at the same time, neither gains self-esteem at the expense of the other and if neither veer away they both die. More generally, a chicken game can describe the actions of agents or countries in negotiations, such as climate change. For example, a country's reputation for defection (veering away) may reduce the likelihood of other countries from defecting (veering away) so as to avoid a potential catastrophe. See *game theory*.

chimera. Organism that has DNA from more than one species.

China syndrome. Term used for the melt-down of a nuclear reactor.

chlorinated hydrocarbons. Commonly called organochlorines, these are synthetic compounds that combine molecules of carbon, hydrogen and chlorine. Many chlorinated hydrocarbons have been used as insecticides, including DDT.

FURTHER READING
Thornton (2000).

chloroacne. A debilitating skin disease caused by contact with some chlorine-based chemicals, such as dioxins and TCDD. See *dioxins* and *TCDD*.

chlorofluorocarbons (CFCs). Compounds (such as freon) that contain chlorine, fluorine and carbon, which were widely used as refrigerants and propellants. Their use has been phased out following the discovery of the "ozone hole" over Antarctica in the 1980s and the confirmation that CFCs were the principal cause of the breakdown of ozone (O_3) in the upper atmosphere at low temperatures. See *Montreal Protocol*.

chlorophyll. Name given to the green pigment in organisms capable of photosynthesis.

choice experiments. Experiments carried out to test assumptions about human behavior and decision making against standard economic precepts.

choice variable. A variable for which value is solved in an optimization problem.

choke price. 1. The upper limit on the price a resource can reach before alternative resources or technologies will be used. 2. The price at which the demand for a good is just zero and for which any diminution in price will result in an at least measurable demand for that good.

cholera. A bacterial disease that can be spread by contaminated water and food. The disease causes diarrhea and dehydration, and is sometimes fatal. The best means of prevention is adequate sanitation and housing, as well as not eating shellfish.

chromosome. A thread of DNA found in the nuclei of plants and animals that carries the genes for inherited characteristics.

Chow test. A statistical test named after Gregory Chow that tests for the equality of coefficients from different sample periods under the assumption of normally distributed errors. The test is frequently used to evaluate whether structural change exists over a given period of time.

FURTHER READING
Chow (1960) and Hill, Griffiths and Judge (2001).

circular-flow economy. 1. A model of an economy which focuses on the flows of income between producers to households (through, for example, the payment of wages and salaries) and from households to producers (through, for example, consumer expenditures). 2. A model of an economic system that incorporates not only the transformation of materials into goods for human consumption, but also the generation of waste and the resulting impact on environmental systems.

CITES. See *Convention on International Trade in Endangered Species.*

civil society. A term used to describe the non-governmental organizations and individuals that contribute their time and resources for the benefit of others, as well as the good of society.

clade. Related species that share an identifiable common ancestor.

cladogram. Diagram that illustrates speciation from a common ancestor plotted against time.

class. (Plural: classes.) A classification of organisms by order. For example, Mammalia are a class of animals that includes the order of Primates. See *order*.

class action suit. A civil law suit most commonly used in the United States whereby a lawyer or a group of lawyers sue for damages on behalf of a group of individuals, each of whom need not be named specifically in the suit.

classical economics. A term used for the body of thought developed mainly by British economists in the late eighteenth to the mid-nineteenth centuries. The best known classical economist was Adam Smith who published his *Wealth of Nations* in 1776.

clathrate gun hypothesis. The theory that rising sea temperatures will trigger methane, one of the most powerful greenhouse gases, to be released from methane clathrate on the seabed into the atmosphere, thus causing rapid and devastating climate change. Of particular concern is the melting of seabed permafrost in the Arctic, the location of vast methane deposits. See *Arctic shrinkage* and *catastrophic climate change*.

clean coal technology. See *carbon capture and storage*.

Clean Development Mechanism (CDM). A flexibility mechanism introduced under the Kyoto Protocol to assist Annex B countries achieve their targeted greenhouse gas emissions reductions by reducing emissions through projects in countries that are not obligated, under the Protocol, to undertake emissions reductions. Despite wide participation, the CDM has been fraught with administrative difficulties pertaining to approval and monitoring of projects. Examples of CDM projects include: nuclear power plant construction, reforestation and hydropower development. Critics have argued that the CDM has financed projects that would have proceeded regardless, and, in some cases, encouraged the production of emissions that are then paid to be removed. See *additionality*, *carbon offset scheme*, and *Kyoto Protocol.*

clearcut.   A forest site in which all the timber has been felled.

clearcutting.   See *clear felling.*

clear felling.   A forestry term used to describe a practice whereby all the trees in a given area are felled when harvesting. Clear felling, especially on steep slopes and soils prone to erosion, can impose significant environmental costs.

climate.   Description of weather patterns over long periods of time (minimum of 30 years) as defined by the mean, range and variability of various weather measures including temperature, precipitation, sunshine hours, wind, and extreme weather events.

climate change.   Past, present and future global change in the Earth's climate. Climate change has occurred as a natural process throughout the Earth's history. Today, a scientific consensus exists that global warming arising from human activity is dramatically altering the climate system in an unprecedented narrow timescale. Anticipated effects extend beyond rising temperatures, including: occurrence of extreme weather events; substantial modification of rainfall patterns; ocean acidification that seriously injures marine ecosystems; and rising sea levels. In the absence of significant and urgent action, environmental and economic damages are expected to be very high, and disproportionately impact upon the livelihoods of the poorest societies and individuals, and future generations. See *climate feedback*, *climate change mitigation*, *greenhouse effect*, and *global warming.*

FURTHER READING
Archer and Rahmstorf (2010) and Ruddiman (2007).

climate change adaptation. Anthropogenic responses towards moderating the potential costs associated with climate change. Potential measures include: construction of barriers against rising sea levels; water-use efficiency programs and expansion of water sources, such as desalination plants; and development of drought resistant crops and improving access to food for vulnerable communities. Adaptation measures are already being implemented in many countries, and governments and the international community are currently mobilizing resources and devising plans to limit damages from future climate change, particularly those occurring in developing countries. See *climate change mitigation*.

climate change commitment. A term referring to the committed level of future climate change caused by past greenhouse gas emissions. As a result of feedback loops, the climate system would take a lengthy period to reach a new equilibrium even if anthropogenic emissions ended immediately. Scientists believe that prior emissions may have already committed the Earth to future warming of 1.5°C. See *climate feedback*.

climate change mitigation. The process by which the production of greenhouse gas emissions is constrained by anthropogenic measures such as fuel switching (including renewable fuel sources), changes in land use, reducing energy consumption and carbon sequestration. Policy instruments for climate change mitigation include: emissions trading schemes, carbon taxes, and investment in research and development. The early consensus of economists concerning cost-efficient mitigation was that it should be a gradual and responsive process. However, more recent advancements in climate science suggest that this strategy is unsuited to the risks posed by catastrophic and irreversible climate change. The Stern Review (2007) altered the discourse concerning the international case for mitigation, and, in subsequent years, the dominant view has become in favor of urgent and substantial mitigation as a form of risk insurance. See *carbon tax*, *emissions trading scheme*, and *Stern Review*.

climate cycle. A naturally occurring pattern of changing climate resulting from the complex dynamics inherent to the Earth's climate system. Such cycles can occur over vastly different time frames, from seasonal variations up to millions of years. Many different cycles have been observed from the paleoclimate record, and each of these may reinforce or weaken the effects of others. Examples include: Milankovitch cycles, the North Atlantic Oscillation and the El Niño Southern Oscillation.

FURTHER READING
Ruddiman (2007).

climate downscaling.   Statistical techniques to project future climatic conditions derived from various global climate models onto specific regions.

climate envelope models.   Techniques to predict what species or ecosystems might be better adapted to future climatic conditions in a region (see climate downscaling), inferring from where those climatic conditions currently exist today and what species and ecosystems are found there.

climate feedback.   Term that refers to the potential of one variable (or set of variables) in a climate system to affect, positively or negatively, another variable (or set of variables) which in turn may affect the original variable. For instance, a positive feedback (that increases the rate of global warming) occurs when initial warming reduces snow and ice cover in high latitudes, thus lowering surface albedo, which, in turn, retains more heat within the climate system. Other examples include: warmer oceans absorbing less carbon dioxide; release of methane gas from melting permafrost; and higher temperatures increasing atmospheric water vapour, the primary cause of the greenhouse effect. See *albedo feedback*, *permafrost*, and *water vapor*.

climate finance.   Financial resources available in the form of grants, subsidies or loans that assist nations, businesses or households to make investments in activities directed at climate change mitigation and adaptation.

climate forcing.   See *radiative forcing*.

climate lag.   The delay that occurs between some factor in the climate system changing and the full effects being realized in the system as a whole. This phenomenon is attributable to feedback loops and inertia present in certain components, as well as the broad complexity of the system. A typical example is the lag in atmospheric greenhouse gas concentration and subsequent global warming.

climate legislation.   Laws instituted by government towards climate change mitigation.

climate modeling.   A mathematical and systems approach to understanding and predicting the climate. The increase in computing power available to researchers since the 1980s has led to the development of complex climate models on a global scale. Global circulation models have been used to predict the potential climate changes associated with increases in the concentration of carbon dioxide in the atmosphere.

climate policy.   Government laws and measures relating to climate change within a national or international context. International policy directed towards climate change mitigation, as well as adaptation, takes place under the auspices of the UNFCCC. Climate policy covers a range of different issues, including: mitigation, adaptation, technology transfer and deforestation. See *bottom-up*, *top-down*, and *United Nations Framework Convention on Climate Change.*

FURTHER READING
Aldy and Stavins (2009).

climate projection.   A projection arising from climate modeling of the response of the Earth's climate system to changes in greenhouse gas emissions and other forms of radiative forcing.

climate refugee.   A person who is displaced from their home or country as a result of changing climate. It is hypothesized that rising sea levels and localized water scarcity may generate the future movement of millions of climate refugees.

climate scenario.   A representation of future or past climate originating from climate models. Multiple scenarios are considered in IPCC reports to account for uncertainty surrounding the complex relationships and assumptions underpinning climate modeling.

climate sensitivity.   A measurement of the climate system's adjustment to radiative forcing from greenhouse gas emissions. Defined specifically as the long-term equilibrium change in mean surface temperature following a doubling of the pre-industrial concentration of carbon dioxide in the atmosphere from 280 parts per million by volume (ppmv) to 560 ppmv. See *radiative forcing.*

climate skeptic.   See *skeptic.*

climate system.   The complex system within the Earth's environment comprising: the atmosphere, cryosphere, hydrosphere, biosphere and land. The dynamic state of this system is subject to the interactions between these components, as well as external forcing from, for example, anthropogenic greenhouse gas emissions, volcanic eruptions and variations in solar radiation.

climate threshold.   A tipping element present in the climate system, which if activated by global warming could initiate unstoppable feedback loops resulting in catastrophic climate change. Events associated with exceeding a climate threshold include: melting of the Arctic, Greenland and West Antarctic ice sheets; disruption of the Atlantic thermohaline circulation, El Niño Southern Oscillation and West African monsoon; and loss of the Amazon and boreal forests. The estimated probable bound for most of these events to occur is a 3 to 5°C global temperature rise above pre-industrial levels. Melting of the Arctic and Greenland ice sheets is predicted to occur with a 0 to 2°C rise.

FURTHER READING
Lenton et al. (2008).

climate variability.   Generally used with reference to natural deviations in long-term weather patterns as a consequence of the internal dynamics of the climate system.

climatology.   The study of climate and its variability over time.

climax community.   An ecological term for the developed or mature community of species that arises after a succession of communities and which is often characterized by a greater species diversity than less mature communities. For example, an old-growth forest is a climax community that harbors a wide variety of species.

clone.   The offspring of an organism that is genetically identical to its parent.

closed area.   Term used in management of natural animal or plant populations whereby harvesting or hunting in an area of land or ocean is prohibited. The use of closed areas can be an important management strategy for ensuring the sustainability of animal populations and can be used, for example, to protect spawning areas.

closed economy.   An economy which has very little or no trade with other countries. See *open economy*.

closed forest.   Forest with at least 30 percent canopy or tree crown cover.

closed loop.   A feedback control system whereby actions respond to changes in variables. See *open loop*.

closed-loop recycling.   The recovery and re-use of residual products as a source of additional material for the same product, as in the use of old newspapers to make newsprint.

closed season.   A period of time when harvesting or hunting is prohibited. Closed seasons are frequently used to ensure the sustainability and viability of animal populations at times when they might be highly vulnerable to capture, such as in the breeding season.

closed system.   A term used to define a physical system where there is an exchange of energy with the environment but no change in mass or matter. On a grand scale, the Earth and its atmosphere is a closed system. See *open system*.

cloud feedback.   A form of climate feedback involving the mutual impacts of changing climate and cloud composition. An example of cloud feedback would be higher temperatures increasing evaporation, thus causing greater cloud formation and greater reflection of solar radiation away from the Earth. See *climate feedback*.

club good.   A good or set of goods the use of which can be restricted at little cost but is congestible such that, past a certain point, the greater the number of users the less satisfaction or utility is derived per individual from its use. The use of the facilities in a health club is a club good.

Club of Rome.   A group of people who brought their collective skills and experience together to examine the state of humanity and the Earth. A famous publication of the group was the book *Limits to Growth* that predicted the medium- and long-term supply and use of natural resources.

FURTHER READING
Meadows et al. (1974).

CO$_2$ equivalent.   See *carbon dioxide equivalent*.

co-adaptation.   The mutual adaptation of at least two different species that is common in many predator–prey and pathogen–host interactions.

coal.   A brown or black sedimentary rock consisting of organic matter that has undergone partial decay in the absence of air. Although the quality of coal covers a continuous range, the four main categories in ascending order are: lignite, sub-bituminous coal, bituminous coal, and anthracite. Coal's high carbon content, combustible properties and relative abundance has made it widely used to produce heating and electricity. However, the carbon dioxide released from its exploitation is one of the largest sources of greenhouse gas emissions. See *black coal* and *brown coal*.

coal bed.   A geological layer of coal that lies parallel and is clearly distinguished from surrounding rock stratification.

coal-bed methane.   Natural gas found in porous coal seams that was trapped during the formation of the coal. Some methods used to enhance extraction involve the use of hydraulic fracturing (or fracking) that uses both water and chemicals to help open underground coal seams. Fracking can, in some locations, impose substantial externalities by reducing both the quality and quantity of water available in aquifers.

coal-fired power generation.   The process whereby coal is combusted to heat water and produce steam for turning an electricity-generating turbine. Although coal-fired power generation is one of the cheapest and easiest ways to produce electricity, it is also one of the most environmentally harmful. See *carbon capture and storage* and *renewable energy*.

coal-seam gas (CSG).   See *coal-bed methane*.

Coase Theorem.   Theorem named after Ronald Coase, a Nobel Laureate in Economics, in recognition of his insights in his paper "The Theory of Social Cost" (1960) in which he examined how the assignment of property rights can be used to overcome the problems of pollution. The Theorem is commonly interpreted as stating that the assignment of a property right can be used to internalize an externality and it does not matter, in terms of economic efficiency, which party (the polluter or persons suffering from the pollution) is assigned the right. The Theorem holds true only when people do not behave strategically, transactions costs are zero, information is perfect and the allocation of rights does not affect the marginal valuations of individuals.

FURTHER READING
Coase (1960).

coastal erosion.   Erosion of shorelines from tidal current and from storms that remove sediments.

coastal management.   The suite of policies and measures for protecting coastal lands and settlements from incursion by the ocean. A high proportion of the Earth's population reside on the coast, and the construction of dykes and breakwater structures, respectively, to prevent flooding and erosion dates back centuries. In anticipation of future sea level rises from global warming, adaptation strategies are being formulated, including: resettlement, vegetation of coastal dunes, and seawall construction.

coastal subsidence. Subsiding shorelines due to the gradual settling of newly deposited sediments.

Cobb–Douglas function. A functional form, for both production and utility, in which at least two factors are combined multiplicatively (e.g., $Y = gK^aL^{1-a}$). It has the characteristic that how the firm or consumer substitutes one good for another is independent of either income or production levels.

co-benefit. An additional benefit arising from an action that is directed at a particular purpose, where the extra benefit does not directly relate to the original purpose. For example, a co-benefit of reforestation for the purposes of climate change mitigation is the creation of animal habitats.

cobweb model. A model named after the shape of the direction of the movement in the price of a good if lags exist between the supply and demand. The dynamic model explains how prices adjust such that if supply exceeds demand, at a given price, the price will fall, which, in turn, will reduce the supply of the good in the next period leading to an increase in price. Under certain conditions, these movements in price will dampen over time and will converge to an equilibrium price where supply equals demand at the existing market price.

Codex. A set of standards on the maximum residue limits in foods devised by a committee established by the FAO and WHO.

coefficient of determination. A measure of the variation in a variable accounted for by other variables. In linear regression analysis, it is called R^2 and is the ratio of the explained sum of squares to the total sum of squares. The coefficient of determination is a frequently used measure of the goodness of fit of a model and is bounded by 0 (the model explains none of the variation in the dependent variable) and 1 (the model explains all the variation in the dependent variable).

coefficient of performance. An efficiency measure for heating and cooling appliances derived from the ratio of energy input to energy output.

coefficient of variation.   A measure of variation of a variable defined as the standard deviation (square root of the variance) divided by arithmetic mean. The higher the value, the greater the variability of the variable.

co-evolution.   See *co-adaptation.*

cogeneration.   The production of both power and heat from a power source or plant. For example, a diesel engine that generates power also generates heat that can then be used for heating purposes. Cogeneration increases the energy efficiency of power plants.

cognitive ability.   The capacity to fully assess all of the benefits and or costs of different choices.

cohort.   Term commonly used to describe individuals of a species of a similar age or maturity. Cohorts are particularly important in fisheries management as one or two cohorts may provide most of the fish that are available for harvest.

cohort analysis.   Analysis in which the focus is on a subset of a population of a particular age. See *cohort.*

cointegration.   A statistical term that describes whether a linear combination of two variables is stationary such that its mean and variance are time invariant. See *stationarity* and *order of integration.*

cold war.   The stand-off that existed between the United States with its allies (primarily in NATO) and the Soviet Union with its allies (primarily in the Warsaw Pact), and that resulted in the global arms race from the 1940s to the 1980s.

coliform bacteria.   Rod-shaped, anaerobic bacteria belonging to the group that includes *Escherichia coli.* Coliform bacteria usually are associated with the large intestines of vertebrates, especially mammals, and are often used as a proxy for possible human pathogens in water. See *coliform index* and *Escherichia coli.*

coliform index.   A measure of water pollution which counts the number of coliform bacteria in a given amount of water.

collective action.   Actions by at least two or more individuals to coordinate their activities. Collective action can be an important means of overcoming externalities. See *community rights*.

collective good. See *public good*.

collective rights. The traditional rights to a resource, area, or practice held by a community or people but which may not formally be recognized. See *community rights*.

colloid. Very small particles dispersed in soils, such as clay.

colony. A group of organisms of the same species that live in a mutually dependent way, such as bees in a hive.

combined sewer. A sewer that includes stormwater and untreated sewage.

combustion. The chemical reaction whereby a substance combines with oxygen to produce energy, also known as burning. Common byproducts of these reactions are greenhouse gases, such as water vapor and carbon dioxide. Combustion is an essential process in deriving energy from fossil fuels.

comet. A small object made of ice and dust found in the solar system.

command and control. An approach to pollution prevention and natural resource management that relies upon direct regulations and standards with enforcement and fines.

command economy. An economy where much of the economic activity is directed by the state, such as was the case in the Soviet Union.

commensalism. Interaction between at least two different organisms where one benefits from, but does not harm, the other.

commercial fisheries. Fisheries, usually large scale, that use modern techniques and gear, and where fishing is done largely for profit.

commercial forestry. The practice of growing and harvesting trees for the purpose of producing wood products.

commercial value. The value of a good or resource only in terms of its market value, ignoring its non-market value. Thus the commercial value of a forest is its value as timber, but this does not include its value in providing amenity and recreation.

commercial wood.   Wood that has commercial value in construction or manufacturing.

Commission for Environmental Cooperation (CEC).   A commission established under the North American Free Trade Agreement by the United States, Canada and Mexico to address environmental issues in North America, and in particular, environmental concerns due to trade liberalization.

commitment period.   The period in which Annex II countries are required to fulfill emissions reductions commitments within the UNFCCC framework. Under the Kyoto Protocol 2008–2012 was designated as the first commitment period.

committed warming.   See *climate change commitment*.

Committee on Trade and the Environment.   A committee of the World Trade Organization (WTO) commissioned to investigate ways of reconciling measures to protect the environment and facilitating trade.

common law.   Law based on judicial precedent rather than upon enactment in legislation. Common law forms the basis of property rights in many English-speaking countries, including the United States.

common-pool resource.   A natural resource which is both rivalrous in use and from which it is difficult or costly to exclude users. If the costs of exclusion are high, or if institutional arrangements do not exist to create or enforce property rights over common-pool resources (CPRs), the exploitation of common-pool resources can result in the "tragedy of the commons" whereby the resource is overexploited in both a biological and economic sense. Despite the potential for overexploitation of CPRs, many examples exist of CPRs which have been sustainably managed, often through the use of community rights.

FURTHER READING
Ostrom (1990) and Ostrom, Gardner, and Walker (1994).

common-property resource.   An often misused term that was incorrectly interpreted as natural resources where users had no property rights over the stock or the flow from the resource. Today, the term is widely accepted as defining a resource over which a community has property rights and where the community can exclude non-members from its use. See *community rights*.

commons. A popular term for common-pool resources. Sometimes the term is interpreted as a common-pool resource (CPR) to which users have no property rights or a CPR over which there exists community rights. See *common-pool resource*.

commons, tragedy of. A term popularized by Garett Hardin in his 1968 paper of the same name. The tragedy of the commons is the biological and economic over-exploitation of common-pool resources (CPRs) when there exists no property rights by users over a CPR, or flow of benefits from a CPR. See *common-pool resource*.

FURTHER READING
Hardin (1968).

communal rights. See *community rights*.

community. A group of animals and plants living together in a defined area.

community development quotas. A program in the Alaska fisheries through which communities are allocated a share of the Bering Sea commercial catch.

community rights. Property rights, defined by a community, that prevent outsiders from using natural resources and that include rules about how the members of the community can use the resources.

FURTHER READING
Ostrom (1990).

comparative advantage (theory of). Theory of trade developed by the nineteenth-century British economist, David Ricardo. It states that countries will export those goods in which they have a relative (rather than absolute) production cost advantage.

comparative statics. A method of analysis whereby changes to a model or a system are analyzed by assessing the changes in one or more variables following a change in another variable.

compensating surplus variation.   A measure of welfare equal to the change in income that will make an individual indifferent between some original state of the world and different state of the world, such as a different set of prices for consumption goods.

compensating wage differentials.   Differences in real wages received by workers in various occupations based on their relative preferences between those occupations.

compensation principle.   The idea that one can distinguish whether the move from one social state to another is an improvement, in terms of social welfare, if the benefits to those made better off by the move outweigh the losses suffered by those made worse off by the move. Determining whether this is the case requires choosing a criterion to determine how those benefits and losses are weighed against one another. See *Kaldor Potential Compensation* and *Scitovsky compensation criterion*.

compensatory growth.   Natural population growth which increases (or compensates) in response to a reduction in the size of a population below its capacity. See *depensatory growth* and *logistic growth*.

competition policy.   Policy, rules and legislation designed to prevent collusive or predatory business practices that are not in the long-term interest of consumers.

competitive exclusion.   Biological competition between plant or animal species that results in the exclusion of one or more species from a resource or geographic area.

complement.   A good or commodity that is consumed or produced with another. For example, cream is a complementary good to coffee as they are often consumed together. See *substitute*.

complements in consumption.   Goods that are combined in consumption. Two goods are complements when an increase in the price of one good causes a reduced demand for both goods.

compliance.   The extent to which individuals or firms conform to environmental regulations.

compliance costs.   Costs of monitoring and enforcing environmental regulations.

compliance market. The market for carbon credits enabling participants in an emissions trading scheme to uphold their reduction commitments.

compost. A natural fertilizer that is a product of the aerobic breakdown of organic material.

compound interest. See *compounding*.

compounding. Financial term to describe a method of paying or receiving interest on both the principal (amount borrowed or loaned) and any accrued interest.

computable general equilibrium (CGE) models. Economic models which link different sectors of the economy and that can be used to evaluate the effect of different policies. For example, CGE models have been used to assess the impact of carbon taxes on employment and income in various sectors of the economy, such as agriculture.

concave function. A function where the area beneath the function is a convex set. See *convex function*.

Concentrated Solar Power (CSP). Term for a class of solar power technologies that use mirrors and lenses to reflect large quantities of sunlight upon a concentrated area. A common design involves heating a liquid to very high temperatures within a column, and the resulting steam turning an electricity producing turbine. See *solar thermal energy*.

concentration. A measurement of the relative quantity of a substance present within a liquid, gas or solid.

concentration ratio. Ratio which is used as an index of the market share of firms. A commonly used concentration ratio is the so-called Herfindahl–Hirschman index which sums the square of the market share of each firm and has a value of 1 if there is only one firm in the market.

concessional debt. Debt arising from terms that are below market rates. Typically, such terms are extended by the creditor with the aim of achieving a certain mandate.

condensation. The transformation of a gas or vapour to a liquid state.

conditional probability. The probability of an event given that another event has already occurred. The conditional probability of an event A occurring if B occurs is written as Pr(A|B).

conduction.   The transfer of energy from one object to another through direct contact.

Conference of the Parties (COP).   The title of the supreme body of the UNFCCC. The approximately 190 countries party to the convention assemble at these sessions on an annual basis to negotiate international climate policy. In addition to government officials from member countries, these conferences attract thousands of participants from United Nations bodies, non-government organizations and the media. The decisions and documents emanating from these meetings form the basis of global action to mitigate against and adapt to climate change. For example, COP-3 in 1997 led to the adoption of the Kyoto Protocol and COP-15 produced the Copenhagen Accord. See United Nations Framework Convention on Climate Change.

confidence interval.   An interval defined by two numbers such that a certain probability is assigned to the true value of an estimated parameter being within the interval. Confidence intervals are calculated by using sample data from a population under the assumption that the sample is representative of the population, except for some random error. A commonly used confidence interval is a 95 percent interval where, using the sample data, there exists a 95 percent probability that the true value of parameter lies between the lower and upper bound of the interval.

FURTHER READING
Greene (2003).

congestible.   Term used to describe a good or asset where beyond a certain threshold of use, further use or consumption detracts from the enjoyment of others using it. For example, at low rates of visitation to a national park an extra visitor does not affect the enjoyment of other visitors, but at high rates of visitation (such as to Yellowstone National Park in the summer), extra visitors detract from the enjoyment people receive from being there. See club good.

congestion.   Crowding. See *congestible*.

congestion externality.   A negative externality that arises when there are too many users who impede each other and raise harvesting costs when they simultaneously try to harvest a resource.

coniferous. Plants, especially trees, that bear cones to reproduce. Many evergreen trees such as pines, cedars and firs are coniferous trees and are important species in terms of timber production in several temperate countries such as the United States, Canada, Russia and New Zealand.

conjectures. See *game theory*.

conjoint analysis. Methodologies used to model and evaluate how individuals choose bundles of goods, services and assets (including aspects of the environment) by using a set of attributes or characteristics of the choices available. The approach is based on the assumption that the choices made by individuals are undertaken by a set of decisions that identifies whether a good, service or asset meets a need, and an information-gathering exercise about the choices available to form preferences about the choices.

FURTHER READING
Alberini and Kahn (2006).

connectance. A measure of the interconnectedness within a food web, defined as the number of direct species interactions divided by the total potential number of species interactions.

conservancy system. Term used to describe a sewage system that returns human excreta to the land for decomposition. The system can be effective if the level of wastes are relatively low, such that the land has sufficient assimilative capacity.

conservation. A general term for a mix of preservation, restoration, use and management of natural and human-made resources for the long-term benefit of societies.

conservation easement. A legally binding restriction on the use of a parcel of land by its owner. Frequently, conservation easements by property owners are compensated, to some extent, by reductions in property taxes.

conservation of matter. The incorrect notion that the total amount of matter or mass in a physical process must always be the same before and after the reaction or process. The idea is a useful approximation in chemical reactions and originated with the observations of the nineteenth-century scientist Antoine Lavoisier.

consistency.   A statistical property of an estimator in a model. An estimator is consistent if its bias converges to zero as the number of observations tends to infinity. See *biased estimator*.

constant dollars.   The value of a dollar (or other currency) in terms of what it would have been worth in real terms in a particular year or base period. For example, 1996 dollars refers to the value or price of a good in terms of how much one would have to pay for the good after accounting for the general change in the price level between 1996 and the actual period. Conversion of prices into constant dollars requires an index of changes in the general price level over time such that if a general index of prices of goods in 1997 is 3 percent higher than in 1996, the 1997 prices can be converted to 1996 constant dollars by dividing by 1.03.

constant escapement strategies.   A management strategy for a natural or wild population whereby a minimum proportion of the estimated population is left unharvested so as to ensure its sustainability. Such a strategy is commonly followed in salmon fisheries where a given number of salmon are allowed to "escape", or swim upstream, so as to breed and ensure future generations of salmon.

constant harvest strategies.   A management strategy for a natural or wild population whereby a fixed proportion of the estimated population is harvested over a given period of time.

constant prices.   See *constant dollars*.

constant returns to scale.   A situation where if all the inputs in a production process are increased by the same percentage, the output will also increase by the same percentage. See *decreasing returns to scale* and *increasing returns to scale*.

constrained optimization.   Optimization (maximization or minimization) of an objective function subject to at least one constraint.

FURTHER READING
Simon and Blume (1994).

constructed wetlands.   Wetlands purposely built for the treatment of waste water.

consumer price index (CPI). An index that tracks the expenditures, over time, associated with a defined "basket" of goods and services.

consumer species. Biological species that serve as net consumers of resources that are scarce, or subject to biological competition.

consumer surplus. The difference between the maximum that a consumer is prepared to pay for a good or service and the amount that the consumer actually pays.

consumption. The amount of goods and services used or consumed in a given period of time. Given that different goods are in different units, such as apples and oranges, consumption is measured by expenditures.

consumption efficiency. The ratio of the energy used at a given stage in the food web to energy produced at a lower level of the food web. For example, a consumption efficiency of 20 percent by herbivores implies that they ingest, or are able to utilize, one fifth of the available energy produced by plants.

consumption function. A relationship between consumption and income.

consumption pattern. A term used to describe the consumptive behavior of an individual, households or communities. Often used with reference to utility services such as electricity and water.

consumptive use. Use of the environment (such as irrigation or mineral extraction) which removes or processes natural resources and so may detract from amenity and other values.

consumptive value. The value of a good or asset that is ascribed to its consumption. For example, the price of salmon sold in the supermarket represents its consumptive value, but salmon in the wild also has a non-consumptive value in terms of its value to the ecosystem and the value it may give people who like to watch salmon swim up river.

contaminant. Any substance (abiotic or biotic) that reduces the quality, quantity or longevity of another substance or material.

continental drift. See *plate tectonics*.

continental ice sheet. See *ice sheet*.

continental shelf.   A relatively shallow and gently sloped area off a coastal land mass. The continental shelf in some places may stretch hundreds of kilometers from land (such as in north-east Australia) but, in other locations, it may end abruptly close to shore.

contingent ranking.   A method for environmental valuation in which respondents rank different goods or environmental alternatives by preference. When non-market goods are ranked against market goods, the analyst can derive upper and lower bounds for the value of the non-market good.

contingent valuation.   A method, usually in the form of a survey questionnaire, of eliciting values for hypothetical goods or services. The contingent valuation method (CVM) is used to estimate the value for certain environmental goods and may be the only means of estimating the passive or non-use values for goods.

FURTHER READING
Alberini and Kahn (2006).

continuous emission monitors.   Mechanisms that track all the emissions from a particular source, rather than taking periodic samples, to calculate emissions attributable to that source.

contract curve.   Based on all feasible allocations, the distribution of goods between all individuals in an exchange economy where the distribution is Pareto efficient and cannot be redistributed without making any individual worse off.

contraction and convergence.   A proposed framework for international climate policy. Under this proposal an overall target for global carbon emissions would be set, and national per capita emissions would converge to an equal level across time, in accordance with the global goal. See *climate change mitigation.*

contributory value.   The value of a species, resource, ecosystem, or habitat in terms of the value it provides to other aspects of the environment. For example, the contributory value of a bat species could be the value it provides in pollinating flowers for commercial fruit production.

controlled waters.   Rivers or streams with dams or diversionary channels to control flooding and in-stream water flow.

control variable.   A variable in an optimal control problem which is under the control of the planner or programmer and which helps to determine or affects the state variable(s). For example, in an optimal control problem of a fishery, the control variable could be the level of harvest and the state variable would be the fish population. See *state variable* and *optimal control theory*.

conurbation.   An urban area where two or more towns or cities have grown together such that it is difficult to discern where one city ends and another begins. Where a large conurbation exists between big cities the conurbation is sometimes called a megalopolis; for example, the conurbation between Boston, New York and Washington DC is such a megalopolis and has been called Bos-Ny-Wash.

convection.   The transfer of heat through a gas or liquid as the warm liquid or gas heats up, becomes less dense and rises. See *conduction.*

Convention on Biological Diversity (CBD).   A 1992 convention which allows for the free trade of genetic resources while providing ways for wealthy countries to finance biodiversity conservation.

Convention on International Trade in Endangered Species (CITES).   The convention was initiated in 1973 and seeks to protect endangered plant and animal species from extinction. CITES has been signed by over 100 countries and has been an important mechanism in preventing (or at least restricting) the trade in animal parts, such as ivory.

Convention to Combat Desertification (CCD).   A convention that was adopted in 1994, and ratified in 1996, that recognizes societal, economic, biological and physical processes that influence desertification and focuses upon technology transfer, particularly in Africa, to address desertification problems.

convergent evolution.   Evolution that leads to very different organisms that resemble each other by superficial appearance, or by behavior.

convex function.   Function where the area above the function is a convex set. See *convex set* and *concave function.*

convex set.   A set where if a line is drawn between any two points in the set, all the points on the line lie within the set.

cooperative game theory.   Models in game theory in which the participants can make binding commitments that allow for the possibility of co-operation.

co-ordination games.   An example of a common property game in which there exist two Nash equilibria: one in which players cooperate and another in which they do not.

COP. See *Conference of the Parties*.

Copenhagen Accord.   The official UNFCCC pronouncement emanating from the COP-15 held in Copenhagen, December 2009. Despite great public expectation prior to this meeting, a binding international agreement on climate change mitigation was not formulated. However, the Copenhagen Accord provided a formal target for limiting global warming to 2°C, and acknowledgment that large global cuts in emissions were required. Although accompanying national emissions reductions pledges at the time were not commensurate with the overall goal, the Copenhagen Accord paved the way for further advancement of international negotiations. See *Conference of the Parties* and *United Nations Framework Convention on Climate Change*.

Copenhagen Amendment.   A 1992 amendment to the 1987 Montreal Protocol that accelerated the phase-out of the production of chlorofluorocarbons (CFCs) which break down stratospheric ozone. See *Montreal Protocol*.

Copenhagen Conference.   Common name for the 2009 United Nations Climate Change Conference that took place in Copenhagen between 7 and 18 December that year. The conference acted as the venue for COP-15 and the fifth meeting of the parties to the Kyoto Protocol. Besides the usual government and UN officials, an unusually large number of activists and other parties attended in hope of a binding deal to emerge from negotiations as anticipated by the Bali Road Map. See *Copenhagen Accord*.

coral.   A subdivision of colonial cnidarians or coelenterates characterized by the existence of symbiotic algae (zooxanthellae) and the tendency of many species to create habitat through calcareous, reef-forming excretions.

coral bleaching.   The process whereby corals are placed under stress and lose their color. This phenomenon is caused by the expulsion of symbiotic energy providing micro-organisms (zooxanthellae) living in the coral's tissue. Although the coral may recover, prolonged bleaching leads to the eventual death of coral reefs. As the primary cause of such stress is a change in ocean temperature, scientists fear that global warming will cause massive bleaching and destroy existing coral reefs.

coram populo. Latin term for an open or public process.

core. A set of possible outcomes between individuals in a transaction where if the transaction were made all individuals would be at least as well off as if they had not made the transaction. As the number of individuals in the transaction increases, the core converges to the outcomes that would arise with perfect competition.

FURTHER READING
Cornes and Sandler (1996).

core accounts. The traditional accounts kept by countries in order to measure economic activity, including gross domestic product. See *satellite accounts.*

Coriolis effect. First discovered by the French physicist Gaspard Gustave de Coriolis, and this is the apparent force experienced by an object as it moves over the surface of a rotating object. It explains why ocean currents and wind move in an apparent clockwise (anti-clockwise) direction in the northern (southern) hemispheres. This occurs because points on the equator on a rotating sphere must travel faster than points at higher latitudes. As a result, if air at the equator travelling at the same speed as the surface moves north to a higher latitude it will, initially, move faster than the air it encounters there. Thus it will appear to be moving faster eastward (in a clockwise direction) than the surface beneath it.

cornucopian. An optimistic view of the future in which resource scarcity and environmental problems do not arise and potential problems can be solved by technological progress. See *Malthusian.*

corporate social responsibility. The notion that companies should account for the social impact of business decisions, commonly used in reference to the environment. See *blended value.*

Corpus Juris Civilis. Latin term for the body of civil law.

corrective tax. See *Pigouvian tax.*

correlation. The degree to which variables move together. Negative (positive) correlation between two variables implies that a rise in one is associated with a decline (rise) in the other. See *correlation coefficient.*

correlation coefficient. A "normalized" covariance that can take on any value between –1 and +1 where a value of +1 (–1) indicates the variables are perfectly positively correlated (negatively correlated) such that they move in exactly the same (opposite) direction. See *covariance.*

correlation matrix. A symmetric matrix that provides the correlation coefficients between different variables.

corridor. An area of undisturbed forest or habitat that permits the movement of flora and fauna to ensure biological diversity and ecological integrity.

co-state variable. Sometimes called auxiliary variable, it is analogous to a Lagrangian multiplier and represents the shadow price of the dynamic constraint in an optimal control problem. See *optimal control theory*.

FURTHER READING
Chiang (2000).

cost–benefit analysis. A methodology for determining whether a project or activity generates a positive net benefit for society by evaluating all the costs and benefits over time.

FURTHER READING
Hanley and Barbier (2009).

cost-effective abatement criterion. Term for methods of pollution control that result, at equilibrium, in equal marginal costs of abatement across firms or polluters. See *marginal abatement cost*.

cost-effective policies. Policies and practices that achieve the desired outcome at the least cost.

cost function. A functional relationship between the level of output and the prices of inputs with the costs of production, and which may be defined as $C(y,w)$ where y is the vector of outputs and w is a vector of input prices.

FURTHER READING
Simon and Blume (1994).

cost minimization. The optimization process by which costs are minimized for a given level of output. Cost minimization is a necessary condition for profit maximization.

cost shifting. When the burden (costs) of an action (say, economic development) is shifted from the responsible actor to others who usually are separated from the actor by time, distance, or scale. When the burden is shifted on to a party that cannot demand compensation or otherwise hold the actor liable, the cost shifting results in the creation of an externality.

countervailing duty. A tariff imposed on imported goods by a country to offset an unfair competitive advantage of the exporting firm, such as a subsidy.

Cournot competiton. Competition between oligopolists where firms independently select the amount of their product that they produce. See *Bertrand competition.*

Cournot–Nash conjectures. A concept used in game theory. It holds that economic agents choose a strategy or set of actions which they believe will lead to the most favorable outcome based on assumptions they make about how other agents will behave.

covariance. A measure of the joint association between two variables. If the covariance is positive (negative) the variables tend to move in the same (opposite) direction.

CPR. See *common-pool resource.*

"cradle-to-grave" management system. A system of managing throughput of a natural resource from initial extraction, through production and consumption, and finally to the ultimate emission of residuals to the environment.

Cretaceous. A geological period lasting from 144 million to 66.4 million years ago and encompassing both the rise and fall of the biological dominance of dinosaurs on Earth. The Cretaceous period also encompassed the arrival of flowering plants.

Creutzfeldt–Jakob disease (CJD). A rare degenerative brain disease in humans. A new form of the disease nvCJD has been shown to be linked to bovine spongiform encephalopathy (BSE), or "mad cow" disease. See *bovine spongiform encephalopathy*.

criteria pollutants.   Term used in the United States to define pollutants which are harmful to health and the environment.

critical depensation.   Description of population growth models, where the population growth rate is negative for population levels close to zero.

critical loads.   See *threshold effect*.

critical mass.   The smallest amount of matter required to ensure a nuclear chain reaction. See *chain reaction*.

critical point.   Threshold point beyond which further use, or increased concentrations of a substance, will significantly affect the environment. See *stationary point*.

critical region.   The region or set of values that will lead to the rejection of the null hypothesis in a statistical test, its size being determined by the probability of Type I error. See *Type I error* and *Type II error*.

crop rotation.   Widely used practice whereby different crops are rotated around the various parcels of land on a farm so as to maintain fertility of soils and to reduce the incidence of crop diseases and pests.

cross-price elasticity.   The percentage change in consumption (or production) of a good in response to a change in price of another good.

cross-sectional data.   Data where observations of the variables are available for individual units of observation, but only for a given point in time. See *time-series data*.

crowding-out effect.   A reduction in investment that occurs because of increased public expenditures.

crown cover.   The percentage of forest floor covered by tree canopy.

crown land.   Term used for public land in commonwealth countries where the head of state is the British sovereign.

crude oil.   See *petroleum*.

crust. The outer layer of the Earth that lies above the mantle.

crustal abundance. The total amount of a mineral that exists on Earth.

cryptosporidium. A water-borne parasite that may survive chlorination and is becoming an increasing problem in developed countries. A recent outbreak of the pathogen in Milwaukee, Wisconsin infected over 400 000 people in 1993 and was responsible the deaths of 104 individuals.

cryosphere. The Earth's ice, snow, permafrost and portion of the planet which has an average temperature of less than $0°C$.

CTE. See *Committee on Trade and the Environment*.

cull. Timber that is not fit for sale due to defects.

culling. Management practice of removing individual animals or trees to improve the overall welfare or well-being of the population or forest. Culling is especially important for animals confined to small areas of habitat, such as national parks, where further increases in the numbers of individuals could be damaging to other species and ecosystems.

cultural control. Methods or processes for controlling natural populations by changing aspects of the environment. For example, better drainage and the removal of stagnant pools are effective cultural controls for reducing the number of mosquitoes in a city.

cumulative density function (CDF). Also known as cumulative distribution function, or simply distribution function. The CDF for a continuous random variable X that takes on the value x is
$$F(x) = Pr(X £ x)$$
and defines the probability that X takes on a value that does not exceed x.

cumulative distribution function. See *cumulative density function*.

cumulative effect. The overall impact or effect of dosages of a substance over time.

currency depreciation. The decrease in value of a currency relative to other currencies.

current annual increment (CAI).   The increase in the volume of wood from a stand of trees over the previous 12 months.

current value (dollars).   Value of past returns or expected future returns in terms of today's value of money. See *constant dollars* and *discounting*.

current-value Hamiltonian.   A function that represents a dynamic optimization problem in continuous time and includes the original objective function and dynamic constraints, but is free of the discount factor. See *Hamiltonian*.

cyanobacteria.   Formerly classified as blue-green algae, but are now classified as a type of bacteria that produce oxygen as a byproduct of photosynthesis.

cyclone.   Powerful low pressure systems renowned for the destructive capacity of high winds associated with their structure. Although cyclones may form across all of the Earth's regions, the term is most commonly used with reference to tropical cyclones, or hurricanes. See *hurricane*.

D

Daisyworld. A hypothetical world described by James Lovelock to illustrate the Gaia hypothesis and how the Earth can be an effective self-regulating environment for plants and animals. In Daisyworld, the temperature can be maintained between critically low and critically high levels, despite significant changes in solar radiation. Thus, by changes in the composition of black daisies (which absorb more solar radiation and do best in colder temperatures) and white daisies (which absorb less solar radiation and do best in warmer temperatures), Daisyworld's temperature remains in the range that can maintain life. See *Gaia hypothesis*.

FURTHER READING
Lovelock (1990) and Lovelock (2000).

dams. Concrete or earth structures that block or collect water flows, forming a reservoir used for water storage or hydroelectricity generation. See *hydropower*.

damage function. A function that relates the level of emissions or discharges to the associated environmental or social costs.

Dansgaard–Oescheger events. Warm–cold oscillations in the Earth's climate during the last glacial (15000–80000 years ago). Sometimes referred to as Dansgaard–Oeschger.

dark green technologies. Processes that are used to directly address pollution; for example, technologies used to remove surface oil on water after an oil spill. See *light green technologies*.

Darwinian evolution. Named after Charles Darwin and refers to the gradual evolution of species due to natural selection whereby individuals with characteristics that do not favor reproductive success tend to die out over time. See *natural selection*.

FURTHER READING
Darwin (1859) and Freeman and Herron (2007).

Darwinian fitness. Named after Charles Darwin and also known as adaptive value. It refers to the relative ability of an individual, with a given genotype, to pass on its genes to future generations. See *Darwinian evolution*.

data envelopment analysis.   A mathematical programming method used to obtain a measure of efficiency and capacity output, also referred to as DEA. DEA models are commonly used to measure capacity in a fishery, but a principal drawback is that they do not normally account for random effects.

data fouling.   Process by which unrepresentative or misleading data is used in decision making because respondents deliberately supply false information.

data mining.   An approach to estimation of statistical models by which a large number of data are used or "mined" to ensure that a given model achieves an acceptable goodness of fit.

DDT.   Acronym for dichloro-diphenyl-trichloroethane. DDT is an insecticide that was once widely used, especially for the control of malarial mosquitoes, and which is very persistent within the environment. DDT accumulates in the fatty tissues of animals. The highest concentrations are often found in animals at the top of the food chain. Its persistence and ability to accumulate in animals, causing deleterious health effects, has led to its ban in a number of countries, including the United States and the European Union, although it is still used in some developing countries.

deadweight loss.   Net economic loss to society from a policy or action that alters a competitive outcome.

death rate.   The number of deaths per year divided by the estimated total population, usually expressed as a rate per 1000 individuals.

debt-for-nature swaps.   Agreements whereby countries or conservation groups buy some of the existing debt owed by a developing country which, in return, agrees to set aside a given area of land for conservation purposes.

debt overhang.   A situation whereby the current debt of a country exceeds its net present value of future revenues and, hence, the countries capacity to repay.

decarbonize.   A term used for the mitigation of carbon dioxide emissions from a particular process or entity. For example, a decarbonized economy and decarbonized business practices.

decentralization. Process by which activities and decisions formerly made at one, or a few locations, become more widely distributed over a greater number of locations.

decentralized power generation. The generation of electricity by small facilities close to or at usage locations. For example, roof-top solar panels and micro-wind turbines. Also termed distributed power generation.

decentralized outcome. An outcome, such as a market equilibrium, which arises from many individuals making their decisions independently.

decibel (dB). A logarithmic measure of sound intensity. The sound of traffic is about 80 dB while the rustling of leaves may be as little as 5 dB or less.

deciduous plants. Plants and trees that at some period during the year lose their leaves.

decision rule. A rule whereby an event triggers a decision or response. For example, a farmer may have a decision rule to plant tomato seeds on the first full moon in May.

decision tree. A representation of the subsequent decisions that can be made and the possible outcomes of each possible decision.

declining block pricing. A pricing mechanism where increments of a good or service, such as electrical power, are priced in blocks, and prices decline for additional increments as the total quantity purchased increases.

decommissioning. The process of terminating and eliminating a productive capital asset, such as a boat or nuclear power station. Decommissioning can be an expensive process because it often involves disposal of environmentally harmful materials.

decomposition. The breakdown of organic material by microbial activity.

decreasing block tariff. A pricing structure whereby the consumer pays a lower price per unit beyond a threshold level of consumption.

decreasing returns to scale. A situation where, if all the inputs in a production process are increased by the same percentage, the output will increase by a smaller percentage. See *constant returns to scale* and *increasing returns to scale*.

deduction.   See *deductive method.*

deductive method.   A method of reasoning by which logic and abstract theory are used to explain or predict actual phenomena. See *inductive method.*

deep ecology.   A philosophical view of life on our planet that places no special or unique importance on human beings and has an holistic or even spiritual view of living systems. The term originated in the 1970s with the work of Arne Naess. See *shallow ecologists.*

FURTHER READING
Devall and Sessions (1985).

de facto.   Latin term meaning "existing in fact" or in practice.

de facto property rights.   Property rights that have no legal recognition, but may nevertheless exist in practice. See *property rights* and *de jure property rights.*

defaunation.   The removal of animal species from a geographic area.

defensive environmental costs.   Expenditures associated with overcoming the potential harmful effects of a deterioration in the environment. For example, if water quality declines, the associated increased expenditures on bottled or filtered water are defensive environmental costs.

deficit.   See *economic deficit.*

deficit forest.   Forest where the current annual cut exceeds the long-run sustainable yield. See *surplus forest.*

deficit irrigation.   Under provision of water to crops, relative to the amount that they can use. This may be done inadvertently or deliberately as part of a strategy to maximize crop water productivity when water availability is restricted.

deflation.   A general fall in prices of goods and services in an economy which may be measured by a reduction in the consumer price index.

defoliant.   A herbicide that destroys the leaves of plants and trees. See *Agent Orange*.

defoliation.   The loss of leaves, usually by chemical or incendiary means. Agent Orange was a chemical used during the Vietnam war by the United States military to cause the defoliation of tropical forests in order to deprive opposing forces of cover.

deforestation.   The transformation of land use from growing trees to other uses, such as for growing crops or to pasture land. Much of the world's deforestation is occurring in tropical and subtropical regions. As trees absorb atmospheric carbon dioxide, deforestation makes a substantial contribution to global warming. Hence mechanisms such as REDD have become an important element of international climate policy. In technical terms, deforestation occurs when crown cover, or the percentage of forest covered by trees, is reduced below 10 percent. See *climate policy* and *Reducing Emissions from Deforestation and Forest Degradation*.

FURTHER READING
Food and Agriculture Organization of the United Nations (2009).

deglaciation.   The decline or removal of land ice from a region, usually caused by higher temperatures.

degradation.   The reduction in value or output. Environmental degradation is the state in which the environment has been altered in some way to make it less productive or valuable. Also used in reference to the geological weathering of land or the breakdown of an organic compound. See *forest degradation* and *land degradation*.

degrees of freedom.   A statistical term that refers to the number of independent units of information (usually number of observations) that are left after estimating the parameters of a statistical model.

de jure.   Latin term for existing or having a basis in law.

de jure property rights.   Property rights that have legal recognition. See *property rights* and *de facto property rights*.

delta.   Deposit of sediments at the mouth of a river or stream.

demand.   The willingness or desire for a good or service at a given price. See *demand function.*

demand curve.   See *demand function.*

demand function.   A relationship between the amount people are willing to spend on a good or service and the price of the good, holding everything else constant.

demand side management.   Policies and measures implemented by utilities and governments to control the amount of energy used in businesses and households, particularly during peak usage periods. These actions may include: differential pricing at peak times, education campaigns and smart metering systems which provide greater information concerning energy consumption.

demersal species.   Fish commonly found at or near the bottom of the sea, such as orange roughy.

de minimis risk.   A trivial or minimal risk associated with an activity.

demographic transition.   The notion that with improvements in the standards of living and health care, death rates drop before an appreciable fall in birth rates thus increasing the rate of population increase. Eventually, economic development leads to increased job opportunities for persons with specialized skills and training, and especially for women, and the birth rate falls until the population growth stabilizes.

demography.   The study of change, over time, in the structure and size of human populations.

dendochronology.   The scientific practice of analyzing tree rings for dating purposes and using subsequent information to ascertain historical data, such as past climate.

dengue fever.   A viral disease that can sometimes be fatal. It is transmitted by mosquitoes that are endemic in most tropical countries. Control is through the spraying of pesticides to reduce mosquito populations, as well as improvements in urban infrastructure to reduce areas of surface water that facilitate the breeding of mosquitoes.

denitrification. The breakdown of nitrogen compounds by micro-organisms.

density. Mass per volume and which is usually defined as grams per cubic centimetre.

density-dependent growth. Growth that is regulated by the density of a population. For example, as the population increases, the rate of mortality and morbidity may also rise, which reduces the growth in the population. In some species, such as sardines, increased density at low levels may actually increase the growth in the population as it enables fish to school together and can be an effective strategy against some predators.

density independence. The notion that, over a certain range of population density, the growth of a population is independent of its density.

deoxyribonucleic acid (DNA). An organic molecule consisting of two strands that form a double helix and which contains genetic information.

FURTHER READING
Frank-Kamenetskii (1997).

dependent variable. The variable in a model or optimization problem which depends on (i.e., is a function of) the other variables. In regression analysis, the dependent variable is found on the left-hand side of the regression equation.

depensation. In population growth models, the case in which the proportional population growth rate is an increasing function of population size.

depensatory growth. Natural population growth which declines in response to a reduction in the size of a population below a critical level. See *compensatory growth* and *logistic growth*.

depletable resources. Those resources that cannot be reasonably expected to be replenished by natural processes.

depleted. The point at which extraction of a non-renewable resource from a deposit ceases. If it is the case that there is no mineral deposit left, then the mine is physically depleted. If it is uneconomic to extract any more, then the mine has been economically depleted although there may still be some mineral deposit left.

depletion costs.   The reduction in value of a non-renewable resource that results from a fall in the total quantity available due to extraction.

depletion indices.   Measures of how quickly a non-renewable resource is being used; typically measured in terms of the years remaining of known reserves based upon either current or projected rates of consumption.

deposit-refund system.   An increasingly common method of encouraging the proper disposal of waste whereby the price of a good includes a deposit that is refundable, provided that the residual of the good in question is returned to a center for appropriate disposal or for recycling.

depreciation.   The physical loss and/or financial loss over time associated with the use of an asset.

derby fishery.   A fishery where the total catch is fixed and fishers compete for a share of the harvest. A derby fishery is characterized by "the race to fish" that dissipates the potential rents. See *common-pool resource*.

deregulation.   Term used to describe the removal or reduction in the rules governing individual or firm behavior.

derived demand.   A demand for a good or service that is derived from the demand for another good or service. For example, an increase in the demand for paper products may lead to an increase in the derived demand for tropical timber.

derived savanna.   Savanna vegetation created by the anthropogenic alteration of another type of vegetation (e.g., forest). Most commonly used to describe human-maintained savanna in sub-Saharan Africa.

DES.   See *diethylstilbestrol*.

desalination.   A term for the removal of salt from ocean water or soil. In the case of water, desalination may provide an important source of water in large population centers with low rainfall. As climate change projections indicate greater future incidence of water scarcity, industrial desalination plants have become recognized as a potential solution to the already significant strains placed on water resources from increasing populations. With regards to soil desalination, this process may be used to restore marginal agricultural land or conserve natural areas.

desalinization. See *desalination*.

desert. An arid region characterized by little rainfall.

desertification. Process by which the moisture in the soil over an area of land and vegetative cover declines over time. Desertification can be caused by natural factors, such as non-anthropogenic climate change, or human activities, such as deforestation.

detergents. Synthetic cleaning agents used for a variety of purposes that can sometimes lead to environmental problems. In the past, phosphates were included in detergents, causing eutrophication of lakes.

determinant. The sum of the product of the elements of any column or row in a square matrix multiplied by their cofactors. The determinant of a square matrix A is denoted by |A|.

deterministic. Term used to describe a model or process where all future values can be explained or "determined" by the model or the process itself such that there is no uncertainty about the future.

deterministic production frontier. A relationship between inputs and an output that represents the maximum possible amount of output that can be produced from any quantity or combination of inputs, assuming no uncertainty or randomness in production.

detrending. Any method that tries to remove a trend from time-series data. A common procedure is to regress the variable of interest against a constant term and regressors involving a measure of time. The residuals from the regression are "detrended" and can then be used to test hypotheses about fluctuations around the trend.

detritus. Fragments composed of partially decomposed organisms.

devaluation. Where countries fix the value of their currency relative to other currencies, a reduction in the value of the domestic currency relative to a foreign currency is called a devaluation.

developed countries. A term used for countries with a high level of economic development. As development is a multifaceted concept it is difficult to define countries in this manner; however, developed countries are often described as being: Japan, Canada, United States, New Zealand,

Australia, and most European countries. See *development* and *developing countries*.

developing countries.   There is no established convention for defining "developing countries"; however, they are typically considered as those that are not developed countries. In general, they are countries with low per capita incomes and which suffer from relatively high rates of morbidity and mortality, such as Bangladesh or Bolivia. See *developed countries*.

FURTHER READING
Perkins, Radelet and Lindauer (2006).

developing world.   See *developing countries*.

development.   A process by which individuals, communities and societies become empowered to better achieve their human potential. For example, improvements in education and health care and increases in real disposable income are often considered as factors that contribute to development. Also used in the context of describing the level to which such factors are present or achieved, for example, a country with a high level of development.

development economics.   Study of the process of economic development on both a macro and microeconomic perspective and with a focus on developing countries. See *developing countries.*

FURTHER READING
Perkins, Radelet and Lindauer (2006).

development pathway.   The progression to higher levels of economic development that constitutes many different technological, economic, political and social factors. This term is commonly used with reference to the volume of greenhouse gas emissions produced as a country develops, and, in particular, the notion that developing countries must follow an alternative development pathway to mitigate climate change.

development rights.   The right to exploit or develop a resource or area.

Devonian.   Geological period from 408 to 360 million years ago.

dewatering.   Removal of naturally occuring water from sediments by drainage.

diadromous. Animals that migrate between salt and fresh water, such as some species of eels.

diameter. Measurement of the size of the tree trunk. Sometimes called d.b.h. for diameter at breast height.

diamond water paradox. The observation by Adam Smith that although water was essential to human life and diamonds were not, the price of water was essentially zero while the price of diamonds was high. This discrepancy between the apparent value and price of the two goods was explained by the fact that there were abundant supplies of water while diamonds were in short supply.

diatom. Type of algae that exist in aquatic and marine environments.

dichotomous choice. Survey questions often used in contingent valuation studies in which the respondent is asked to choose one of two choices. For example, the question "Would you be willing to pay $10 to ensure the preservation of community wetlands?" requires the answer yes or no.

dieback. A widespread death of trees and plants. Diebacks may be caused by pollutants, such as ground-level ozone and acid rain, or by natural factors.

diesel exhaust particulate matter. Particulates made up of chains of carbon matter containing a variety of organic compounds, some of which are carcinogenic.

diethylstilbestrol (DES). A synthetic chemical that acts like the hormone estrogen and which for many years was given to pregnant women to "normalize" their pregnancies. Recent evidence has linked DES to physical abnormalities and other health problems.

FURTHER READING
Colborn, Dumanoski and Myers (1997).

difference equation. A representation of the changes in variables in discrete time. An example of a difference equation is
$$x_t = a + by_{t-1}$$
where a and b are parameters, x_t is the value of the variable x at time period t, and y_{t-1} is the value of the variable y at time period $t - 1$. The solution must be consistent with the difference equation and contain no lagged terms.

FURTHER READING
Simon and Blume (1994).

differencing. A method used to eliminate trends from time-series data whereby an observation of a variable in period t is subtracted from an observation of the variable in period $t + n$. The degree of differencing is determined by n, where if $n = 1$ the resulting observations are called first differences.

differential equation. A representation of the rates of changes in variables in continuous time. An example of a differential equation is
$$dx/dt = a + by$$
where the solution must be consistent with the differential equation and contain no derivative terms.

FURTHER READING
Simon and Blume (1994).

differential rent. See *Ricardian rent*.

differentiated regulation. Standards or requirements that are not applied uniformly and may have the effect of favoring different types of technology or products (such as emission standards placed upon newer model cars that are absent for older cars).

diffusion. 1. Movement of material from areas with higher concentrations to areas with lower concentrations. 2. The adoption of ideas or practices developed elsewhere and then applied in new environments or circumstances.

dilution approach. The method of managing pollution by not treating the waste emitted but rather diluting it as it is discharged through various means (such as discharging raw sewage directly into the sea).

diminishing marginal product. See *diminishing returns*.

diminishing returns. A process where the application of a given input (holding all other inputs constant) results in an increase in output, but at a decreasing rate. This is equivalent to stating that the marginal product of an input is positive, but falling as the amount of the input increases. See *marginal physical product*.

dimorphism. Term for the existence of two distinct types of organisms within a population of a species. The most common form of dimorphism is sexual dimorphism where there exist physiological differences between males and females.

dioxins. Many dioxin compounds exist and are formed in the production of organochlorine compounds (such as herbicides) and in the manufacture of chlorine-bleached wood-pulp. Some dioxins, such as tetrachlorodibenzo-p-dioxin (TCDD), are extremely toxic and are endocrine disrupters.

direct emissions. Greenhouse gas emissions directly attributable to the specific actions of a particular entity. For example, the direct emissions related to car use would be the actual exhaust emissions. See *indirect emissions*.

direct taxes. Taxes imposed on income or wealth. See *indirect taxes*.

direct valuation. The direct measurement of changes in environmental quality attributable to different policies or actions, and their impact on individuals' well-being in financial terms.

discarding. Term used in fisheries to describe the actions of fishers who dump unwanted fish overboard at sea. Discarding can be a potentially serious problem as the discards are not reported in the catches and the dumped fish often have a very high mortality rate.

discounted cash flow (DCF). A valuation method whereby all future net cash flows of a company, asset or project are discounted to the present and aggregated. See also *net present value*.

discount factor. A numeric value that serves to convert a future value to a present value. See *discount rate*.

discounting. Method by which future costs and benefits are converted into current dollars. For example, if the discount rate is r percent per year then a payment of $\$A$ after n years is worth $A/(1+r)^n$ in current dollars, after discounting. The discount rate chosen for an economic analysis can be controversial due to the strong influence of this parameter upon results. For example, debate concerning the timing of cost-effective climate change mitigation has frequently centered on the rate used to discount future damages.

discount rate.   In consumption analysis, a factor by which future welfare or utility is multiplied to indicate that future consumption is less valuable in the present than current consumption. See *time preference*.

discovery cost.   The cost of finding new mineral deposits.

discrete choice.   Models of consumer choice in which the good or alternative chosen by the consumer, is available only in discrete (e.g., integer) units.

disembodied technical change.   Technical change that arises from the development of new techniques and methods of production rather than from the use of new machines and capital goods. See *technological change* and *embodied technological change*.

disequilibrium.   A biological or mechanical process that is not in equilibrium or non-stationary, but which may be in the process of converging to an equilibrium.

Dismal Theorem.   An influential critique of traditional cost–benefit analysis (CBA) applied to climate change mitigation presented by economist Martin Weitzman. He argues that CBA cannot consider appropriately the possibility of catastrophic climate change because of the uncertainty about the probability and impact of its occurrence. CBA is therefore faced with small probability, high impact adverse events (fat-tailed uncertainties) that will not yield a definitive conclusion on the amount of climate change action to take. Climate change action should therefore be treated as a form of risk insurance. This approach is often referred to as an application of a strong form of the precautionary principle. See *catastrophic climate change*, *cost–benefit analysis*, and *Stern Review*.

FURTHER READING
Weitzman (2009).

dispatchable capacity.   The output capacity of an electricity generation facility that is available on average to supply electricity to meet peak demand (also known as effective capacity).

dissolved oxygen (DO).   The amount of oxygen dissolved in water, usually specified by parts per million. Low levels of DO are deleterious to fish as well as other aquatic life and may indicate the presence of anthropogenic pollutants.

distance-related distortions.   A failure to internalize the costs of an action or behavior due to a distance over time, space, or scale between the agent and the environmental impact.

distributed generation.   Electricity generation facilities such as small-scale gas-fired, solar and wind turbine plants that are located near to where the electricity is consumed. Benefits include the avoidance of significant capital investment in large, centralized electricity generation facilities, improved peak demand management and the reduction in electricity transmission losses.

distributional effects.   See *distributional impacts*.

distributional impacts.   The economic costs imposed upon, and benefits received by, different individuals because of policy decisions (for example, a program to provide a price floor for wheat may benefit farmers but harm consumers through higher prices).

disturbance term.   See *residual*.

diurnal temperature range.   The difference between the maximum and minimum temperature in a 24-hour period.

diurnal variation.   The difference in sunlight, and hence energy available to organisms, and its impact on natural processes during the course of a day.

divide.   The boundary between two drainage basins.

divisibility.   A property right characteristic that represents the ability to divide a resource or asset into sub-units, such as denominating pollution permits into tons.

division of labor.   Term used to describe how, in general, individuals specializing in particular activities can become very good at that activity and thus can increase their productivity over time. Provided that people can exchange their goods and services with others at low cost, the division of labor can benefit everybody.

DNA.   See *deoxyribonucleic acid*.

Dobson unit.   Named after Gordon Dobson, a Dobson unit is a measure of the level of ozone concentration in the atmosphere.

domesticated. Animals and plants that have been bred for human use. Almost all the plants and animals used in agriculture are domesticated species.

dominant groups. Species that have the most influence within an ecosystem.

dose–response relationship. The measurable effect of different levels of pollution on some aspect of the environment, including people's health.

double dividend. Term used to describe the potential benefits from increasing "green taxes" on pollution or emissions with a corresponding reduction in other taxes, especially direct taxes, such that the net effect is revenue neutral. The so-called double dividend is an improvement in environmental quality and an increase in output as workers have a greater incentive to work more with reduced direct taxes.

dracunculiasis. A parasitic disease, also known as the Guinea Worm disease, found in sub-Saharan Africa and the Indian sub-continent.

drainage basin. The area of land from which all surface water flows into a single river and/or its tributaries.

drawdown. Term used to describe the reduction in the annual allowable cut of forest to its long-term sustainable level. Also known as falldown.

dredging. The removal of sediment from a water body to deepen channels or harbors.

drift net fishing. A method of fishing in which large nets attached to buoys are left in the sea for a period of time and then brought on board the fishing vessel. The practice is particularly destructive as all fish that are larger than the mesh size are caught in the net. Large-scale drift nets are banned by the United Nations.

drought. An extended period of abnormally reduced rainfall causing water shortages that may negatively impact environmental, economic and/ or social welfare. One of the most important impacts of droughts upon human welfare is the fall in agricultural production, and hence food availability. Scientists predict that the incidence of drought may become more pronounced in many regions, such as equatorial areas and south east Australia, with the onset of climate change.

dryland farming.   The production of crops on land without using irrigation.

dryland management.   An umbrella term for the management of arid and semi-arid lands, primarily for agriculture.

dual-fuel vehicle.   A vehicle that can run on traditional gasoline mixed with ethanol or methanol in the same tank.

dummy variables.   Variables that take on value of 1 or 0 for categorical variables such as male (dummy variable = 1) and female (dummy variable = 0).

dumping.   1. The sale of goods in a market at below the cost of production. 2. The improper disposal of wastes.

dung.   See *animal feces*.

duopoly.   Term given to where there exist only two sellers of a good or service in a market.

durable goods.   Goods, such as refrigerators, that provide a service over an extended period of time.

duration.   A characteristic of property rights that represents the expected longevity of the asset, or the expected time a user may be allowed to use the asset.

Durbin–Watson statistic.   A statistical test used to determine whether the residuals in a multiple regression are correlated. See *autocorrelation*.

FURTHER READING
Greene (2003).

dustbowl.   A large area of land in the midwest and southwest of the United States where air-borne soil erosion due to drought and poor agricultural practices became a huge environmental problem in the 1930s.

dyke.   See *levee*.

dynamic control.   A system in which the control variable can respond to changes in the state of the system.

dynamic efficiency.   A mode of production where firms continuously improve their practices over time. Methods of pollution control are dynamically efficient if they encourage firms to reduce emissions over time.

dynamic equilibrium.   See *steady-state*.

dynamic game.   A repeated game in which a player's strategy can span the multiple iterations of the game.

dynamic model.   A model whose variables change with respect to time.

dynamic programming.   A recursive method of optimization for problems defined in discrete units of time. See *Bellman's equation*.

dynamic stability.   The ability of a system or process to return to an equilibrium following a disruption or shock to the system.

E

E-4 scheme. Coined by Maurice Strong as the four variables that best describe sustainable development programs: equity, economics, ecological integrity, and empowerment.

Earth sciences. A term for the many branches of science involving the study of the Earth, which includes geology, oceanography, glaciology and atmospheric sciences.

Earth Summit. See *United Nations Conference on Environment and Development.*

Earth's energy budget. The balance of incoming and outgoing energy within the Earth's entire physical system.

earthworm biotechnology. The treatment of livestock effluent by first preconditioning the slurry in tanks and then applying the same to land already enriched with earthworms. Earthworm technology is thought to reduce odorous emissions from effluents and to speed the decomposition of the slurry.

earmarked revenues. State income linked to a particular set of state expenditures. It currently plays a major role in the financing of resource extraction in the developing world. Also termed hypothecated revenues.

easement. A legal right between a landowner and another to use the land for a particular purpose, such as the passage of telephone lines. See *conservation easement.*

Ebola. A viral disease named after a place in Zaire where it was first discovered in 1976. The disease is extremely infectious and often fatal. The host of the disease is most likely endemic to the equatorial forests of central Africa, and thus increased human activity in the forest, such as deforestation, may increase the number of outbreaks.

eco-feminism. A school of thought that places the exploitation of women by men as the fundamental explanation for the misuse and abuse of the environment. See *social ecology.*

eco-labeling. See *green labeling.*

E. coli.   See *Escherichia coli*.

ecological balance.   The condition in which a community of organisms or an ecosystem is in dynamic equilibrium, within which genetic species, and ecosystem diversity, remain relatively stable over time.

ecological carrying capacity.   The ecologically determined maximum sustainable stock (e.g., animals, humans) that can be supported by a given unit of habitat.

ecological corridor.   See *corridor*.

ecological debt.   A concept describing the balance of resources within an ecosystem. It occurs when the consumption of resources exceeds the ecosystem's regenerative capacity.

ecological economics.   A branch of economics that explicitly recognizes environmental limits, and the interactions between human activities and the environment. Ecological economics places a priority on the understanding of ecological processes, systems, causes and effects in addressing economic and environmental problems.

ecological fallacy.   The notion that analyses based on different levels of aggregation of data may give very different results.

ecological footprint.   A measure of the appropriated ecosystem area that theoretically would be required to sustain a given population. Ecological footprinting is sometimes used as an indicator of the degree to which a population's consumption is ecologically sustainable.

ecological growth efficiency.   The proportion of biomass energy produced at a trophic level that is used by the next highest trophic level.

ecological integrity.   A term used to describe the general well-being of ecosystems in terms of habitat, species diversity and other factors.

ecological sensitivity.   The potential carying capacity of an ecosystem for a given pollutant. For example, lakes found in areas where the soil is alkaline are less affected by acid rain than those found in areas where the substrate is acidic.

ecology.   Study of life and interactions between organisms, and between organisms and their biotic and abiotic environment.

econometrics.   A branch of economics that uses statistics to test hypotheses about economic models and to estimate model parameters.

economic bad.   A good that is a useless byproduct of the production process. Many waste products are economic bads.

economic convergence.   The notion that, over time, the per capita *levels* of income of different countries will converge to a common value. Conditional convergence is the notion that, over time, the *growth rates* in per capita income converge to a common rate.

FURTHER READING
Perkins, Radelet and Lindauer (2006).

economic cost.   The full cost of producing a good or service including all explicit and implicit costs of the producer, and the costs or benefits absorbed by the rest of society.

economic deficit.   The amount by which current spending exceeds current revenues.

economic development.   Process by which individuals, communities and societies are able to become empowered so as to meet their needs in an equitable and efficient manner.

economic efficiency.   A state where all available inputs are used in the correct amounts and proportions to produce the maximum amount of output at the lowest cost of production. It is calculated as the product of allocative efficiency and technical efficiency. See *allocative efficiency*, *scale efficiency*, and *technical efficiency*.

economic growth.   A process that is usually defined as the sustained increase in the per capita real value of the production of goods and services in an economy.

economic impact.   Effect of a policy or event upon the economy.

economic instruments.   Approaches to pollution control that involve economic incentives and include effluent charges, deposit-refund systems for recyclables, and market-based instruments. See *market-based instruments*.

economic profit. The difference between total cost and total revenue, where total cost includes the opportunity cost or the value of any single input's next best alternative use. For example, the opportunity cost of a skipper who owns a boat but does not receive a salary would thus be included in total costs, measured by the amount the skipper would receive as a hired-skipper on another boat. By contrast, accounting profit is simply the difference between total revenues and explicit cash outlays.

economic rent. The return over and above that necessary to ensure the supply of a factor of production.

economic reserves. Those resources that have been identified (which includes actual deposits that have been measured as well as those that have been inferred) and can be extracted without prohibitive costs.

economics. The science of choice, or a discipline that studies the allocation of scarce resources, with limited and alternatives uses, for the production and distribution of goods and services.

economic surplus. The amount by which current revenues exceed current spending.

economic sustainability. A concept related to sustainable development, where economic integration, participation and equity are stressed. See also *environmental sustainability* and *sociopolitical sustainability*.

economic water scarcity. A lack of water caused primarily by institutional or financial factors which hinder sufficient water resource development. See *physical water scarcity* and *water scarcity*.

economies of scale. The reduction in long-run average cost of production due to an increase in total output. Thus, if a firm triples its production but only doubles its costs, it benefits from economies of scale.

economies of scope. A reduction in the total average costs of a firm by producing a range of related products rather than producing just one type of product.

economies in transition. See *transition economies*.

ecophysiological process. The physical response of organisms to variations in the natural environment.

economy. Term given for the collective economic activities and interrelationships of a defined geographical region or country.

ecosphere. The portion of the Earth that includes the biosphere, hydrosphere, atmosphere, and soils and minerals.

FURTHER READING
Holmberg, Robert and Erikson (1996).

ecosystem. A discrete or distinct unit or community of organisms with particular characteristics.

ecosystem approach. An approach to environmental management that focuses upon the whole, and linkages among components within an ecosystem, including human activities, so as to ensure ecological integrity.

ecosystem services. Term given to the services that the environment can provide as part of its natural functions. For example, the water cycle provides an invaluable service by helping to restore water contaminated with pollutants.

ecotechnological productivity. The use of biotechnology to enhance the primary productivity of naturally occurring ecosystems for appropriation by the human economy.

ecotone. A clearly defined zone that demarcates the boundary between two or more different communities.

ecotourism. Tourism, travel, and adventure focusing on outdoor activities or outings that have as their primary destination the enjoyment of nature. Often ecotourism is associated with attempts to promote tourism that is environmentally focused and environmentally sensitive. See also *wilderness tourism*, *adventure tourism*, and *sustainable tourism*.

ecotoxicity. The negative effects of a substance on ecological (and in some cases geophysical) systems.

ecotoxicology. Study of the effect of chemicals on the functioning of an ecosystem, including all of its individual components.

ecotype. A locally adapted variant of a population.

ecozones. Method of classifying broadly similar terrestrial and marine ecosystems.

ecumene. The area of the planet where there exists permanent human habitation.

edaphic. The classification system in which vegetation type is determined primarily by soil and groundwater conditions.

effective capacity. See *dispatchable capacity.*

effective fishing effort. The adjusted value of nominal fishing effort to account for technological change or the actual change in fishing power over time.

effective irrigation. A way to measure irrigation efficiency at a basin-scale that takes into account the efficient quantity of water delivered from and returned to a basin's water supply, which is potentially reused by other irrigators downstream. See *irrigation efficiency.*

efficient equilibrium. An equilibrium where the marginal social cost of production equals the marginal social benefit. See *marginal social cost* and *marginal social benefit.*

efficiency. See *Pareto efficiency*, *economic efficiency*, and *statistical efficiency*.

efficiency wage. The notion that firms may wish to pay more than the existing wage so as to elicit greater productivity from their workers.

Effort–yield curve. The production or output of renewable resources, such as the amount of fish caught, based upon the effort expended in trying to harvest the resource. In some cases, total yield first rises as effort increases and then falls despite increasing effort, such as when increased fishing efforts reduces stocks making it more difficult to harvest fish.

eigenvalues. Also called the characteristic roots. See *characteristic equation* and *polynomial equation.*

8.2 ka event. A sudden decrease in global temperature approximately

8200 years ago. Scientists believe this event was caused by melting North American ice sheets releasing freshwater that disrupted the thermohaline circulation of the Atlantic Ocean. See *thermohaline circulation*.

El Niño Southern Oscillation (ENSO). An irregular warming of the surface waters of the eastern tropical Pacific which suppresses the cold ocean upwelling of the Humboldt current off Chile and Peru. An El Niño event can last up to 2 years and results in significant global changes in weather patterns including greater rainfall on the west coast of South America, droughts in Australia and New Zealand and drier and hotter summers in North America. See *La Niña* and *modes of climate variability*.

FURTHER READING
Ruddiman (2007).

elastic. Term used to describe the consumer or producer response to a change in price. A good is elastic if the percentage change in demand (or supply) equals or exceeds the percentage change in the price. For example, if the price of a good fell by 1 percent and demand increased by more than 1 percent, the good would have an elastic demand at that price. See *inelastic*.

elasticity. A numerical measure of how responsive one variable is to changes in another variable. It is calculated as the percentage change in one variable divided by the percentage change in the other variable.

electric power grid. The system connecting electric power generators and consumers via infrastructure such as transmission and distribution lines. The incorporation of renewable technologies and smart metering systems as part of future climate change mitigation will require extensive investment in power grids.

electric vehicle. A vehicle powered by batteries that are charged by connection to an electric power grid.

electricity generation. The process of producing electricity to satisfy industrial, commercial and retail consumer demand. Traditional sources of electricity generation include fossil fuel combustion, nuclear and hydropower. Renewable sources include wind, tidal, geothermal, biomass and biogas, and solar.

electricity grid. See *electric power grid*.

elfin thicket.   A densely vegetated grove composed of gnarled and stunted woody vegetation, usually found at high altitudes.

embedding.   Term used to describe the apparent insensitivity in contingent valuation surveys of respondents' willingness to pay measures for changes in environmental assets that may be "embedded" in each other. For example, the willingness to pay to ensure the continued existence of wolves in a national park may be very similar to the willingness to pay measures to ensure the continued existence of the park itself.

embodied technological change.   Technological change that arises from the use of new machines or capital goods. See *technological change* and *disembodied technical change*.

embodied water.   See *virtual water*.

eminent domain.   The notion that the state has the right to appropriate private property for the common good provided that the owners are paid a fair market value for their property.

emission.   The discharge of pollutants into the atmosphere. Most commonly used in the context of greenhouse gas emissions.

emission coefficient.   A unit of measurement for expressing emissions in terms of a quantity of carbon dioxide or carbon dioxide equivalent emissions per unit of a specific activity. May relate to the ratio of emissions to a unit of a factor input, such as a barrel of oil, or an output, such as a megawatt hour of electricity generated.

emission/effluent charge.   A pollution charge or tax levied on an emitter based on the amount of pollution emitted.

emission factor.   See *emission coefficient*.

emission inventory.   A listing and description of the amounts of pollutants by particular sources emitted within a defined geographic area over a specified period of time. Commonly used for environmental compliance activities and international climate policy.

emission permit.   A permit or property right which allows the holder to emit or release a given quantity of a pollutant in a defined period of time. See *emissions trading scheme*.

emission permit rental charge.   A charge payable to a regulator that is based on the market price of emission permits.

FURTHER READING
Grafton and Devlin (1996).

emissions reduction unit (ERU).   An emissions credit for reductions achieved under the Joint Implementation mechanism of the Kyoto Protocol. See *emissions credit*.

emissions banking.   The saving of emissions or pollution permits for future use. See *pollution permits*.

emissions concentration.   The level or amount of pollutants discharged into the environment measured in terms of relative volume. In the case of greenhouse gases, emissions concentration is normally measured in terms of carbon dioxide equivalent parts per million.

emissions credit.   A generic term for either: a certified emissions reduction credit within a carbon offset scheme, such as the Clean Development Mechanism, or an emissions permit within an emissions trading scheme. Most commonly used in the context of climate change mitigation and expressed in units of metric tons of carbon dioxide equivalent.

emissions intensity.   The volume of pollutants emitted from a particular source as a proportion of some other activity. For example, tons of carbon dioxide emitted per unit of economic growth. This concept is important in the context of climate change mitigation because limiting greenhouse gas concentrations in the context of continued economic growth will require sharp reductions in the emissions intensity of the global economy. See *Kaya Identity*.

emissions offset scheme.   See *carbon offset scheme*.

emissions projection.   Forecasts of future greenhouse gas emissions derived from computer modeling of assumptions concerning an array of economic, political and social factors. See *Special Report on Emissions Scenarios*.

emissions quota.   The fraction of permitted emissions for an individual entity in the context of a broader national or global limit. See *cap-and-trade*.

emissions reductions.   A decrease in greenhouse gas emissions attributable to the activities of a nation, state, business or individual. Most commonly used for reductions recognized and traded under a formal arrangement, such as the European Union Emissions Trading Scheme. See *climate change mitigation.*

emissions reduction credit (ERC).   A formal collective term used for emissions credits such as certified emissions reductions (CERs), emissions reduction units (ERUs), and removal units (RMUs), generated under the Kyoto flexibility mechanisms.

emissions scenario.   See *emissions projection.*

emissions standard.   A legal limit on emissions which, if exceeded, may result in a fine or other costs for the offending party.

emissions trading (ET).   The process of trading the right to pollute in a formal scheme. Officially recognized as a flexibility mechanism for countries to meet their commitments under the Kyoto Protocol.

emissions trading scheme.   A system whereby firms or countries are allowed to purchase the right to emit or pollute from others so that they are able to meet their obligations, caps or limits on their own emissions. Participants buy and sell permits such that those who can meet the emission limits at low cost can sell excess permits to others who are only able to meet their obligations at a high cost. This trade involves mutual gains and the market for permits creates a price for pollution that theoretically can achieve emissions reductions at least cost. Emissions trading schemes are a common feature of climate change mitigation policy. For example, the Kyoto Protocol and the European Union Emissions Trading Scheme. See *cap-and-trade* and *baseline-and-credit.*

emitter.   An entity that is a source of emissions, such as a coal-fired power plant.

endangered species.   Plants and animals threatened with extinction including at least 3000 animals and 24000 plants listed under CITES. See *extinction* and *Convention on International Trade in Endangered Species.*

Endangered Species Act.   Legislation passed in the United States in 1973 requiring the US government to identify species that are in danger of extinction. Species that are identified are placed upon a list, at which point public and private landowners are prohibited from any activities that may harm the endangered species.

endemic disease. Disease that is commonly found and continuously present in a particular region or location.

endemicity. The degree to which a tax or species is found only within the boundaries of a given area.

endemics. See *endemic species.*

endemic species. Species found in a defined geographical area.

endocrine. Name given to the system, and its hormones, that is used to regulate body function in complex organisms.

endocrine disruptors. Chemicals that affect the endocrine hormone system of animals, including organochlorines such as PCBs, DDT and many pesticides. Such chemicals can lead to physical abnormalities and a variety of health problems.

FURTHER READING
Colborn, Dumanoski and Myers (1997).

endogenous growth. Name given to economic models that try to explain economic growth endogenously such that growth is not explained by something exogenous such as technological progress. Such models have emphasized, among other things, the importance of human capital in economic growth.

FURTHER READING
Barro and Sala-i-Martin (2004).

endogenous technological change. Innovation that occurs as a result of market signals and consumer demand, rather than government subsidies. In the context of climate change, this term may refer to the emergence of carbon emissions mitigating technology as a market response to the effects of climate change. See *autonomous technological change*, *induced technological change*, and *technological change.*

endogenous variable. A variable in a model or an equation which is determined by the model. See *exogenous variable.*

end-of-pipe treatment.   Methods of pollution control that focus on treatment of pollutants at the end of the production process rather than the prevention of pollution.

endothermic.   A process or reaction by which heat is absorbed from the surrounding environment.

energy.   In physics, the capacity for a system to do work on or change other systems. A common measurement of household electrical energy consumption is the kilowatt-hour (kWh).

energy budget.   A measure of the energy flows within an economic or ecological system.

energy carrier.   A secondary or intermediate form of energy comprising hydrocarbons, hydrogen and electricity that can be used either immediately or stored for the future that is obtained from processing or converting a primary energy source. See *energy conversion chain* and *primary energy*.

energy conservation.   Methods and practices to reduce the levels of energy consumed.

energy conversion chain.   A conceptual model, that is also called an energy pathway, used to track a primary energy source to an energy carrier and then to its final end-use service that accounts for processing, storage, end-use conversion, and energy losses and emissions of pollutants. The comprehensive approach used to create an energy conversion chain provides for a complete assessment of an energy system through an energy conversion chain analysis. See *primary energy* and *energy carrier.*

energy crisis.   Term used to describe the economic downturn and concern over the scarcity of non-renewable resources that followed the rapid increases in oil prices in 1973.

energy efficiency.   The proportion of energy input into a system or process that can be transformed into useful work rather than dissipated. For example, miles per gallon of gasoline.

energy independence.   The state of being whereby a nation is not reliant on external resources to satisfy the energy requirements of its economy. See *energy security*.

energy intensity. Ratio of expenditures or use of energy to the total value of production in an economic process or system. The energy intensity of global production is significant to climate change mitigation if energy is produced from greenhouse gas emitting fuel sources. In that case total emissions can be cut back from reductions in energy intensity. See *Kaya Identity*.

energy pathway analysis. An alternative term used for an energy conversion chain analysis. See *energy conversion chain*.

energy recovery. The process whereby energy is created from the waste materials or byproducts of an industrial process.

energy reserves. Stocks of non-renewable resources for energy production, such as natural gas and oil.

energy savings. Reductions in energy consumption arising from adoption of energy efficient technologies or techniques.

energy security. A broad term relating to the security of a nation's energy supply. As the majority of fossil fuel reserves are concentrated in localized areas, many countries rely on imports. The economies of these importing nations are highly susceptible to variability in supply. It is argued that energy security may be obtained if countries importing fossil fuels convert to renewable energy, such as solar power, which can be sourced within national borders.

energy storage. Converting actual energy into potential energy that can be recovered later. For example, using windmills to pump water into a reservoir when the wind is blowing so that the water can later be released to create power when there is no wind.

enhanced adaptive capacity. The notion that states and communities have a relatively higher or improved ability to adapt to climate change. Adaptive capacity may be enhanced by higher incomes, improved environmental management and forward planning.

enhanced greenhouse effect. The idea that the Earth's natural greenhouse effect has been amplified by anthropogenic greenhouse gas emissions. See *greenhouse effect*.

enforceability. The ability to protect and maintain the property rights of individuals, communities or firms.

Engel's law.   So-called economic law observed by Ernst Engel that states that as income rises, the proportion of income spent on food declines.

ENGOs.   Acronym for environmental non-governmental organizations.

entropy.   The level of disorder in a system that, in a closed system, represents the amount of unavailable or bound energy. Living organisms (including human beings) try to maintain, or at least slow down, an increase in their own entropy by using low entropy materials in their environment. Economic activity can be broadly characterized as the transformation of low entropy natural resources (such as oil) into higher entropy forms, such as wastes. See *second law of thermodynamics.*

FURTHER READING:
Georgescu-Roegen (1973).

enumerator.   One who administers survey questionnaires.

envelope theorem.   A mathematical theorem widely used in economics which states that the partial derivative of the optimal value function by a parameter is identical to the partial derivative of the associated Lagrangian function, holding all variables at their optimal values.

FURTHER READING
Simon and Blume (1994).

environment.   The abiotic and biotic elements that form our surroundings.

environmental accounting.   A broad term referring to the measurement of environmental costs and benefits, as well as the incorporation of such valuations into business decisions.

environmental amenities.   Utility enhancing attributes of the natural environment. Usually, amenities refer to non-extractive, consumption attributes (e.g., site attributes, air quality, views).

environmental assessment.   An assessment of the environmental impacts from a proposed policy or action. In the United States, this term is used in the context of the formal procedures concerning planned federal agency operations.

environmental assets.   Elements of the natural environment which provide environmental functions or services; such services may have direct use value (such as floodplains which regulate water levels) or non-use value (such as the existence of biodiversity).

environmental audit. A formal investigation of the environmental performance of a business or facility, particularly with respect to environmental regulations.

environmental bond. A bond made by a firm (or an individual) against possible environmental damage that could result from that firm's actions. The firm can reclaim its bond by employing environmentally sensitive practices or through environmental remediation that leaves the environment at a level of quality agreed upon in the bond.

environmental capital. The renewable and non-renewable resources present in the natural environment.

environmental covenant. A formal agreement between two or more parties in which the owner of an environmental asset or resource, in return for some consideration, agrees to restrict who has access and/or how the asset or resource is used. Covenants often arise between owners of land and governments interested in conserving the habitat or species found on private land.

environmental economics. The application of economic principles to the management of environmental quality. Environmental economics differs from resource economics in that it focuses on the valuation and management of environmental quality and not just resource stocks.

environmental effectiveness. The degree to which a policy or action achieves a desired environmental outcome.

environmental equity. The degree to which environmental amenities and risks are distributed evenly within society.

environmental federalism. Term used to describe how in federal states (such as the United States, Canada, Mexico, Germany and Australia), policies have been developed to assign responsibilities to the appropriate level of government.

environmental flows. The quantity, quality and timing of water flows in streams and rivers necessary to maintain ecosystem services.

Environmental Impact Assessment (EIA). A detailed accounting of the impacts on the environment that would be expected given a proposed development. The findings of the EIA are presented in an environmental impact statement (EIS).

Environmental Impact Statement (EIS).   The formal document that details the likely impacts on the environment due to proposed development. See *environmental assessment*.

environmental justice.   The principle that poor people and future generations should not suffer the burden of environmental pollution and contamination disproportionately.

environmental Kuznets curve.   An inverted U-shaped functional form relating levels of environmental degradation with income. The curve suggests that environmental degradation first increases as countries start to develop, reaches a turning point, and then subsequently declines as countries become wealthier.

FURTHER READING
de Bruyn (2000) and Ekins (2000).

environmental policy.   A stated policy, or suite of policies, pertaining to environmental protection or improvement adopted by a government, political party or organization.

Environmental Protection Agency (EPA).   The US federal agency charged with controlling pollution.

environmental risk.   Risks to human, ecological, or material well-being that emanate from environmental sources.

environmental services.   The services provided by the environment which include the ability to assimilate and dispose of waste, provision of natural resources (such as fish and trees), and aesthetic and recreational value.

environmental services flows.   A series of beneficial services produced by an environmental asset over time.

environmental sex lability.   A characteristic of some plants and animals in which extremes in environmental factors (e.g., temperature or humidity) can determine sex. Environmental sex lability is known to occur in certain amphibians, reptiles and eels.

environmental shirking. When an individual or firm fails to comply with environmental regulations.

environmental sustainability. A concept related to sustainable development. It implies the idea that an environment's natural capital stock remains constant over time. Differences in what constitutes the natural capital stock and what exactly remains constant over time have led to a number of different definitions of sustainability. See also *economic sustainability* and *sociopolitical sustainability*.

environmentally sustainable growth (path). A path of economic development in which per capita social welfare is monotonically increasing and environmental quality is non-decreasing over time.

environmental values. Human perception and satisfaction derived from the physical attributes or functions of a landscape or ecosystem.

environmental valuation. The valuation of environmental goods and services through environmental economic techniques, such as contingent valuation, hedonic pricing, travel-cost method, choice modeling and benefits transfer.

FURTHER READING
Grafton et al. (2004).

enzyme. A set of proteins used by plants and animals to accelerate the speed of chemical reactions.

EPA. See *Environmental Protection Agency*.

Eocene. Geological epoch from 57.8 to 36.6 million years ago.

ephemeral. Short lived. Ephemeral refers to plants that thrive only for a short time, usually after heavy rain.

epidemic. A widespread infection or outbreak of a disease involving many individuals.

epidemiology. The study of human diseases using data on the distribution and size of diseases, and their causal factors.

epilimnion. A warmer and upper layer of water in a lake.

epiphyte.   Plant that depends on other plants for physical support but not for nutrition.

epoch.   A subdivision of a geological time period.

equality.   A term used to describe the belief that consideration should be given to the distribution of income, living standards and opportunity within a society, between countries, and across generations.

Equator Principles.   A voluntary set of standards for financial institutions to determine and manage the environmental and social risks associated with the projects they finance. See *corporate social responsibility*.

equatorial currents.   Large surface currents found near the equator that flow in a westerly direction.

equilibrium.   A state of balance in a system. In the static case, it represents a balance between competing forces (e.g., supply and demand). In the dynamic case, equilibrium represents the state of a system in which stocks tend to remain at constant levels over time.

equilibrium line.   The border on a glacier between the region where there is a net annual gain in ice mass, and the area below it where there is a net loss.

equitability.   A measure of the evenness with which a resource or its associated benefit stream is distributed among members of society.

equivalent carbon dioxide concentration.   See *carbon dioxide equivalent*.

equivalent variation.   A measure of consumer surplus based on the amount a consumer would be willing to accept to forgo a change in economic circumstances due to a proposed policy. The difference in consumer surplus is measured by subtracting how much someone would have in the current situation, without the policy change, from what she would have under the proposed policy. This tells us what the individual would need to make her as well off without the policy change as she would have been with it. See *compensating surplus variation*.

ericoid.   Plants that demonstrate a shrub-like growth habit.

erosion.   The transfer of soil, sediment, or rock caused by the movement of wind, water bodies, or ice. Although erosion occurs naturally, human activities can significantly exacerbate this process. Examples include: removal of trees and other vegetation, and dam reservoir construction. Erosion can cause significant environmental and social damages, particularly on agricultural and coastal land. For example, the rising sea levels associated with anthropogenic climate change threaten erosion of coastal landforms, potentially causing inundation of seaside communities and ecosystems. See *breakwater* and *sea level rise*.

erratum.   Latin term for error.

error term.   See *residual*.

ESAM.   Expanded social accounting matrix. In addition to standard income and capital accounts, the ESAM also includes environmental stocks and flows.

escapement fishery.   A fishery that is managed by ensuring a certain number of fish are allowed to live or "escape" free from exploitation. Salmon fisheries are escapement fisheries as a certain number of salmon are needed to swim upstream to spawn and ensure the future of the fishery.

Escherichia coli.   Bacteria that live in the digestive tracts of vertebrates, including humans, and that provide a useful function in the absorption of nutrients. See *Escherichia coli O157:H7*.

Escherichia coli O157:H7.   A recently discovered pathogen that can be found in the feces and tissues of some animals and can contaminate water supplies. The pathogen has been called "hamburger disease" because it was first observed in the United States and, in the initial outbreak, was transmitted from the sale of undercooked hamburgers. Outbreaks have occurred in several countries including Japan, England and Canada.

estuary.   The lower part of a river or stream that enters the ocean and is influenced by tidal waters.

ethane.   See *ethylene*.

ethanol.   Common name for ethyl alcohol. It is produced from the fermentation of plants such as sugar cane and is commonly used as a substitute for, or additive to, petroleum. See *biofuel*.

ethnobotany.   The study of the relationship between plants and people.

ethno-medicinal.   Traditional knowledge related to the collection and processing of naturally occurring medicines.

ethylene.   An important plant hormone that is emitted by ripe fruit and that is used commercially to hasten the ripening of stored fruit.

ETS.   See *emissions trading scheme*.

et sequens (et seq.).   Latin term meaning "and the following".

eugenics.   The notion that the human species can be "improved" genetically by selective mating.

eukaryote.   Multicellular (including humans) and unicellular life forms whose cells are characterized by a nucleus and membrane and other features.

Euler's equation.   An equation that, under certain conditions, yields the necessary condition to maximize the value of the objective function in a dynamic optimization problem.

FURTHER READING
Chiang (2000).

euphotic zone.   Upper levels of a lake or the ocean where light is sufficient for photosynthesis.

European Union (EU).   A group of more than 25 European countries which share a common free trade area, a central parliament and justice system. The majority of members have also adopted a single currency, the euro. The history of the EU has been characterized by increasing integration and acceptance of new members. Although primarily an economic union, members share rules governing a range of policy areas, as well as cooperating closely on foreign policy, military affairs and humanitarian aid. The European Union has long been a strong advocate of climate change mitigation, as evidenced by its proposal of the 2° global warming target now central to international climate policy.

European Union Emissions Trading Scheme (EU ETS).   The largest transnational emissions trading scheme in the world, and the central element of

EU climate policy. Operational since 2005, the EU ETS covers around half of all member state emissions, as well as providing a marketplace for emissions credits from international mitigation instruments such as the Clean Development Mechanism.

eustatic sea level change.   Alteration in mean global sea level due to change in the total volume of water in ocean basins. See *local mean sea level change* and *sea level rise*.

eutrophic.   A body of fresh water which is highly productive because of high nutrient concentrations.

eutrophication.   A natural process that can be hastened by human activity if it increases the level of nutrients and reduces the level of oxygen in the water. The runoff from farms which have received applications of phosphorous and nitrogen, in particular, can hasten eutrophication of small lakes and ponds.

evaporation.   Process by which liquid in bodies (such as water in soil and oceans) is transformed into vapor or a gaseous state (such as water vapor).

evapotranspiration.   An important part of the water cycle whereby evaporation of moisture at the Earth's surface and transpiration from plants discharges water into the atmosphere. See *evaporation* and *transpiration*.

even-aged forest.   Forest where the trees are of a similar age. Many commercial and artificial forests have even-aged stands of trees and it is claimed that this reduces ecosystem and species diversity relative to uneven-age and old-growth forests.

even-aged stand.   A forest in which all stands are approximately the same age (within 20 years.) See *uneven-aged stand*.

evergreen trees.   Tree species, such as cedar, that retain their leaves all year round.

evolution.   The notion that changes in species occur incrementally over succeeding generations. See *Darwinian evolution*.

evolutionary facilitation.   The notion that, over time, some species are becoming better able to evolve or adapt.

evolutionary paradigm.   A mechanism for economic adaptation and learning using three basic processes: (1) information storage and transmission, (2) generation of new alternatives, and (3) the selection of superior alternatives according to performance criteria.

ex ante.   Latin term meaning before the event. Often used in the context of predictions or forecasts. For example, *ex-ante* scientific analysis of future climate indicates a high probability of anthropogenic climate change. See *ex post*.

ex-ante crediting.   Issuance of emission credits in anticipation of future reductions. See *ex-post crediting*.

exceptionally unlikely.   See *likelihood*.

excess capacity.   The difference between the output level a firm or industry could produce if production was efficient and inputs were fully utilized, and the output level actually produced for a given set of input and output prices and a given technology. See *capacity*.

excess demand.   Demand that exceeds the current level of supply at the existing market price.

excess supply.   Supply that exceeds the current level of demand at the existing market price.

exchange controls.   Controls imposed on individuals and companies to limit the exchange of one currency for another.

exchange rate.   The price of a currency in terms of other countries' currencies, or the rate at which one can exchange one currency for another.

excludability.   The ability to deny others access to, or use of, a good or asset.

exclusive economic zone (EEZ).   A zone of extended jurisdiction up to 200 nautical miles from the shores of coastal states.

exclusivity.   A characteristic that refers to how able a society, or the holder of a property right, is to exclude others from infringing upon recognized property rights. Without some degree of exclusivity, property rights do not exist.

exhaustible resource. See *non-renewable resource*.

existence value. The value placed on a good or resource solely because that good or resource exists (exclusive of any use value). For example, many people wish to ensure the continued existence of blue whales despite the fact they will almost certainly never see one or benefit from its "use". See *consumptive value* and *non-use value*.

ex officio. Latin term for office or position (of an individual).

exogenous technological change. See *autonomous technological change*.

exogenous variable. A variable in a model or an equation that is pre-determined. That is, its value is determined outside of the model.

exothermic. A process or chemical reaction in which heat is emitted into the environment.

exotics. See *exotic species*.

exotic species. Species that are not native to an environment. The introduction of exotic species, especially in isolated island habitats, has been a major cause of species extinction. See *native species.*

expansion path. The path traced in terms of relative input usage from an expansion of output of a firm.

expected value. The expected value of variable X is denoted by $E(X)$ and is a weighted average of the variable, with the weights corresponding to the probability that the variable will take on a particular value. Thus if the probability that the variable X takes on a given value x_i is

$$Pr(X = x_i) = f(x_i)$$

and if X takes on n possible discrete values then

$$E(X) = x_1 f(x_1) + x_2 f(x_2) + \ldots + x_n f(x_n).$$

expected value maximization. A choice rule for consumption or resource allocation models in which the consumer or planner chooses among several uncertain (random) allocations of goods or resources in order to maximize the expectation of utilities from each possible outcome. The expectation of utilities is calculated as the utility from each possible allocation times the probability of occurrence of that allocation.

expenditure function.   Function that represents the minimum level of expenditure necessary to achieve a defined level of utility.

exploitation rate.   The ratio of the weight of the fish harvested in a period of time to the total biomass of fish at that time.

explicit costs.   Costs that directly affect the cash flow of firms or individuals. See *implicit costs*.

exponent.   The power to which a variable is raised. For example, in the equation $b = a^3$, the variable a has an exponent of 3.

exponential function.   A function of the form $f(x) = b^x$ where b is the base and x is the exponent.

exponential growth.   Growth or increase which is a fixed percentage per time period. For example, a population which increases by 1.5 percent per year exhibits exponential growth.

exposure analysis.   A method of analysis that begins at the source of contamination and assesses its transmission and potential impacts on the environment, especially human health.

ex post.   Latin term for referring to past events. For example, *ex-post* analysis of recent climate indicates a high probability of anthropogenic climate change having already occurred. See *ex-ante*.

ex-post crediting.   Issuance of emission credits for verified and measured reductions that have already occurred. See *ex-ante crediting*.

extant.   Species that currently exist.

extended jurisdiction.   The extension of national sovereignty outside of national borders, such as the creation of 200-nautical-mile exclusive economic zones. See *exclusive economic zone*.

extensive form.   See *game tree*.

extensive margin.   The term for units of land where the value of production exactly equals the costs of production. At the extensive margin, no differential rent is earned. See *Ricardian rent*.

FURTHER READING
Van Kooten and Bulte (2000).

external benefit. The benefit that accrues to others but which is not accounted for in the actions of consumers and producers.

external cost. The cost imposed on others and the environment that is not accounted for in the actions of consumers and producers.

external economies. Positive externalities that arise in a region following an increase in production or output, such as reduced transportation costs due to a greater density of population.

external debt. The debt owed by a country (both private and public debt) to foreigners.

external diseconomies. Negative externalities arising in a region or area from increases in production or output, such as traffic congestion.

external forcing. The process whereby changes to the climate system originate from outside the system. For example, variation in solar radiation, volcanic dust and anthropogenic greenhouse gas emissions. See *radiative forcing*.

externality. A situation where the actions of an individual, company or community has an unintended effect on others or the environment, and these effects are not taken into account in the decision making of the parties causing them. For example, if a firm pollutes a river without considering the costs it imposes on downstream users, this represents an externality.

FURTHER READING
Baumol and Oates (1988).

extinction. Shortened form of species extinction, which has occurred throughout life on Earth including several mass extinctions such as that which occurred at the end of the Permian period some 245 million years ago. The current rate of species extinction is considered by many observers to be several times higher than in the recent past. Contributing factors to the high rate of species extinction include habitat destruction through land clearing and introduction of non-native species. Scientists predict that anthropogenic global warming will accelerate extinction rates, as species may struggle to adapt to changing climatic conditions and damages to natural habitats.

FURTHER READING
Kaufman and Mallory (1993).

extirpation.   The eradication of a species in a particular area or region.

extractive use value.   The consumptive value of a good when removed from nature.

extrapolation.   Terms used for predicting future values of variables from past values.

extrinsic value.   The value placed by a consumer upon an object derived from either the benefits the consumer receives or from interaction with the object, rather than from its intrinsic value. For example, the pleasure received from viewing wildlife rather than the inherent value such wildlife possesses. See *intrinsic value*.

extreme weather events.   Severe weather that may cause large environmental and economic damage. For example, floods, hurricanes, droughts and heat waves.

Exxon Valdez.   An oil tanker that ruptured its hull in Prince William Sound, Alaska, in 1989. The oil spill led to billions of dollars in clean-up costs and damages.

FURTHER READING
National Geographic (1999).

F

$F_{0.1}$.   A constant rate of exploitation of a fishery that is slightly less than that which maximizes the yield per recruit. $F_{0.1}$ is a widely used but ad hoc method of determining the total allowable catch in a fishery.

factorial design.   A method of design in experiments that allows for statistical comparisons between different "treatments" or applications. For example, a factorial design may be used to assess the performance of different varieties of wheat in field trials where each wheat variety represents a treatment.

factors of production.   The inputs in the production process which, collectively, include land (natural resources and the environment), labor (physical, human and social capital), and capital (physical capital).

Fahrenheit scale.   Named after the seventeenth-century scientist G.D. Fahrenheit and still used in the United States. Under the Fahrenheit scale, 32 degrees is the temperature at which ice melts and 212 degrees is the boiling point of water. To convert a temperature from the Fahrenheit scale to the internationally used Celsius scale subtract 32 from the Fahrenheit temperature and then multiply by 5/9. See *Celsius*.

fair compensation.   The notion that appropriation of assets or resources by the state or another authority requires that those affected receive a compensation that reflects the value of the property. See *eminent domain*.

fair game.   A gamble or game where the cost of playing the game equals the expected return. See *expected value*.

fair market value.   The price of an asset or resource should it be sold in the current market. The term is often applied in real estate when assessing the value of property.

fallacy of composition.   The mistaken assumption that what might be true on a small scale must be true on a much larger scale. For example, an individual fisher may feel that increasing the catch slightly would have no effect on the fish population. However, if all fishers make the same decision, the aggregate effect might be enough to reduce the stock level, if not eliminate it.

falldown effect.   The decline in timber harvest associated with the transition from a first cut on a mature natural stand of timber to a sustainable harvest-level with a balanced age–class structure.

fallow.   Land that is left uncultivated that may or may not have been previously farmed.

false-negative rate.   The proportion of results in a diagnostic test that fails to identify what is being tested for but is, nevertheless, present.

false-positive rate.   The proportion of results in a diagnostic test that incorrectly identifies what is being tested.

falsifiable.   Principle used in model-building that states that all the hypotheses derived from a model should be testable and can be proven false.

family.   A classification of organisms by genera. For example, the family Hominidae includes the genus Homo, of which *Homo sapiens* (modern humans) is the only surviving species. See *genus*.

famine.   A severe shortage of food that may lead to mass starvation and widespread mortality. See *food security*.

FAO.   See *Food and Agricultural Organization of the United Nations*.

fauna.   The animal life found in an ecosystem. See *flora*.

Faustmann rotation.   The optimal time to cut a stand of uniformly aged trees assuming the land on which the trees are growing has no other value than the return it generates from timber and that the landowner plans to have an infinite number of rotations. The Faustmann rotation is shorter than the single-period rotation or Fisher rotation. See *Fisher rotation* and *Hartman rotation*.

FURTHER READING
van Kooten (1993).

FDA.   See *Food and Drug Administration*.

fecal coliform (FC).   See *coliform bacteria*.

fecundity.   A measure of fertility commonly defined by the number of offspring per period of time.

feedback. An occurrence that arises when one component in a system changes in response to another component. A negative feedback arises when the response to a change in a component in a system tends to move the system back to its former state while a positive feedback tends to move the system away from the former state. See *albedo feedback* and *climate feedback*.

feedback loops. Processes in which the secondary effects tend to either reinforce the process (a positive feedback loop) or tend to limit the process (a negative feedback loop). An increase in the average birth rate in an animal population might tend to lead to a rapidly rising population at first as more adults lead to more breeding (a positive loop) until increased competition for food may lead to an increased mortality (a negative loop).

feedback mechanisms. Cyclical processes in system and climate change modeling in which process flows are indirectly self-influencing through an intermediary process. Feedback is considered positive (negative) when it increases (decreases) the level of flow in question. See *climate feedback*.

feedback rules. Explicit rules that relate current decisions to the current state of nature. See *feedback*.

FURTHER READING
Grafton, Sandal and Steinshamn (2000).

feedlot. Pens or structures in which livestock are enclosed and fed in order to increase their rate of growth. Large feedlots can impose significant environmental costs if the animal wastes are not disposed of appropriately.

fee-simple. Land-ownership that is freehold and without encumbrances on transfer or inheritance.

fen. English term for a wetland, swamp or bog.

feral. Domesticated animals that have subsequently returned to living in the wild, such as dingoes in Australia.

fertility rate. The number of live births in a year per 1000 women of child-bearing age in the population.

fertilizers. Substances added to soil to aid in plant growth. Fertilizers include animal manure and compost as well as artificial substances like phosphates and nitrates. Too heavy a treatment of fertilizers can impose environmental consequences as the material may be washed or leached away and can contaminate water sources.

f-gas. See *fluorinated greenhouse gases*.

finfish. Vertebrate fish species.

fire suppression. A common practice in forest and park management in which forest fires are suppressed. A potential problem of fire suppression in forestry is that it increases the chances of very large fires by allowing combustible matter to accumulate. In parks, it can change the nature of habitats and may be detrimental to some species.

first fundamental theorem of welfare economics. A theorem stating: in a perfectly competitive market, with no externalities or public goods, the market equilibrium is Pareto efficient. The result provides much of the justification for the arguments in favor of "free markets". See *Pareto efficiency* and *second fundamental theorem of welfare economics*.

first law efficiency. A measurement of the energy efficiency of an apparatus, which is expressed as the ratio of the amount of input energy to the amount of useable output energy. See *second law efficiency.*

first law of thermodynamics. See *law of conservation of energy*.

first-order conditions. A set of necessary conditions that requires the first-order partial derivatives of the choice variables to be equal to zero if the objective function is maximized or minimized.

FURTHER READING
Simon and Blume (1994).

fiscal policy. Any policy of the government that is related to its revenues and expenditures.

Fisher rotation. Single-period rotation in forestry where the optimal time to cut a stand of uniformly aged trees is when the rate of growth in the value of the trees equals the rate of interest. See *Faustmann rotation.*

FURTHER READING
van Kooten (1993).

fishers' individual salmon harvesting rights (FISHRs). Property rights for salmon fishers that allow them to bid for the privilege of participating at salmon fishing openings.

FURTHER READING
Grafton and Nelson (1997).

fishery management plans (FMPs). Plans developed by Regional Fishery Management Councils in the USA that set the broad objectives and strategies for managing specific fisheries.

FURTHER READING
National Research Council (1999).

fishing effort. A single measure that tries to aggregate the total effort or inputs used in fishing. Fishing effort may be defined as days at sea, fuel consumption, the number of fishing boats actively used, or any measure that is reflective of fishing inputs. Standardized fishing effort is a measure of fishing effort after making adjustments for differences in the gear and vessels used.

fishing mortality. Mortality, or death, of fish that is attributable to harvesting or capture. Fishing mortality and natural mortality together represent total mortality in a fishery.

fishing power. A measure of the ability of a fishing vessel to catch fish; may be a function of vessel characteristics, gear used, the skipper, or the crew.

fission. See *nuclear fission*.

fitness. A measure of how well adapted an individual or species is to its environment.

550 ppm. The limit on atmospheric greenhouse gas concentration proposed by the Stern Review (2007). Estimated to be commensurate with 3°C of global warming. Although many early critics of Stern argued that this increase in temperature was too high, the stated target for international action has since been reduced to 450 ppm, thus reflecting more recent scientific research on climate change. See *Stern Review*.

fixed costs. Costs of production that do not vary with the level of output.

fixed effects.   Term applied to statistical models where differences across units of observations can be identified by dummy variables or different constant terms. See *random effects*.

FURTHER READING
Greene (2003).

fixed exchange rate.   A term that refers to when a country's exchange rate is fixed or pegged to that of other currencies. See *floating exchange rate*.

fixed factor.   A factor of production that can only be changed in the long run, such as the size of a factory.

fixed input.   See *fixed factor*.

flexibility.   A property right characteristic that refers to the ability of the holder of the property right to accommodate changes in the asset and individual circumstances.

Flexibility Mechanisms.   Policy instruments under the Kyoto Protocol for reducing the cost of mitigating greenhouse gas emissions in developed countries, including: Clean Development Mechanism, Joint Implementation, and Emissions Trading. See *Clean Development Mechanism*, *Joint Implementation*, and *Kyoto Protocol*.

flexible functional forms.   Functions used to estimate costs, profits and output from data, that do not impose arbitrary restrictions on the elasticities of substitution among inputs or outputs. A commonly used flexible functional form is the Translog.

FURTHER READING
Takayama (1985).

floating exchange rate.   A term that refers to when a country's exchange rate is able to adjust freely to changes in currency markets. See *fixed exchange rate*.

flocculation.   Term used to describe the aggregation of fine sediments suspended in a solution. Flocculation is used in sewage treatment to help remove fine particles of waste from treated water.

flood plain.   The area of land surrounding a water channel that is subject to flooding.

flora.   The plant life found in an ecosystem. See *fauna*.

flow.   The input or output from a stock that can vary over time. The annual harvest of fish from a fishery is a flow. See *stock*.

flow pollutant.   A pollutant that imposes social costs at the time of its discharge, but does not accumulate in the environment, such as noise pollution.

flowback.   Term used for the process or water returned to the surface after hydraulic fracturing is complete. See *hydraulic fracturing*.

flower.   The part of an angiosperm used for reproduction. See *angiosperm*.

flue gas desulfurization.   Method for removing some of the sulfur dioxide produced in the combustion of coal by the spraying of the fumes in a discharge chamber with a lime-rich liquid so as to produce calcium sulphate.

fluidized bed combustion.   Method for reducing sulfur dioxide emissions by burning coal with a mix of limestone and sand.

fluoridation.   See *fluoride*.

fluoride.   A form of the element fluorine that, in small doses, can be beneficial to the development of teeth and bones. Many urban communities add small quantities of sodium fluoride to drinking water for its potential health benefits, but this procedure is becoming increasingly controversial as some individuals are unable to tolerate even small doses (one part per million).

fluorinated greenhouse gases (f-gas).   A broad term for hydroflurocarbon, perfluorocarbon and sulfur hexafluoride gases. These gases were introduced as an alternative to ozone-depleting chlorofluorocarbons in refrigeration during the 1990s. However, these gases have very high global warming potential and a long atmospheric lifetime. Although much less prevalent than carbon dioxide, f-gases are subject to considerable mitigation efforts because they are highly damaging. See *global warming potential* and *lifetime*.

fluorocarbons.   The term applied to carbon–fluorine compounds, often also contain hydrogen and other elements. See *chlorofluorocarbons*, *hydrochlorofluorocarbons*, and *hydroflurocarbons*.

F_{max}.   A constant rate of exploitation of a fishery that maximizes the yield per recruit.

folivorous.   Leaf-eating.

folk theorem.   In game theory, the hypothesis that as the time horizon approaches infinity, the incentive to "free-ride" will diminish.

Food and Agriculture Organization (FAO).   An organization of the United Nations based in Rome, Italy, that is mandated to improve the living standards and nutrition of rural populations.

Food and Drug Administration (FDA).   US federal agency charged with ensuring the safety of food and drugs consumed within the United States.

food chain.   The chain of events in the food consumption of an ecosystem in which plants are eaten by animals which in turn are eaten by larger animals.

food security.   This term relates to the availability, access and utilization of food resources, and can be considered in the context of households and communities or, more broadly, at a national and global level. Food security can be threatened by: crop destruction through extreme weather, poor institutional infrastructure, unsustainable agricultural practices and use of environmental resources, and land-use changes such as biofuel production.

food web.   See *food chain*.

foot.   Unit of length equal to 0.3049 of a meter.

foehn.   Dry wind that blows on the eastern side of a range of mountains.

forage.   Vegetation, including grasses, herbs, shrubs, and leaves that can be used to feed livestock.

forecast.   A prediction of future events based upon an assessment of current trends and future probabilities.

Fordism.   Term describing the large-scale mass production of goods, often characterized by assembly lines.

forest.   An area of land densely covered by trees. In addition to wood production, forests are an important source of natural resources and support large ecosystems containing other vegetation and animals. As trees consume carbon dioxide, forests place a vital role in the carbon cycle. See *deforestation*.

forest carbon. A term used in the context of climate change mitigation referring to carbon that is stored in the vegetation and soil of forests.

forest clearing. See *deforestation*.

forest degradation. 1. Tree felling such that crown cover is reduced below 20 percent. 2. Degradation of forest ecosystems, non-tree vegetation and soil. See *deforestation*.

forest management. The organized management of forest stands to achieve stated outcomes. Forest management often refers to a plan of use and stewardship, as implemented by a private owner or public agency proprietor.

forest principles. Principles of forest management agreed to at the 1992 United Nations Conference on Environment and Development.

forest profile. A description of the range of forest conditions that exist across a given landscape. A forest profile may contain information on timber species, condition, age, location, elevation, and so on.

forestry. The science and practice of managing forest resources for human benefit.

Forestry Stewardship Council (FSC). A not-for-profit organization that issues sustainability certificates and engages in other activities promoting sustainable forest management.

formaldehyde. A poisonous gas that is widely used in plastics and that can be a major source of indoor air pollution as it is slowly emitted from such products as laminated woods, upholstery and even permanent-press fabrics.

formation. See *geological formation*.

forward crediting. The sale of *ex-ante* credits in a carbon market.

forward market.   Any market in which buyers and sellers contract to buy or sell goods in the future at fixed prices.

fossil.   The geological remains or evidence of past life.

fossil fuels.   Sources of energy, such as oil, natural gas and coal that are obtained from organic matter that was laid down millions of years ago. As fossil fuel combustion is a major source of carbon dioxide emissions, one of the most significant greenhouse gases, their usage is a primary cause of anthropogenic climate change. A large proportion of climate change mitigation activities focus upon switching from fossil fuels to renewable sources of energy. See *climate change*.

4° Celsius.   A worst case, yet plausible, scenario for global temperature rise by 2060 without significant mitigation of carbon emissions in the near future. The negative environmental, social and economic consequences of this situation are predicted to be much worse than a 2° rise.

FURTHER READING
New et al. (2011).

450 ppm.   The concentration of atmospheric greenhouse gases estimated to be commensurate with a 2° temperature rise above pre-industrial levels and, therefore, the implicit target of international mitigation efforts. Although early economic research concluded that reaching such a target would be too costly, more recent analysis suggests that such costs are acceptable as a form of insurance against the risk of catastrophic climate change. See *350 ppm* and *550 ppm*.

Fourth Assessment Report (AR4).   The fourth report of the IPCC. Released in 2007, the AR4 significantly upgraded the assessment of risks posed by climate change from the Third Assessment Report. See *assessment report* and *Intergovernmental Panel on Climate Change*.

fracking.   See *hydraulic fracturing*.

fractals.   Seemingly complicated patterns where the degree of irregularity is the same at any scale. Fractals can be formed from equations that exhibit chaos. See *chaos*.

fragility.   The opposite of resilience. See *resilience*.

free access.   See *open access.*

free good.   Goods that are so abundant they have no market price. Paradoxically, free goods can be the most essential, such as the air we breathe.

free radical.   Name given to any atom or molecule that has a "free" electron and is thus highly reactive.

free-rider.   An individual who benefits from the actions and efforts of others without contributing to the costs incurred in generating the benefits.

free trade.   The unimpeded international exchange of goods and services without trade restrictions, such as tariffs, quotas and non-tariff barriers.

freon.   Common name for a chlorofluorocarbon. See *chlorofluorocarbons.*

frequency distribution.   An approach to representing data where a variable is classified into appropriate groupings, levels or classes with the frequency of the number of observations in each grouping.

fresh water.   Water (H_2O) that contains salt in quantities of less than 0.05 percent.

FURTHER READING
Pielou (1998).

freshwater ecology.   Limnology. The study of aquatic systems in fresh waters such as lakes, ponds, rivers, streams, and other water bodies.

front-end charge.   Charge paid by consumers upon purchasing a product. The charge may be refundable if the waste product is disposed of appropriately.

frontier.   A term for the edge of a forest that is threatened by encroachment from human activity.

frugivorous.   Fruit-eating.

F-test.   A frequently used statistical test for the equality of variances. In regression analysis, an F-test is a joint test used to test the null hypothesis that all coefficients are equal to zero with $K-1$ and $n-1$ degrees of freedom, where K is the number of regressors and n is the number of observations.

FURTHER READING
Greene (2003).

fuel.   A substance that is consumed during the production of energy.

fuel cell.   A device that produces an electrical current from a chemical reaction, such as the oxidization of hydrogen.

fuel economy.   A term used to describe the efficiency of fuel usage by automobiles. Often measured in units of miles per gallon.

fuel loads.   Accumulated dead wood, fallen timber and other combustible material found within a forest stand. More of these materials means a higher fuel load, and an increased likelihood that fires will be severe.

fuel switching.   The switching or substitution of one fuel for another. For example, fuel switching from the use of coal to natural gas in the generation of electricity reduces emissions of carbon dioxide for a given amount of electricity produced.

fuelwood.   Wood that is harvested and consumed for heating and cooking purposes rather than being converted into other products.

fugitive resource.   A natural resource that is non-stationary, such as a fishery.

Fukushima.   The nuclear complex operated by Toyko Electric Power severely damaged by a tsunami following an earthquake on March 11, 2011. The force of the tsunami triggered a series of equipment failures leading to ongoing releases of radioactive materials and Japan's worst nuclear power crisis to date.

full-cost pricing.   The pricing of goods and services such that the price the final consumer pays includes both the private and social costs of its production. See *social cost* and *externality*.

full-tree harvesting.   A method of timber harvesting where the trunk, branches, and sometimes roots are removed at the time of harvest. Full-tree harvesting is sometimes employed to control root disease.

fully regulated forest management. Management of a forest where the forest is divided into a number of areas each capable of producing an equal volume of timber and where the total number is equal to the length of time in years in which it takes the trees to reach the age at which they can be harvested. Each year, one area or stand is harvested.

function. 1. A relationship whereby the independent variables determine unique values for the dependent variable. 2. Activity of an organism.

functional extinction. A decline in a species' population from which it is not possible to recover and extinction is inevitable.

fundamental equation of renewable resources. The equation that defines an optimal steady-state for the level or size of a renewable resource. In its general form, the fundamental equation may be defined as,

$$\rho = F'(x) + \frac{\partial \pi / \partial x}{\partial \pi / \partial h}$$

where ρ is the discount rate, x is the level or size of the renewable resource, h is the harvest, π is the current rent from the renewable resource, $F(x)$ is the growth function of the renewable resource, $F'(x)$ is the change in growth from a change in the size of the renewable resource and the second term on the right-hand side of the equation is called the marginal stock effect. The left-hand side of the equation may be interpreted as the external rate of return (return on assets separate from the renewable resource) while the right-hand side may be interpreted as the internal rate of return.

FURTHER READING
Conrad (1995).

fundamental equation of non-renewable resources. See *Hotelling's rule*.

fund pollutant. See *stock pollutant*.

fungi. Micro-organisms that develop spores to reproduce and include mushrooms, yeasts and molds. Unlike plants or algae, fungi cannot photosynthesize and must absorb their nutrients from their surrounding environment.

furans. Chlorine-based chemical compounds that have similar health effects to dioxins. See *dioxins*.

fusion. See *nuclear fusion*.

futures.   Contracts to deliver or buy goods in a futures market at a defined price.

future value.   Value of returns assessed for a particular period in the future.

fuzzy logic.   Methodology used to model problems that involve fuzzy sets. See *fuzzy sets*.

fuzzy sets.   Sets whose membership or elements are uncertain. Fuzzy sets are used to model situations where the degree, or the extent to which an event occurs is uncertain.

G

G8. See *Group of Eight*.

G20. See Group of Twenty.

G77. See Group of 77.

Gaia hypothesis. Idea popularized by James Lovelock, an atmospheric scientist, who hypothesized that the Earth could be viewed as a single organism where its constituent elements coexist in symbiosis. Thus animals and plants collectively help to manage a world, through positive and negative feedbacks, and ensure the planet maintains an equilibrium that is desirable for all life. See *Daisyworld*.

FURTHER READING
Lovelock (2000).

gallery forest. Tropical forests that grow at high densities and tend to follow riverine courses. See also *riparian*.

gambler's fallacy. A belief that because an event has not occurred for a long period of time then it is more likely to occur in the near future.

game ranching. Raising wildlife under controlled conditions within pens.

game theory. An area of study and research that deals with the strategic interactions of (utility) maximizing agents under specified sets of constraints.

game tree. Also known as extensive form of a game, this is a structure that provides information about a game where the nodes represent potential moves in the game and the edges or lines or branches that move from the nodes represent the possible strategies or actions available at a particular node. Each node is labeled for the player making the move or, if the node has no edges or branches, it is labeled with the pay-offs for the chosen actions. See *game theory*.

FURTHER READING
Binmore (2007).

gamma radiation. The most dangerous form of the three main types of nuclear radiation. It consists of a stream of particles released from radioactive isotopes. See *alpha radiation* and *beta radiation*.

gap analysis (GA).   Analysis designed to bridge or fill a gap in either knowledge or a defined objective. For example, in the United States, gap analysis includes various techniques to identify areas of importance in terms of biodiversity and then to give priority for conservation of areas that are not currently under protection.

garbage in, garbage out.   The notion that if the data used in an analysis is of poor quality then the results will also be of poor quality.

gas.   The state of matter distinct from solids and liquids by virtue of low density, expansion and contraction potential, as well as diffusion properties.

gas flaring.   The practice of burning off natural gas found at oil well sites.

gasohol.   A mix of gasoline, ethanol and methanol.

gasoline.   A flammable mixture of hydrocarbons derived from crude oil and used in combustion engines, such as those powering automobiles. The combustion of gasoline creates emissions of greenhouse gases and other harmful atmospheric pollutants.

Gaussian distribution.   See *normal distribution*.

Gauss–Markov theorem.   A theorem that proves that the least squares estimator of the parameters of a multiple regression is the best linear unbiased estimator (BLUE) with lowest variance, provided that the residuals are identically distributed with a mean of zero.

FURTHER READING
Greene (2003).

GDP.   See *gross domestic product*.

gear restrictions.   Limits placed on the type of gear that may be employed in the harvesting or capture of fish and other animals. Gear restrictions are a form of input control and are widely used in fisheries management.

Geiger counter.   Shortened name for the Geiger–Muller counter that measures ionizing radiation by assessing the electrical field disruption of the argon gas in the instrument.

GEMS. See *global environmental monitoring system.*

gene. A unit hereditary factor comprised of a segment of deoxyribonucleic or ribonucleic acid that codes for one or more related functions of an organism.

gene bank. A repository of plants, seeds and genetic material stored for the purpose of preserving genetic diversity.

gene pool. The total number of genes possessed by the reproductive members of a population.

General Agreement on Tariffs and Trade (GATT). A trade agreement which resulted due to the failed negotiation over the ITO. It was in effect from 1949 to 1993 and was replaced by the WTO in 1995. See *International Trade Organization (ITO)* and *World Trade Organisation (WTO)*.

general circulation. The planetary-scale movements of the atmosphere and oceans resulting from the differential heating caused by the Earth's rotation.

general circulation model (GCM). A computer model of the general circulation of the climate system that incorporates a wide variety of physical and chemical processes. GCMs are used to provide information pertaining to past and future climate, particularly with respect to projections of anthropogenic climate change.

general equilibrium analysis. A method of economic analysis which attempts to simultaneously link all sectors and markets of an economy. See *partial equilibrium analysis.*

generalized least squares (GLS). A method of estimating parameters of a statistical model where there exists autocorrelation or heteroscedasticity and where ordinary least squares (OLS) would lead to inconsistent estimators. See *ordinary least squares.*

FURTHER READING
Greene (2003).

generation. 1. The predetermined "life span" of an agent in an economic or biological model. 2. A cohort of a population of a similar age.

generational equity. The notion that the welfare of future generations should be a central concern in current environmental decision making. A frequently discussed issue in the context of anthropogenic climate change due to: the potentially large scale of future damages; inertia in the climate system that elongates the impact of current greenhouse gas emissions; and the discount rate used in economic analysis of the timing and extent of climate change mitigation.

genetically modified (GM). Term usually applied to plants where one or more of their genes have been altered by genetic engineering for a particular purpose. Genetically modified foods have caused a great deal of controversy due to concerns over the potential long-term health hazards.

genetically modified organisms (GMOs). See *genetically modified.*

genetic diversity. Variability in genotypes found within a species or population. See *biodiversity.*

genetic drift. Chance fluctuations, independent of natural selection, in the genetic make-up of an isolated population.

genetic engineering. The anthropogenic change in genetic structure for a desired purpose brought about by the removal of, or a change in, chromosomes.

genetics. The science and study of how genetic influences affect organisms and the inheritance of characteristics.

genome. The genetic material of an organism contained in its chromosomes.

genomics. The study of genomes.

genotype. The specific genetic composition of an organism. See *phenotype.*

gentrification. The purchase and renovation of homes in run-down urban neighborhoods, often displacing lower-income households.

genuine progress index (GPI). An index of overall societal well-being that includes attempts to take into account changes in social, human, physical and natural capital to reflect the progress, or lack thereof, in a society.

genuine savings. A sustainability measurement which aggregates assets related to development, such as: physical, natural and human capital.

genus. (Plural: genera). A classification of species with common characteristics. For example, modern humans (*Homo sapiens*) are the only surviving species of the genus *Homo*. See *species*.

geochemically scarce metals. Those metallic elements the relative abundance of which falls while costs of extraction rise because of a mineralogical threshold.

geoengineering. Anthropogenic attempts to change global environmental systems, particularly in response to climate change. The increasing interest in geoengineering has followed greater comprehension of the risks posed by catastrophic climate change and the absence of substantial international mitigation. Proposals have included: release of reflective material in the atmosphere; mirrors placed in the Earth's orbit; ships spraying saltwater to promote cloud formation; air scrubbers; and iron fertilization of the oceans to encourage growth of carbon dioxide consuming plankton. Critics argue that many of these options may have unintended consequences and resources should instead be directed towards mitigation.

FURTHER READING
Schneider (2008).

geographic information system (GIS). Digitalized approaches to combining and analyzing information and data to visualize existing data.

geological formation. A distinctive body of earth material that has homogenous characteristics.

geology. The study of the earth, its structures and processes.

geometric growth. Growth or increase where the ratio of the next value in the series to the current value is the same as the ratio of the current value to the previous value. For example, the series 3, 6, 12, 24 . . . exhibits geometric growth where the ratio of the next number in the series to the current number is always 2.

geometric mean.   A summary statistic defined as the nth root of the product of a series of n numbers. For example, the geometric mean of the numbers 3 and 48 is $(3 \times 48)^{1/2} = 12$.

geometric series.   A series of numbers that can be written as $a + ac + ac^2 + \ldots + ac^n$, given c does not equal 1. If $-1 < c < 1$ then the sum of an infinite geometric series is $a/(1-c)$.

geomorphology.   The study of Earth surface processes, landforms and events.

geosequestration.   Capture and storage of greenhouse gas emissions by injection into underground geological formations such as caves. See *carbon capture and storage*.

geostationary.   Description of a satellite that does not appear to move its position relative to the planet around which it orbits.

geothermal energy.   Energy derived from heat beneath the Earth's surface. One of the renewable energy sources for generating electricity. A common method is the injection of water into high temperature areas underground, creating steam that turns a turbine.

ghost fishing.   The capture of fish by fishing gear that has been lost or abandoned. For some gear, such as driftnets, ghost fishing can be a major cause of fishing mortality.

Giardia.   A water-borne parasite that infects animals and humans.

giga.   Term signifying a billion or 1×10^9.

gigalitre.   A common unit of measurement for water flows, equal to one thousand million litres.

gill net.   A fishing net attached to buoys that ensnares the heads of fish.

Gini coefficient.   A common measure of income and wealth inequality. The Gini coefficient measures the area between the Lorenz curve and the 45° line relative to the entire area below the 45° line, where a measure of 0 indicates that all individuals have identical incomes. See *Lorenz curve*.

FURTHER READING
Perkins, Radelet and Lindauer (2006).

GIS. See *geographic information system.*

glacial. A period of extreme cooling during an ice age that is marked by the advance of ice sheets towards the equator. See *ice age, interglacial,* and *Milankovitch Theory.*

glacial period. 1. A geological period embracing an ice age. 2. An epoch during which large continental ice sheets have moved towards the equator, such as during a glacial.

glaciation. The covering of an area of land by glacier ice.

glacier. A land-bound body of ice formed by the accumulation of snow at high altitudes and subsequent downhill ice flow. During the coldest periods of past ice ages, glaciers covered significant areas of most continents.

glacier retreat. The reduction in area and mass of a glacier. Across the paleoclimate record glacier retreat has been a natural event occurring during warming periods within and outside of ice ages. However, global warming from anthropogenic climate change is now causing widespread glacier melt and withdrawal. This process is raising sea levels and reinforcing warming through feedback loops. See *albedo feedback* and *ice age.*

global commons. Aspects of the environment that are global in nature and are common-pool resources, such as the Earth's atmosphere. See *common-pool resource.*

global dimming. A decline in the intensity of sunlight from atmospheric pollution particles experienced during the second half of the twentieth century. It is thought that this dimming masked the temperature change associated with past greenhouse gas emissions, and, ironically, pollution controls may partially increase global warming as a result. See *aerosol.*

global ecological services. Global benefits provided by aspects of the environment. For example, growing forests provide a global benefit by sequestering carbon.

global environmental facility (GEF). A fund established by the United Nations but held at the World Bank. The GEF is used for projects and initiatives in developing countries to prevent or mitigate global environmental problems.

global environmental monitoring system (GEMS).   A system administered by the United Nations Environment Program. The GEMS monitors pollution, particularly airborne pollutants.

global financial crisis (GFC).   A crisis that occureed in 2008–09 but had it origins several years before began with the marketization of finacial instruments that included high-risk housing mortgages in the United States. The spread of these financial instruments globally, but especially in American and British financial markets, triggered a global loss of confidence in financial instititions towards the end of 2008 and precipitated the world's greatest financial crisis. The immediate crisis (2008–2010) was resolved by providing government guarantees and loans to major financial institutions. The GFC trigerred recessions in many developed countries and in some countries, such as Iceland and Ireland, severe economic downturns.

globalization.   The process of economic, social and cultural integration of the world's population. In the context of environmental sustainability, globalization can have both positive and negative effects. For example, international trade can encourage damaging activities such as logging and mining. Alternatively, enhanced distribution of information supports dissemination of sustainable practices and technologies, and awareness of environmental issues.

global surface temperature.   An aggregate measure of global air temperature combining land and sea surface measurements. Also referred to as mean surface temperature.

global positioning system (GPS).   A system whereby a receiver can be used to track particular satellites in the Earth's orbit to determine a spot on the Earth's surface within meters of where the receiver is located.

Global Reporting Initiative (GRI).   A network organization composed of both public and private members which pioneered one of the most widely used standards for sustainable reporting.

global village.   Term used to describe a set of ideas that represents the interconnectedness of people across the globe for good (communications, transport, etc.) and ill (spread of disease, global pollution, etc.).

global warming. The rise in mean global temperature due to anthropogenic emissions of greenhouse gases. Scientists believe there is a high probability that committed future warming from past emissions may already be 1.5°C above pre-industrial levels. Without significant and urgent mitigation it is thought that warming will exceed the international target of 2°C, and potentially 3°C, whereupon tipping points in the climate system may initiate catastrophic climate change. Inertia and feedbacks within the climate system delays the full extent of warming from current emissions being realized for decades, and even centuries. See *2° Celsius*, *4° Celsius*, and *greenhouse effect*.

FURTHER READING
Archer and Rahmstorf (2010).

global warming potential (GWP). An index which values the relative contribution of greenhouse gases (GHGs) to global warming. This index is always expressed with respect to time, reflecting the variable atmospheric lifetime of different gases. Carbon dioxide is used as a baseline and assigned a value of 1. Gases with a value greater (less) than 1 are expected to have a greater (lower) relative contribution for an equivalent amount emitted. For example, nitrous oxide has a GWP of 289 across 20 years. See *lifetime*.

GNE. See *gross national expenditure*.

GNI. See *gross national income*.

GNP. See *gross national product*.

golden rule. Term used to describe a time path for an economy where the level of savings and investment maximizes long-term per capita consumption.

Gondwanaland. A supercontinent that broke off from the even larger supercontinent Pangea and eventually split and formed into Antartica, Africa, India, Australia, New Zealand, Sri Lanka and Madagascar.

goodness of fit. A measure of how well the predicted values from a statistical model "fit" the observed values from sample data. A commonly used measure is the R-squared of a model.

Gordon–Schaefer model. A bioeconomic model of a fishery that combines a surplus production model, or biomass dynamic model, for the population dynamics of the resource, and a static profit maximizing model of fisher behavior. A fundamental result of the model is that the maximum economic yield occurs at a higher level of the resource stock than the maximum sustainable yield.

governance. The structure and practice of decision making in an organization or society.

government failures. Errors of commission and omission for which the government is directly responsible. For example, a nation that is unable to provide security of property rights and personal security for its citizens suffers from government failure. See *market failure*.

GPI. See *genuine progress index.*

grandfathering. Allocation of emission permits on the basis of historical production or consumption levels, usually free of charge.

Granger test for causality. Test used to help determine the direction of causality between two variables. If a variable X "Granger causes" a variable Y, then past values of X are useful in predicting future values of Y.

FURTHER READING
Granger (1969).

grassland. Biomes, such as prairies or savanna, that consist primarily of grasses. Most grasslands have been modified by human activity.

gravity model. Model based on Newton's Law where the gravitational force between two masses is proportional to the product of their masses and inversely proportional to the square of the distance between the masses. Modified forms of the model have been used to help explain other phenomena such as trade between countries and migration between geographically separated populations.

Gray (Gy). The internationally accepted measure for absorbed ionizing radiation where $100 \text{ rad} = 1 \text{ Gy}$.

gray water. Industrial or household wastewater which does not contain excreta, normally disposed of through the stormwater system. As a result of global water shortages, recycling of gray water is becoming an increasingly common practice.

grazing land.   Grassed areas of land used for grazing livestock. Much of the world's grazing land has been artificially created by forest clearing. This practice has degraded soil quality and has been a major source of past greenhouse gas emissions. See *deforestation*.

grazing lease.   A lease on land for the purpose of grazing livestock.

grazing pressure.   The reduction in vegetative cover on a land parcel due to grazing by livestock.

Great Acceleration.   The current period in human history beginning after the end of World War II. It is characterised by dramatic increases in population, economic activity, resource use, scientific knowledge and social liberalization.

Great Lakes Water Quality Agreement (GLWQA).   A 1972 agreement (updated in 1978) between the United States and Canada to reduce the pollution entering the Great Lakes (Superior, Michigan, Huron, Erie and Ontario) which are shared between the two countries. See *International Joint Commission*.

green accounting.   Approaches to national income accounting in which an attempt is made to record changes in the quality of the environment in physical and/or value measures.

FURTHER READING
Nordhaus and Kokkelenberg (1999).

greenbelt.   An area of natural habitat, largely free from development, set aside by easement or by community designation, as a buffer to future development or as a conservation area.

greenfield site.   Agricultural land that is highly desirable for conversion into a developed site for industry or residential purposes.

green GDP.   The conventional measure of gross domestic product adjusted for changes in values of the environment.

greenhouse effect.   Refers to the process whereby solar radiation passes through the atmosphere, warming the Earth's surface and causing infrared radiation to be reflected back up to the atmosphere. Greenhouse gases (such as carbon dioxide) in the atmosphere absorb some of this radiation, thereby preventing the heat from leaving the planet. The greenhouse effect occurs naturally and is a necessary prerequisite for life. However, amplification of this natural process by anthropogenic greenhouse gas emissions is causing global warming. See *greenhouse gases*, *global warming*, and *climate change*.

FURTHER READING
Archer and Rahmstorf (2010).

greenhouse gas concentration.　Term for the total concentration of greenhouse gases in the Earth's atmosphere, usually measured in carbon dioxide equivalent (CO_2-e) parts per million (ppm). The level of the Earth's outgoing heat trapped in the atmosphere is determined by this concentration. Pre-industrial levels of carbon dioxide are estimated to have been 280 ppm, and have now risen above 390 ppm. See *carbon dioxide equivalent* and *450 ppm*.

greenhouse gases (GHGs).　Atmospheric gases which absorb thermal radiation emitted from the Earth. The most important greenhouse gases include carbon dioxide, water vapour, methane, nitrous oxides and hydroflurocarbons. See *climate change* and *greenhouse effect*.

green jobs.　Occupations in industries providing environmentally sustainable activities or environmental improvement, such as construction of wind turbines.

green labeling.　Information placed on products concerning the environmental outcomes associated with production and consumption. This information may be required by government regulation or offered voluntarily as a form of marketing. For example, refrigerators in many countries are required by law to display a rating of energy consumption.

green marketing.　Advertisement of the environmental attributes of a company's products or activities as a major selling point. See *greenwashing*.

green power.　Electricity generated from environmentally sustainable sources such as solar, wind and geothermal energy.

green revolution.　Period beginning in the late 1940s whereby plant scientists were able to use plant breeding to develop much higher-yielding varieties (HYVs) of various grains such as rice and wheat. The subsequent diffusion of the HYVs in countries such as India and Pakistan, along with the increased use of fertilizers and pesticides, significantly increased world grain production.

green tax. A tax that promotes environmental sustainability by altering consumer behavior or providing funds for environmental improvement. Often such taxes are introduced to correct the failure of markets to account for environmental damages. For example, carbon taxes incorporate the social cost of climate change in greenhouse gas emissions intensive goods or services.

green technology. Machinery that contributes to improving the environmental outcomes from human activity. Commonly used in the context of climate change mitigation.

green-tree retention. A method of harvest in which live trees of specific species and size are reserved (spared) during harvesting.

greenwashing. The false presentation of a company's activities as being environmentally friendly or sustainable through misrepresentation of their operations or fabrication of information. For example, an oil company may advertise their investment in renewable technologies, although this forms a very small component of their business.

green water. Fresh water located in soils that is available for plant growth but not directly accessible for extraction or pumping. See *blue water*.

grid. See *electric power grid*.

grim strategy. The behavior or set of actions in a game chosen by a participant to impose the worst possible outcome upon the other participants, regardless of the effect upon themselves.

gross domestic product (GDP). A measure of the total value of goods and services produced in an economy in a year.

gross national expenditure (GNE). A measure of the annual total expenditure in an economy, including consumption, investment and government expenditure.

gross national happiness (GNH). Developed in Bhutan, it is a substitute for the GDP, insomuch as it attempts to measure social progress and quality of life in an holistic manner. See also *quality of life*.

gross national income (GNI).   See *gross national product*.

gross national product (GNP).   A measure of the total value of goods and services produced in an economy in a year, including net income paid to foreigners less net income paid to domestic nationals.

gross primary production.   The gross energy produced by organisms before respiration.

gross registered tons (GRT).   A measure of the size of a vessel that is listed on its registration certificate.

gross world product.   The total value of goods and services produced by all economies over one year.

groundfish.   Term usually applied to demersal fish species (such as cod) that are commonly found near the sea bottom.

ground-level ozone.   See *tropospheric ozone*.

ground resolution element.   The smallest unit of the Earth's surface that can be detected from a remote sensing device.

groundtruthing.   Checks on the reliability of data obtained from secondary sources, or from remote sensing, by verifying the original sources of data or through direct observation.

groundwater.   Water contained in rocks and soil beneath the Earth's surface and which has precipitation as its source of replenishment. The rate of withdrawal of many groundwater supplies or aquifers exceeds the rate of recharge.

Group of Eight (G8).   A forum of the world's richest economies originating in 1975 and currently comprising the United States, Canada, United Kingdom, France, Germany, Italy, Russia and Japan. The principal global economic forum until superseded by the G20 in 2009.

Group of Twenty (G20).   A council of finance ministers and heads of central banks from the world's twenty wealthiest nations. The G20 has replaced the G8 in importance, reflecting the rising economic power of nations such as China, India and Brazil. Participation in G20 meetings extends to heads of multilateral organizations such as the World Bank.

The G20 acts as a forum for discussion of the global economy and financial system.

Group of 77 (G77). A loose coalition of developing countries within the United Nations system, also often referred to as the G77+China. The G77 act as a negotiating body in international climate negotiations.

Groves–Ledyard mechanism. A scheme for eliciting a level of a public good that is Pareto efficient whereby individuals report to a planner the increment (which can be positive or negative) of the public good they would like, in addition to the total amount demanded by everyone else.

FURTHER READING
Cornes and Sandler (1996).

growing season. The span of time during the year in which plants are actively increasing in size.

growing stock. The volume of merchantable trees that are still actively growing within an area.

growth overfishing. Term used to describe how a reduction in the total fishing effort can increase the total future harvests by increasing the average weight of fish.

growth rate. A measure of change in a variable. Using discrete units of time for a variable x, the growth rate between period $t-1$ and t is
$$[(x_t - x_{t-1})/x_t] \times 100.$$

groyne. See *breakwater*.

Gulf Stream. A warm ocean current named after its place of origin, the Gulf of Mexico, which passes up the coast of eastern North America past Iceland, and eventually reaches the waters around Scandinavia. Without the Gulf Stream, Western and Northern Europe would be much colder.

gyre. Planetary-scale rotating systems of ocean currents, usually driven by the Earth's winds.

H

habitat. The environment in which a given plant or animal is usually found. For example, gazelles are usually found in a savanna habitat.

Habitat Conservation Plan (HCP). In the United States, a plan prepared by landowners that specifies how they will deal with an endangered species found on their land. HCPs are submitted by landowners who hold permits for the "incidental taking" of endangered or threatened species on non-federal lands during otherwise lawful activities. The permits and HCP form an agreement between the landowner and the federal government that allows the landowner to engage in activities that might result in the landowner affecting the endangered species or modifying its habitat in exchange for measures that will enhance the survival of the species. See *incidental taking* and *Endangered Species Act.*

habitat diversity. Variations in the ecology of a site that increase the number of niches for different species.

Hadean era. A geological era that started at the formation of the Earth 4550 million years ago and continued until life first began some 3800 million years ago.

Hadley cell. A circulation system whereby rising warm air near the equator moves towards higher latitudes where the air cools and sinks back towards the subtropics, eventually returning to the equator.

half-life. The time it takes for half of the material of a radioactive substance to decay.

halocarbons. A family of compounds containing carbon and either bromine, chlorine or fluorine. These can act as greenhouse gases (GHGs) in the atmosphere. Chlorofluorocarbons (CFCs) are halocarbons and can damage stratospheric ozone. See *chlorofluorocarbons.*

halogens. Term for a group of highly reactive elements that includes chlorine, fluorine, bromine and iodine.

halons. Organic compounds, formed from bromines and fluorine, which have been found to damage stratospheric ozone.

halophyte.   Plant adapted to growing in a salt-rich environment.

Hamiltonian.   An expression that includes the function to be optimized over time in a dynamic optimization problem plus the dynamic constraint multiplied by the co-state or auxiliary variable. Using the Hamiltonian, the necessary conditions for a maximum can be derived with optimal control theory.

FURTHER READING
Chiang (2000).

Hammer–Aitoff Conic Equal Area Projection.   A type of map projection used in AVHRR and NDVI data. See *AVHRR* and *Normalized Difference Vegetation Index*.

hantavirus.   An often fatal viral disease spread by rodents and exposure to their excreta. The spread of the disease to different regions has been aided by the unwitting transport of rodents in cargo ships.

hard energy.   Energy from large-scale production facilities using non-renewable resources such as uranium, coal or oil. See *soft energy*.

hardpan(s).   A hard, usually impermeable, sub-surface soil layer often created by the precipitation of iron and aluminum.

hard water.   Water that contains a high proportion of alkaline compounds derived from calcareous sources. See *soft water*.

hardwoods.   Any member of the class of deciduous, broad-leaved trees (such as oak or maple), especially those that are suitable for timbering.

harmonic mean.   A summary statistic defined as the arithmetic mean of the reciprocal of a series of numbers. For example, the harmonic mean of the numbers 3 and 6 is $\frac{1}{2}(1/3 + 1/6) = \frac{1}{4}$. See *arithmetic mean*.

Hartman rotation.   The optimal time to cut a uniformly aged stand of trees assuming the trees provide both timber and non-timber values. The Hartman rotation is a modification of the Faustmann rotation and includes amenity and non-timber values. Provided that non-timber values increase with the age of the trees, the Hartman rotation is longer than the Faustmann rotation. See *Faustmann rotation*.

FURTHER READING
Hartman (1976) and van Kooten (1993).

Hartwick's rule.   The rule requires that the rents from the extraction of non-renewable resources be invested in man-made capital, rather than consumed, so as to ensure a sustainable level of consumption over time.

FURTHER READING
Hartwick (1977).

harvesting.   The anthropogenic removal of biomass from an ecosystem. Harvesting of renewable resources without well-defined property rights or controls can lead to biological and economic overexploitation.

harvesting capacity.   The maximum output level of a resource (or harvest) that can be produced from a given set of inputs, or the minimum level of inputs used to produce a given harvest or output level.

harvesting costs.   The fixed and variable costs associated with harvesting a natural resource. See *natural resources*.

hazardous pollutants.   Material discharged into the environment which can cause harm to humans within localized areas.

haze.   A collection of fine atmospheric particles that obscures visibility in the sky, usually as a result of air pollution and weather conditions that prevent its dispersal. Haze often causes respiratory problems and, as a result, is the focus of environmental regulation in both developed and developing countries.

HCP.   See *Habitat Conservation Plan*.

healing.   The act of improving an individual's health (which could include physical, mental, or spiritual well-being).

health capital.   A measure of health or health status that can depreciate over time and can be maintained or improved over time through investment (e.g., exercise).

FURTHER READING
Grossman (1972).

health production function.   A numerical production function in which the health of an individual is the product of a combination of exogenous factors (e.g., environmental variables) and choice variables (e.g., health care, preventative medicine).

Healthy worker effect. A widely observed phenomenon whereby employed workers often have lower mortality and morbidity rates than other groups in a population.

heat island. See *urban heat island effect*.

heat pump. A device providing heating or cooling by transferring heat from one source to another.

heat wave. An extended period of unusually hot weather.

heating, ventilation and air condition (HVAC). All the components available for climate control in a building, or a specific internal work or living space.

heavy metals. Metals of a high atomic weight including lead, mercury and cadmium. These metals, and in their associated soluble organic forms, can pose significant environmental and human health hazards.

heavy water. Water in which normal hydrogen atoms have been replaced by hydrogen atoms with twice the atomic weight.

heavy-water reactor. A nuclear reactor that uses deuterium oxide (also called heavy water) as both moderator and the primary coolant. Ordinary water is used as the secondary coolant which through this process produces steam for turning a turbine. The efficiency of heavy water as a moderator allows natural uranium which has a low percentage of fissionable isotope U^{235} to be used as fuel.

Heckscher–Ohlin model. Model of international trade in which countries export goods whose factors of production are locally abundant.

hectare. Unit of area equal to 10 000 square meters or 2.47 acres.

hedonic analysis. Valuing activities or uses based on the observable characteristics or attributes of a good or service, such as valuing the effects of noise pollution by using property prices near and far away from an airport. See *hedonic property model*.

hedonic methods.   Empirical methods used to estimate the implicit or hedonic prices of goods, or attributes of goods, using the explicit or market prices of goods. For example, the implicit price of air quality in a neighbourhood may be valued using data on property prices, house characteristics and environmental characteristics.

FURTHER READING
Rosen (1974) and Callan and Thomas (2007).

hedonic pricing.   A method of evaluating the economic value of goods based on econometric techniques that identify and quantify the utility of various attributes of the good.

hedonic property model.   A (hedonic) model of preferences for property characteristics in which the (resale) value of property is modeled as a separable function of the attributes of that property.

hedonic travel cost model.   A travel cost method, first developed by Brown and Mendelsohn (1984), in which the implicit value of recreational site attributes are thought to contribute to the total revealed willingness to pay to gain access to a recreational site, as evidenced by an individual's choice to pay for the travel costs to a recreational site.

hedonic wage model.   A hedonic model of preferences for job attributes, including job-related risks, in which the wage is modeled as a separable function of job and worker attributes.

height–diameter relationships.   The correlation between the diameter and height of different tree species in terms of either growth and/or its eventual mature size.

Helsinki Agreement.   A 1989 modification to the 1987 Montreal Protocol in which parties agreed to end the production of chlorofluorocarbons (CFCs) by the year 2000 so as to prevent further deterioration of stratospheric ozone.

Helsinki Process.   A meeting between European countries called the Second Ministerial Conference on the Protection of Forests in Europe, held in Helsinki in 1993, and which identified the general guidelines for sustainable forestry management in Europe and the development of a series of criteria and indicators. See *Montreal Process*.

hepatitis.   A viral disease classified by three types A, B and C. Hepatitis A is transmitted through contaminated water and food that is often a result of poor hygiene and inadequate sanitation.

herb.   A seed-bearing plant that does not have, and does not develop, wood tissues.

herbicides.   Chemicals used to inhibit growth or destroy undesirable plants. Herbicides are widely used, and their misuse can lead to serious environmental problems including groundwater contamination.

herbivores.   Animals, such as cows and horses, whose diet consists primarily of plants.

heredity.   Transfer of genetic traits to offspring. The degree to which offspring exhibit the physical traits of their parents depends on the inheritance of the characteristic, environmental factors and random factors.

Hessian matrix.   A square matrix that consists of the second-order partial derivatives of a function, $f(x_1, x_2, \ldots, x_n)$, where the element in row i and column j is the second-order partial derivative f_{ij}. The Hessian matrix can be used to determine the curvature of a function.

heteroscedasticity.   A statistical term for error terms or residuals that are correlated with one or more of the variables in the model. Heteroscedasticity commonly occurs with cross-sectional data and, because it affects the standard errors of estimators, it invalidates hypothesis tests about the estimators. If heteroscedasticity is present, generalized least squares (GLS) may be used.

heterotroph.   An organism that obtains its nutrition from other organisms. See *autotrophy*.

hexachlorobutadiene (HCBD).   A byproduct of the manufacture of some organochlorines which can bioaccumulate in the environment.

Hicksian demand.   Sometimes called compensated demand. The Hicksian demand for a good or service is a function of the prices of goods and the level of utility to the consumer. By contrast, the normal or Marshallian demand is a function of prices and the level of income of the consumer.

Hicksian income. Named after the Nobel Laureate economist, Sir John Hicks, it is the national income that can be consumed while maintaining a non-decreasing level of the capital stock.

Hicks neutral technological change. Disembodied technological change where the ratio of capital to labor in production remains unchanged over time. See *technological change* and *disembodied technical change*.

high forest. Old-growth, closed-canopy forest.

High Forest Cover Low Deforestation (HFLD). Term used in the context of climate change mitigation for countries that retain large amounts of forest due to a historical lack of deforestation, such as Guyana and Suriname.

High Latitude Mode. See *annular modes*.

highgrading. Removal of the best or most desired of a species from a renewable resource. Highgrading occurs in fisheries where smaller and less-valued fish may be dumped at sea so as to maximize the hold capacity with more valuable fish. In forestry, highgrading involves the removal of the best trees leaving behind low-quality stands of trees.

high-level radioactive waste. See *radioactive waste*.

high-sulfur coal. Coal which contains a high proportion of sulfur compounds that can be released as emissions when burnt. Sulfur concentrations vary substantially across coal deposits.

high yielding varieties (HYVs). Domesticated plant varieties that, with the appropriate application of fertilizers, can provide higher yields than traditional varieties. See *green revolution*.

HIV. See *human immunodeficiency virus*.

Holocene. The current geological epoch which began at the end of the last glacial, 10–12 000 years ago.

homeostasis. The tendency of a system to maintain an equilibrium through negative feedbacks.

hominids. A family of primates called Hominidae from which modern humans (*Homo sapiens*) are the last remaining species. Modern humans are believed to have evolved in Africa some 100 000 years ago.

homogeneous function.   A set of functions widely used in economic analysis that have the following property where if f(x) is the function then
$$f(ax) = a^k f(x)$$
for all values of a greater than zero, and where k is defined as the degree of homogeneity.

Homo sapiens.   See *hominids*.

homoscedasticity.   A term used in statistical models to refer to residuals or error terms that have equal variance around a mean of zero. See *heteroscedasticity.*

horizontal drilling.   The process of drilling a portion of a well at a horizontal angle to increase the exposure of a geologic formation to the well. This is a common procedure in hydraulic fracturing operations.

horizontal equity.   Outcomes in which similarly situated people are treated equally. For example, people with the same income are taxed at the same rate. See *vertical equity.*

horizontal integration.   Economic term for the agglomeration of companies at the same level of production, such as a merger of oil companies.

hormone.   An organic compound used to control various functions in plants and animals, including growth and sexual development.

hormone disrupter.   See *endocrine disruptors.*

host.   Organisms that sustain parasites. See *parasites.*

host country.   Term for a country that is the site of a CDM, JI or REDD project. See *Clean Development Mechanism, Joint Implementation*, and *Reducing Emissions from Deforestation and Forest Degradation.*

hot air.   Term used to describe the issue where some participants, such as Russia, to the Kyoto Protocol were allocated an emissions allowance well in excess of their actual or feasible emissions. Hence these countries were able to trade a right to pollute greenhouse gas emissions that did not originate from mitigation activities.

Hotelling model. Model that explains the spatial location of firms and suggests that firms in a similar business are likely to locate close to each other.

Hotelling rent. Named after the economist Harold Hotelling, this is a method of valuing non-renewable resources defined as the marginal net return on the last unit of the resource extracted.

Hotelling's lemma. A result that states that the output supply and input demand functions can be derived from a profit function by taking its partial derivative with respect to the output price and input prices.

Hotelling's rule. Named after the economist Harold Hotelling, it is a rate of extraction of a non-renewable resource such that the net price of the extracted resource rises at a rate equal to the rate of interest in an economy. Hotelling's rule is an equilibrium condition and will hold true only under special conditions such as perfect foresight and information, and competitive markets.

human capital. The learned and inherent abilities that are embodied in individuals.

human ecology. The transdisciplinary study of the interaction between humans and their surrounding environment.

human genome project. An international project involving thousands of researchers that has mapped the human genome.

human immunodeficiency virus (HIV). A virus contracted through a transfer of bodily fluids. The four main transmission pathways are unsafe sex, contaminated needles, breast milk and birth. The World Health Organization considers the HIV infection in humans as pandemic. It causes acquired immune deficiency syndrome (AIDS). See *acquired immune deficiency syndrome*.

Humboldt current. Also known as the Peru current, it is a cold ocean current that travels north along the western coast of South America. Upwelling of nutrients from the current provides sustenance for an abundance of marine life and one of the world's largest fisheries, the Peruvian anchoveta.

humidity. A measure of the amount of water in a given volume of air. The higher the temperature, in general, the more water the air can hold.

humus.   Partially decomposed plant and animal material found in soils. It is important for retaining moisture and maintaining soil fertility.

hunter-gatherer.   Term used to describe a pre-agricultural mode of living whereby animals are hunted, and plants and fruits are gathered to provide sustenance.

hurricane.   A severe tropical cyclone causing strong winds, heavy rain and, in coastal areas, storm surges. As hurricanes are caused by rising condensation of warm oceanic water, it is hypothesized that an increasing frequency of severe hurricanes will accompany global warming.

hybrid.   Term used to describe the progeny or offspring of a union between different breeds or species.

hybrid vehicle.   A vehicle that uses two different sources of energy, usually gasoline supplemented by electricity. As hybrid vehicles generally involve lower greenhouse gas emissions than traditional types, they are seen as an important early step in mitigating climate change.

hydraulic fracturing (fracking).   A procedure where high pressure is used to pump a fluid (comprised of water, sand and chemicals) down a well and into underground rock formations. This causes the rock to fracture and increases the amount of gas or oil that can be extracted. Often used for marginal gas deposits for which it is not economical to use conventional methods. There is large concern that the increasing use of fracking will cause major damage to groundwater supplies from: depletion of groundwater to meet the large volumes of water required by the process, accidental release, or incorrect disposal of the water that flows back after fracturing occurs. See *coal-bed methane*.

hydraulic fracturing water life cycle.   The entire life cycle of water throughout the process of hydraulic fracturing. This includes: obtaining water from a source; mixing water with chemicals to create the fluid for fracturing; inserting fluid into rock formations with pressure; removing fluid from well after fracturing; treatment of wastewater.

hydrocarbons.   Compounds of hydrogen and carbon formed from organic material, including fossil fuels. Their combustion contributes to the emission of greenhouse gases, such as carbon dioxide and methane.

hydrochlorofluorocarbons (HCFCs).   A widely used substitute for chlorofluorocarbons (CFCs) used to decrease damage to stratospheric ozone. For equivalent weight, HCFCs are at least 10 times less damaging to stratospheric ozone than CFCs.

hydro-electric.   See *hydropower*.

hydrofluorocarbons (HFCs).   Organic carbon compounds that have been used to replace chlorofluorocarbons (CFCs) and hydrochlorofluorocarbons (HCFCs) in many uses, especially in refrigerants. The global warming potential of HFCs can be over a thousand times greater than carbon dioxide and, hence, they are a focus of international mitigation efforts.

hydrogen oxides.   Naturally occurring compounds that contribute to stratospheric ozone depletion.

hydrogen sulfide (H_2S).   A poisonous and flammable gas that smells like rotten eggs and is often produced in oil production and in oil refineries.

hydrological cycle.   The physical processes by which water is recycled, whereby precipitation on land is followed by infiltration of water tables and run-off, transpiration from plants, and evaporation to complete the cycle. At sea, precipitation is followed by evaporation.

hydrological drought.   A prolonged instance of below average rainfall that causes water deficiencies, as measured by low water levels in the physical elements of the hydrological system, such as rivers. See *drought*.

hydrological system.   The components and dynamic processes of the hydrological cycle.

hydrology.   The science and study of the hydrological cycle, groundwater and water management.

hydrophyte.   Plant adapted to growing in water or in very wet environments, such as a water lily.

hydroponics.   The practice of growing plants in aquatic media under controlled conditions.

hydropower.   The production of electricity from turbines propelled by moving water from either in-stream flows or the diversion of dammed water through channels. Despite harnessing natural energy and involving few greenhouse gas emissions, critics argue that large hydropower dams cause extensive environmental and social damages. For example, flooding of upstream areas and depletion of downstream flows, both of which can cause riverbank erosion and harm ecosystems. Furthermore, dam walls block natural fish migration and sediment transportation. Proponents of dam construction contend that they strengthen water management through flood prevention, and as a storage facility that can be utilized during dry spells.

hydrosphere.   The component of the Earth's climate system comprising surface and underground liquid water, such as rivers, oceans, lakes and aquifers.

hydrothermal vents.   Heated water from geothermal activity that emerges in the deep ocean.

hyperdepletion.   A relationship between catch per unit of effort (CPUE) in a fishery and total abundance where the CPUE drops much more rapidly than does overall abundance. See *hyperstability.*

FURTHER READING
Grafton et al. (2010).

hyperstability.   A relationship between catch per unit of effort (CPUE) in a fishery and total abundance where the CPUE drops much less rapidly than does overall abundance. See *hyperdepletion.*

hyperthermophiles.   Bacteria that live at extremely high temperature environments that range between 80°C and 150°C. Some scientists have hypothesized that hyperthermophiles may have been some of the earliest life on our planet.

FURTHER READING
Davis (2000).

hypolimnion.   A cooler and lower layer of water in a lake.

hypothesis.   A statement used to explain a phenomenon. A hypothesis should be able to be tested and falsified. See *null hypothesis.*

hypothetical bias.   The so-called bias that arises when persons are asked a hypothetical question in which they have limited ability to provide a meaningful response.

hypothetical resources. Deposits that have not yet been discovered but that can reasonably be expected to be discovered based on a similar geological structure to that in which current deposits are found.

hypoxia. Severe oxygen deficiency in an aquatic environment. This phenomenon is generally caused by the excessive presence of land-based nutrients causing overgrowth of photoplankton. Can lead to devastation of aquatic life forms in affected areas. See *anoxic event.*

I

ice age.   A major interval of time where continental ice sheets are present on the Earth's surface, as well as mountainous inland glaciers. Technically, the climate system is currently experiencing an interglacial (warmer) stage of an ice age. The term ice age is also used to describe the extremely cold glacial stage of an ice age, the last of which ended approximately 10 000 to 15 000 years ago. See *Milankovitch Theory*.

FURTHER READING
Ruddiman (2007).

ice cap.   A mass of ice that covers a large portion of the Earth's surface. Smaller than an ice sheet.

ice-core samples.   A cylindrical section of ice sampled from a glacier or ice sheet. Chemical analysis of the air inside these ice samples allows scientists to determine the historical composition of the Earth's atmosphere.

icehouse effect.   Term used to describe the factors that can combine to produce an ice age.

ice sheet.   A very large mass of land ice, usually defined as covering an area greater than 20 000 square miles. Although ice sheets can only be found today in Antarctica and Greenland, during the last glacial they covered much of North America, northern Europe and southern South America. Global warming is gradually melting modern ice sheets, thus causing rising sea levels. See *albedo feedback*.

ice shelf.   A thick platform of ice at the end of a glacier that floats upon the ocean.

ice stream.   A section of an ice sheet that moves faster than surrounding ice.

identity matrix.   A square matrix whose elements along its main diagonal (elements a_{ij} where $i = j$) are all equal to one and every other element equals zero.

IGOs.   Acronym for intergovernmental organizations, such as the World Trade Organization (WTO).

illite.   A type of clay formed by the weathering of feldspars and mica.

imission.   Term used in Germany to describe emissions from distant sources that have negative local effects.

impact analysis.   An analysis where decision makers weigh the consequences of different actions or policies, as measured by changes in variables stemming from the different actions.

impact area/zone.   The area in which the environment is likely to be affected by a specific policy.

implementation.   A legal term for the actions taken by a party to meet their obligations under a treaty. Also used in reference to the execution of a policy or plan.

implicit costs.   Costs that are not directly considered by firms or consumers but are, nevertheless, a cost of production imposed on either the firm itself or others. See *explicit costs*.

import quotas.   Legal limits placed on the quantity of a good that can be imported into a country.

impoundment.   Used in the context of water storage or water-based pollutant storage to describe a physical barrier, such as a dam, that confines its contents.

improvement cutting.   The removal of (economically) undesirable species in order to promote forest health and growth.

in-breed.   To reproduce from members of a species that are very similar genetically.

in camera.   Latin term meaning "in private".

incentive compatibility.   Term used to describe a situation where a personal or individual interest coincides with the intentions of a regulator or the public interest. For example, the deductible in an insurance contract provides the person buying the insurance with an added incentive to prevent the loss.

inch.   Unit of length equal to 2.54 centimeters.

incidental taking.   Under the Endangered Species Act in the United States, an incidental taking is defined as when a landowner is engaged in otherwise lawful activities but indirectly harms an endangered species. See *taking*.

income effect.   The change in consumption that results from a change in the price of good in terms of its effect on the real income of an individual. See *substitution effect*.

income elasticity.   The change in the consumption of a good or service following a change in income. If the change in demand is in the same (opposite) direction as the change in income, the good in question has a positive (negative) income elasticity.

increasing block tariff.   Pricing structure where the consumer pays a higher per unit price above a threshold. It is a common method of charging for water to residential consumers.

increasing returns to scale.   A situation where if all the inputs in a production process are increased by the same percentage, the output will increase by a higher percentage. Increasing returns to scale tend to lead to large-scale plants, as the greater the amount produced the lower the per unit cost. See *decreasing returns to scale* and *constant returns to scale*.

incremental effects.   Changes that occur in a small period of time. Often used in reference to small and gradual change that leads to larger long-term effects.

incrementalism.   Making gradual rather than abrupt changes in policy.

independence of irrelevant alternatives.   A concept used in welfare economics which states that the introduction of an alternative choice should not alter the relative ranking of two other choices. Also used as a criterion to determine the outcomes of certain games in game theory.

independent variable.   Variable in a regression that is set exogenously from the model and helps determine the dependent variable.

indexation.   Automatic increases in wages, the value of pensions and other benefits that follow increases in the rate of inflation.

indicated resources.   Deposits that have been discovered and the size of which estimated based on sampling and projections derived from similar types of deposits.

indicator species.   A species that can be used to assess the quality of the environment. Indicator species often have a low level of tolerance for a particular aspect of the environment and thus are most vulnerable to changes in ecosystems.

indifference curve.   A graphical representation of the different combinations of goods that yield the same level of utility for an individual.

indigenous people.   Peoples, communities and nations who claim a historical continuity to endemic societies that existed prior to integration with foreigners. They compose a non-dominant sector of the prevailing society and contest their sovereignty and self-determination. Indigenous populations play a major role in environmental discussions throughout the world due to their land right and land use claims.

indigenous knowledge.   The body of knowledge of local inhabitants regarding the physical functions, uses and interdependence of resources and living organisms found within their environment. See also *traditional knowledge*.

indigenous species.   A plant or animal species native to a particular region or area.

indirect emissions.   Greenhouse gas emissions indirectly attributable to the activities of an individual, household or business. Indirect emissions often arise from the purchase or use of a product. For example, the indirect emissions associated with car use would be those involved in the manufacture of the car or, particularly relevant to electric vehicles, the emissions associated with the electricity supplied for recharge. See *direct emissions*.

indirect objective function.   See *optimal value function*.

indirect taxes.   Taxes imposed on the sale of goods and services rather than income. See *direct taxes*.

indirect utility function.   A function that provides a measure of the utility of an individual, but is a function of prices and income rather than a function of the quantities of goods and services consumed.

indirect valuation. The measurement of the impact from a change in pollution on some aspect of the environment (such as an increase in air pollution on human health or on plant growth).

Individual Fishing Quotas (IFQs). Term used for individual transferable quotas in the US fisheries. See *Individual Transferable Quotas*.

individual rights. Property rights to an asset or resource held by an individual. See *community rights*.

Individual Transferable Quotas (ITQs). An output or quantitative management tool for regulating fisheries. The total allowable catch is allocated among fishers in the form of harvesting rights that can be traded. Provided that the ITQs are viewed as an exclusive and durable property right, they can help overcome "the race to fish" associated with input-based controls.

FURTHER READING
Grafton (1996).

Individual Vessel Quotas (IVQs). Term used for individual transferable quotas in some Canadian fisheries. See *Individual Transferable Quotas.*

induced technological change. Technological change that occurs as a result of a deliberate government policy or action. For example, the establishment of climate change mitigation technology that occurs as a result of public financing for research and development. See *autonomous technological change*, *endogenous technological change*, and *technological change*.

induction. See *inductive method*.

inductive method. A method of reasoning by which facts and general observations are used to form a generalization or a theory or a scientific principle. See *deductive method.*

industrialized country. See *developed country*.

industrial revolution. Term used to describe the rapid transformation of an essentially agriculture-based economy to one where most of the output and employment is generated by the industrial sector. The first industrial revolution began in the United Kingdom at the beginning of the nineteenth century and coincided with a massive increase in the use of natural resources, a social transformation and a major change in the environment and landscape.

industrial sector.   The segment of an economy involved in the manufacture and construction of finished products. The industrial sector is a major source of greenhouse gas emissions.

industry concentration.   The share of a market served by the largest firms. The larger the market share for a given number of firms, the greater is the industry concentration. See *concentration ratio*.

inelastic.   Term used to describe the consumer response or producer response to a change in price. A good is inelastic if the percentage change in demand (or supply) is less than the percentage change in the price. For example, if the price of a good fell by 1 percent and demand increased by less than 1 percent, the good would have an inelastic demand at that price. See *elastic*.

inertia.   The propensity of a system or object to maintain its state of motion. Consequently, the application or removal of external force tends to cause only gradual change. For example, if all greenhouse gas emissions ceased immediately, inertia from feedback loops in the climate system would still cause global temperatures to continue rising for many decades.

infant industry argument.   The notion that to enable an industry to be internationally competitive it may first be necessary to protect its domestic market through import quota tariffs.

infant mortality rate.   The annual number of deaths of children under the age of one year per 1000 live births in a population.

inferior good.   A good or service which has a demand that is inversely related to a consumer's income.

inferred resources.   Deposits that may exist contiguous to existing deposits in unexplored areas.

infiltration.   The absorption of water by soil.

inflation rate.   A measure of the change in the price level of a defined set of goods and services over time. See *consumer price index*.

infra.   Latin term meaning "below".

infrared radiation.   Long-wave radiation in the segment of the electromagnetic spectrum between microwaves and visible light. In the physical sciences, a distinction is made between solar infrared radiation that originates from the sun, and thermal infrared radiation emitted by the Earth and its atmosphere. See *thermal infrared radiation*.

infrastructure. The physical communication, delivery and transportation systems within an economy.

initial conditions. The starting values of stocks or state variables in a dynamic model.

inorganic. Substances that either have no carbon, or only carbon combined with elements other than hydrogen.

in perpetuum. Latin term meaning "forever".

in personam. Latin term meaning "in person".

input controls. A type of management that seeks to control the inputs used by resource users so as to control the total effort and/or harvest from common-pool resources. Input controls are widely employed in fisheries, but have frequently failed to prevent excess fishing effort or capital stuffing because fishers are often much better at circumventing input controls than are regulators at devising such regulations. See *output controls*.

Input-orientated measure of technical efficiency. The minimum amount of inputs required to produce a given level of output.

input–output analysis. A method of modeling an economy by specifying the relationship between inputs, intermediate outputs, outputs and final demands in an economy. The methodology can be used to examine the impact on the rest of the economy of a shock in a particular sector.

input substitution. A term that describes the process by which a firm substitutes one input for another. With input controls, it is also typical for fishing vessels to substitute from regulated inputs (such as vessel size) toward unregulated inputs (such as electronic gear).

inputs. Factors of production used to produce intermediate or final outputs. See *factors of production* and *outputs*.

in situ. 1. Latin term for the natural state. 2. Attributes and characteristics of a natural resource or environmental asset that arise from its current state. For example, a growing forest provides *in-situ* value in terms of carbon sequestration.

In-situ price.   See *in situ*.

insolation.   The amount of solar radiation that reaches the surface of an object in a defined period of time.

institutional economics.   A sub-discipline of economics that emphasizes the importance of institutions and social structures and their affect on economic performance.

FURTHER READING
Eggertsson (1990).

institutions.   The anthropogenic rules and norms of behavior that govern all human interactions.

in-stream augmentation.   Supplementing the flow of water within a stream through either increased supply of water to the stream, or practices that increase flows (such as releasing more water from a dam during periods of low stream flow).

in-stream flow protection.   The maintenance of water flows within a stream, which may be diminished through the diversion of water for irrigation purposes, or for power generation.

in-stream flows.   The state in which water remains in its natural course (e.g., streams, creeks, or rivers) as opposed to water that has been diverted artificially for other purposes (e.g., irrigation, reservoirs, drinking, etc.).

in-stream rights.   The right to enjoy a minimum amount of water flow within a stream or river.

in-stream use.   Use of water flowing in a stream or river, such as for fishing or navigation.

instrumental temperature record.   Historical data of atmospheric and ocean temperatures measured by instrumental sensors. Aggregate global records extending back to 1850 show a sharp temperature increase of nearly 1°C after the 1970s.

insulation. Method by which heat is retained in a body by reducing conduction. Insulation of houses and offices can result in significant savings in terms of energy use.

intangible benefits. Benefits that accrue to individuals that cannot be quantified in financial terms.

integrated assessment model (IAM). An environmental model incorporating economic and social factors that is used to analyze policies and provide forecasts. IAMs have been commonly employed in the past for economic analysis of climate change mitigation in a cost–benefit framework. In recent times, the appropriateness of this approach has been widely questioned as it is unable to adequately account for the risks of catastrophic climate change. See *Dismal Theorem* and *Stern Review*.

integrated pest management (IPM). An ecological approach to the management of pests which eschews the use of pesticides. IPM involves a variety of techniques, including biological controls and the planting of mixed crops to reduce, but not necessarily eradicate, pests on crops.

integrated pollution prevention and control. A broad-based approach to the management and control of pollution that considers all potential sources and their possible impacts across the environment.

integrated solid waste management. A combination of recycling, composting, incineration and judicious use of landfills to reduce the amount of solid waste produced by society.

integrated water resource management (IWRM). An inclusive and flexible process for managing water that supports both environmental sustainability and economic development. IWRM has become the international standard for planning and implementation of water policy.

integratibility. In utility or demand theory, the requirement that the Slutsky matrix of substitution terms be symmetric and semi-definite. Integratibility ensures that preferences can be recovered by integrating a system of demand functions to recover the expenditure function, which, in turn, can be used to derive the indirect and direct utility functions.

intellectual property rights (IPRs). Patents or copyrights that provide the originators of creations, innovations, or ideas with a legally enforceable right to exclude others from appropriating their intellectual capital without their prior consent. IPRs exist for a range of intellectual creations that include the written word, inventions, art, techniques, image, and designs but such protection is, in general, not provided for indigenous or traditional knowledge. See *property rights* and *indigenous knowledge*.

intelligent controls.   A broad term for technology used in buildings to increase cost efficient use of electricity, heating and air-conditioning. For example, light activating motion sensors.

intensive agriculture.   Agricultural practices associated with high yields but with concomitant use of high inputs of pesticides, herbicides and artificial fertilizers. Intensive agriculture in terms of livestock often involves the production of large quantities of excreta that can impose a major burden on the environment and can lead to contamination of soil and water.

intensive forest management.   A range of forest practices such as pruning, fertilizing and spacing designed to yield higher timber quality, increase harvest volumes, or reduce rotation periods.

intensive margin.   Term used for a plot of land or stand of timber where the value of the marginal product associated with a strip of the land, or of a tree in the stand, exactly equals the marginal cost of production.

FURTHER READING
Van Kooten and Bulte (2000).

inter alia.   Latin term meaning "among other things".

Inter-American Development Bank (IADB).   Bank established in 1959 by member countries that is mandated to further the economic and social development of nations in the Caribbean and Latin America.

inter-basin transfer.   The transfer of water across watersheds for a particular purpose. Inter-basin transfers, widely practiced in the Western United States, have been highly controversial because of their deleterious effect on ecosystems.

intercropping.   The practice of growing different crops between the rows of another.

Interdecadal Pacific Oscillation (IPO).   A pattern of climate variability that occurs every two to three decades across the Pacific Ocean. First identified in 1997, the IPO is driven by surface oceanic temperature changes and is an aggregate effect from a variety of processes, including the El Niño Southern Oscillation.

intergenerational equity. The notion that the present generation should consider the needs of future generations in current decision making. Intergenerational equity is a fundamental feature of sustainable development. See *equity* and *sustainable development.*

interglacial. A period of time during an ice age, such as the present, of comparatively warmer temperatures. During an interglacial, ice sheets retreat from their greater coverage established during the colder glacials. Over the last million years interglacial periods have generally lasted around 10 000 years, and taken place approximately 100 000 years apart. See *ice age* and *Milankovitch Theory.*

Intergovernmental Panel on Climate Change (IPCC). The scientific advisory body to the UN charged with compiling comprehensive reports that assess academic knowledge concerning climate change science, mitigation and adaptation. Since being established in 1988, the IPCC has provided these updated summaries on a regular basis and is widely viewed as the authoritative source of information relating to climate change. See *assessment report.*

intermediate good. Goods that are used in the production process to produce other goods. For example, wheat is an intermediate good as it is the raw material for bread and other wheat products.

intermittent energy source. An energy source that due to its characteristics is subject to some variability in the quantity of energy it supplies. Renewable energy sources such as solar energy and wind energy are intermittent as they can only generate if the primary energy source (sunlight or wind) is available.

internal combustion engine. Engines operated from burning fuel internally, such as those used in most cars. As the majority of combustion engines rely on fossil fuels, they are a major global source of atmospheric carbon dioxide.

internal rate of return (IRR). The discount rate in cost–benefit analysis at which the present value of benefits equal the present value of costs. See *cost–benefit analysis.*

internalize. Term given for the process by which individuals or firms explicitly consider the costs or benefits that they impose on others from their actions or decisions.

International Bank for Reconstruction and Development (IBRD). More commonly known as the World Bank. It was established in 1944 and its mandate is to lend to member countries to assist them in their social and economic development.

International Council for the Exploration of the Sea (ICES). An international body that is almost 100 years old. Its 19 member countries promote and coordinate fisheries and marine research.

international emissions trading. Sale and purchase between states of the right to emit greenhouse gases in formal frameworks such as the Kyoto Protocol. See *climate policy* and *emissions trading scheme*.

International Energy Agency (IEA). An agency formed in 1973 by developed countries in response to the first oil price shocks. The IEA provides information on global energy production and consumption.

International Joint Commission (IJC). A commission and advisory body formed in 1909 by the United States and Canada to oversee the management of their boundary waters. Since the early 1970s, the IJC has been instrumental in setting pollution guidelines for the Great Lakes and has helped initiate considerable improvements in the quality of the water in the lower Great Lakes, especially Lake Erie.

International Monetary Fund (IMF). An international organization formed by the Bretton Woods Agreement in 1944 that now has over 180 members. Its membership finances special drawing rights that enable countries to deal with temporary balance of payments difficulties. In providing funding to countries in need, the IMF may also specify conditions in terms of fiscal and monetary policy that must be followed before assistance is provided.

International Organization for Standardization (ISO). An interna-tional organization based in Geneva that is mandated to provide international standards. See *ISO 14001*.

International Trade Organization (ITO). Precursor to the WTO suggested at the Bretton Woods Conference in 1944. It was intended to complement the IMF and the World Bank; however, it never passed the US Congress and failed to be created. It was succeeded by GATT, which eventually resulted in the creation of the WTO in 1995. See also *General Agreement on Tariffs and Trade (GATT)* and *World Trade Organization (WTO)*.

International Union for the Conservation of Nature (IUCN).   Also known as the World Conservation Union. It was formed in 1948 by governments and non-governmental organizations and today has over 139 countries represented in its membership. Its mandate is to conserve the integrity and diversity of nature in an equitable and ecologically sustainable way.

International Union of Forest Research Organizations (IUFRO).   Formed in 1892, its membership consists of over 700 non-governmental organisations in 110 countries. They aim to improve the understanding of ecological, economic and social aspects of forests.

International Whaling Convention (IWC).   Convention signed in 1946 that established the International Whaling Commission, mandated to preserve whale stocks. The Commission has been highly controversial and failed to prevent the near extinction of several whale species. More recently, a moratorium on commercial whaling has helped some species to increase in number but controversy still remains as some members of the commission would like to ban all commercial whaling in perpetuity while other countries would like to resume commercial whaling.

interstadial period.   A period of approximately 1000 years of warmer temperatures and glacier retreat during the cooler glacial stage of an ice age. See *interglacial.*

intertemporal efficiency.   See *dynamic efficiency.*

intertidal community.   A community of marine organisms living in coastal areas and along a shoreline that is submerged during high tides and exposed during low tides.

intertropical convergence zone (ITCZ).   A low pressure zone which changes with the seasons but is always found close to the equator where the rising of warm tropical air caused by surface heating creates an area of low pressure. The variability of the ITCZ causes seasonal droughts in Africa and Asia.

interviewer bias.   The bias in responses by respondents in a survey due to the particular behavior of an interviewer.

intrinsic value.   Value of good, asset, or aspect of the environment that is separate from its value in use. See *non-use value*.

FURTHER READING
Fisher and Raucher (1984).

invasive species.   Species of flora or fauna originating from outside an ecosystem that thrive in their new habitat, usually causing damages to native species.

inverse demand function.   A relationship between the quantity of a good consumed and the marginal willingness to pay for the last unit of the good consumed, holding everything else constant.

inversion.   Shortened term for a temperature inversion where air temperature is either unchanged or increases with altitude. The effect is that air pollution can be trapped by warmer air above, as occurs in Los Angeles at certain times of the year.

invertebrates.   Animals without a backbone that include insects and crustaceans.

investment.   The expenditures associated with the purchase of new capital goods. See *capital good*.

invisible hand.   The notion, first described by Adam Smith, that individuals pursuing their own self-interest in their particular trade or profession can provide benefits for society as a whole by ensuring goods and services are supplied that are demanded.

in vitro.   Latin term meaning "existing outside of a living creature".

in-vitro fertilization.   Fertilization of an egg outside of a living organism.

in vivo.   Latin term meaning "existing within a living creature".

involuntary risk.   Risk imposed on individuals without their consent, such as the effects of air pollution.

ion.   An atom which has lost or gained an electron and is either negatively or positively charged.

ionization.   The process by which atoms become ions.

ionizing radiation.   Radiation capable of causing atoms to become negatively or positively charged by the gain or loss of electrons. Ionizing radiation can be caused by various forms of electromagnetic radiation and can lead to various health problems, including cancers.

ionosphere.   The outermost layer of the Earth's atmosphere.

IPRs.   See *intellectual property rights*.

ipso facto.   Latin term meaning "by the fact itself".

irradiation.   Exposure to ionizing radiation. Irradiation can be used to protect food from contamination by micro-organisms but is controversial because of concerns about its possible health effects.

irrigation.   The process of supplying water to dry land, usually for agricultural purposes. As climate change is predicted to cause increased incidence of drought in certain regions, efficient irrigation practices will play a significant role in supporting global food security. See *drought*.

irrigation efficiency.   The ratio of water volume beneficially used by plants to the volume of water delivered through an irrigation system, adjusted for effective rainfall and changes in the water storage in the root zone.

irreversibility.   The notion that many processes and developments are difficult, or may even be impossible, to change. For example, the extinction of a species is irrevocable.

island biogeography.   The study of the positive relationship between habitat size, remoteness and the number of species found within the habitat.

IS–LM model.   A macroeconomic model developed by Sir John Hicks and based upon John Maynard Keynes' book, *General Theory of Employment*. The model includes IS (investment-savings) that represents the combinations of interest rates and savings that generate equilibrium in investment and savings and LM (liquidity-money supply) that ensures that money demand equals money supply for different combinations of interest rates and income.

ISO 14001. A set of environmental standards developed by the International Organization for Standardization that assists organizations to improve their environmental management.

ISO 14064. International standards for greenhouse gas emissions inventories, covering: inventory development; emissions monitoring and reporting; and emissions validation and verification.

isobar. In meteorology, a line drawn on surface-charts that represents points of equal atmospheric pressure.

isocost. Combinations of input quantities for given input prices that generate the same total cost of production.

isohyet. Line on a map joining areas of equal average rainfall.

isoquant. The combinations of input quantities that produce the same level of output.

isotope. Forms of an element that differ in terms of the number of neutrons. Isotopic forms of oxygen and carbon have been used for the dating of past events.

Itai-Itai. A degenerative bone disease caused by excessive amounts of cadmium in the body.

IUCN. See *International Union for the Conservation of Nature*.

ius disponendi. Latin term meaning "the right to sell or transfer". See *transferability*.

ius excludendi. Latin term meaning "the right to prevent interference". See *exclusivity*.

ius fruendi or usufructus. Latin term meaning "the right to withdraw". See *withdrawal right*.

ius possidendi. Latin term meaning "the right to possess or own". See *quality of title*.

ius utile. Latin term meaning "the right to enjoy".

J

Jacobian matrix. A matrix consisting of first-order partial derivatives that is used to test for functional dependence among a system of equations.

FURTHER READING
Simon and Blume (1994).

Jensen's inequality. An inequality where the expected value of a concave function of a random variable, $E[g(x)]$, is equal to or less than the concave function of the expected value of the random variable, $g(E[x])$, if $g(.)$ is a concave function, x is the random variable and $E[]$ is the expectations operator. See *expected value* and *random variable*.

FURTHER READING
Greene (2003).

jet stream. A stream of air found at the upper boundary of the troposphere (8 to 15 kilometers above the Earth's surface depending upon the latitude) which can travel as fast as 300 kilometers an hour.

joint and several liability. Term used to describe a legal standard whereby a single person or party is liable for all damages despite the fact that the party may have been responsible for only a part of the total damages.

Joint Implementation (JI). A flexibility mechanism under the Kyoto Protocol for Annex B countries to meet their emissions reductions obligations by investments in other Annex B countries, usually former Soviet economies in transition. See *Kyoto Protocol* and *Clean Development Mechanism*.

jointness in inputs. A production relationship where all inputs are required to produce all outputs.

joint product. Goods or services that are produced jointly in more or less fixed proportions, such as mutton and wool.

joule. A unit of energy defined as the work from the force of one newton exerted over one meter.

Jurassic. A geological time period that occurred from 208 to 144 million years ago.

just compensation.   The fair value to be paid to an individual or group in compensation for the environmental degradation or harm created by the development or use of the resource by others.

K

Kalahari Sands.   Aeolian sands of Quartenary origin that are found from Southern Africa north into the Republic of Congo.

Kaldor Potential Compensation.   One social state is preferred to another if there exists another third state that could be reached after making the move to that state, and that third state is Pareto superior to the initial state (the potential state that could be reached is preferred by all parties to the initial state). There is no assurance, however, that this third state will be reached and this criterion can lead to contradictory results.

kaolinite.   A clay derived from the weathering of feldspar-rich rocks, usually under acidic conditions in humid tropical soils.

katabatic wind.   A downward air-flow in a valley caused by cooling at higher altitudes at night-time.

Kaya Identity.   An equation describing the determinants of anthropogenic greenhouse gas emissions, commonly used for projecting emissions. The Kaya Identity formulates global emissions, GHG, as a function of world population (p), global per capita GDP (g), energy intensity of global GDP (e), and greenhouse gas emissions intensity of energy (c), such that:
$$\text{GHG} = p \times g \times e \times c$$

kelp.   Large and often fast-growing marine plants that serve as important habitat for marine organisms.

Kelvin scale.   A temperature scale named after the British scientist Lord Kelvin. Each degree Kelvin is equivalent to one degree on the Celsius scale but the scale starts at 0 at absolute zero or –273°C.

keystone species.   A dominant species in an environment such that its removal has an important impact on all other species.

kilogram.   Metric measure of mass that equals 2.204 pounds in the imperial measure.

kilometer.   Metric measure of distance that equals 0.621 of a mile.

kilowatt hour.   The standard measurement of electricity consumption.

For example, an 80 watt light bulb used for 10 hours consumes 0.8 kilowatt hours.

kinetic energy.   The energy of an object in motion. It depends on its mass and velocity.

kingdom.   Broadest classification of organisms that includes the animal, plant, fungi, monera and protista kingdoms.

knot.   One nautical mile per hour.

known deposit.   See *proved reserves*.

Krakatoa.   An Indonesian volcanic island that erupted in 1883, depositing enough material in the atmosphere to affect world temperatures in 1884.

Krebs cycle.   The chemical reactions in cells that provide the energy organisms use to function.

k-strategists.   Term used to describe animals that spend a great deal of time and energy in rearing their young, such as primates.

Kuhn–Tucker conditions.   Set of conditions for nonlinear programming problems that, if satisfied, are the necessary conditions for an optimum.

kurtosis.   A measure of how a unimodal probability distribution differs from a normal distribution in terms of whether it is flatter or more pointed.

Kuznets curve.   A supposed relationship observed by Simon Kuznets between per capita income and the degree of inequality in an economy. According to Kuznets, inequality rises with economic development and then peaks, after which point it declines, even as growth continues. See *environmental Kuznets curve*.

FURTHER READING
Perkins, Radelet and Lindauer (2006).

kwashiorkor.   Human disease associated with a low-protein diet, common in children in many developing countries.

Kyansur forest disease.   A viral disease transmitted by ticks and found in southern India. Deforestation has increased the incidence of the disease as livestock near the forest provide ready hosts for the ticks.

Kyoto mechanisms.   See *Flexibility Mechanisms*.

Kyoto Protocol.   The 1997 agreement by parties to the United Nations Framework Convention on Climate Change for Annex I (or developed) countries to collectively commit to a 5 percent reduction in greenhouse gas emissions by 2008–12, relative to 1990 levels. This reduction was to be achieved through binding national emissions constraints, in conjunction with mechanisms for reducing mitigation costs, such as emissions trading and projects in developing, or non-Annex I, countries. Although the Protocol represented an important first step towards coordinated global action, global emissions rose significantly over this period. A major drawback was the non-participation of the United States, who refused ratification because large developing country emitters, such as China and India, were not subjected to binding commitments. See *Clean Development Mechanism*, *climate policy*, and *emissions trading scheme*.

L

labor-intensive. Goods and services that use more labor, relative to other factors, in their production process.

labor–leisure trade-off. The supposed trade-off that people face when determining how many hours they should work. Greater hours worked means higher income, but the trade-off is reduced leisure hours.

labor productivity. The real value of output per unit of labor input, usually measured per hour or person days.

labor theory of value. The notion, attributed to Karl Marx, that the measure for the value of goods and services should be the amount of labor (such as person hours) devoted to its production.

Labrador current. A cold current that flows southwest from Greenland towards eastern Canada, and the coasts of Labrador and Newfoundland.

Laffer curve. A curve relating tax rates to government revenue. As the tax rate increases on a certain activity, so too should its revenues. However, after some tax level, the activity will be discouraged and revenues will fall.

Lagrangian function. Function formed by the objective function from an optimization problem plus each of the constraints each multiplied by a Lagrangian multiplier. See *Lagrangian multiplier.*

FURTHER READING
Simon and Blume (1994).

Lagrangian multiplier. A variable created to help solve optimization problems with constraints. It is treated like a choice variable in the optimization problem. See *Lagrangian function.*

lag time. The period of time from an initial event before its effects can be observed or realized.

lahars. Landslides of volcanic origin.

Lake Nyos. Site in Cameroon where in 1986 a release of carbon dioxide from the lake asphyxiated almost 2000 people.

Lamarckian evolution.   The discredited theory, named after Jean Baptiste Lamarck, that evolution occurs due to the inheritance of acquired characteristics. For example, Lamarck would have believed that a person with an acquired characteristic, such as a darker skin because of a suntan, could pass this characteristic on to her offspring.

land clearing.   See *deforestation.*

land cover.   The physical materials covering the Earth's surface such as vegetation. The albedo, or reflectivity, of land cover affects the amount of solar radiation that is absorbed as heat by the Earth's surface. See *albedo.*

land degradation.   The decline in the health and productivity of land as a result of human activities such as deforestation and over-grazing. Potential consequences include high salinity levels and desertification. Land degradation contributes to greenhouse gas emissions by rendering soil unfit for vegetation. See *Reducing Emissions from Deforestation and Forest Degradation.*

land desertification.   See *desertification.*

land expectation value (LEV).   A discounted cash flow technique used to value timberland in perpetual production. See *discounted cash flow.*

landfill.   A site where waste is buried under the earth. Landfill waste dumps cause serious environmental damages through: dissemination of toxic chemicals into groundwater and soil, and as a source of methane, a powerful greenhouse gas.

land grant.   Land given by the government to individuals, companies, or institutions.

land rent.   The payment associated with a site that can be attributed to some scarce factor about the land, such as fertility or location, and which can be taxed away without inducing a change in how the land is used.

landslide.   An extensive and sudden movement of rocks or soil down the slope of a landform. Although landslides are commonly associated with natural processes such as heavy rainfall, human activities such as devegetation and groundwater removal can be contributing factors. Scientists believe that erosion from rising sea levels will increase the frequency of coastal landslides in the future, particularly in the vicinity of cliffs.

land surface air temperature. Measurement of the air temperature over land. See *global surface temperature*.

land use. The set of activities, inputs and outputs carried out on an area of land by humans. Land-use change takes place when land is converted from one purpose to another. For example, from forests to agricultural land.

land use, land-use change and forestry (LULUCF). A term used to categorize and evaluate greenhouse gas emissions from land usage, including deforestation.

land-use planning. The practice of managing land for the most desirable purposes. Such planning ensures that green spaces exist in urban environments and prevents undesirable industries from locating in residential neighborhoods.

landings data. Data on catch and fishing vessel characteristics obtained at port, or when the harvest is delivered to port.

La Niña. An irregular event where the waters off the coast of Chile and Peru are colder than average, causing opposite weather effects to its counterpart, El Niño events. See *El Niño Southern Oscillation*.

FURTHER READING
Ruddiman (2007).

large-scale singularities. See *climate threshold*.

Laspeyres index. A commonly used index that employs weights from a base period. For example, a Laspeyres price index may be defined as $p_1 q_0 / p_0 q_0$ where q_0 is the quantity consumed in the base period, p_0 is the price of the good in the base period and p_1 is the price of the good in the current period. See *Paasche index*.

Last Glacial Maximum (LGM). The episode during the last glacial when ice sheets reached their greatest extent, between 19 000 and 27 000 years ago. See *glacial*.

latent heat. Heat that is radiated or absorbed during a change of state that occurs at a constant temperature and pressure, such as melting of ice.

laterite.   A tropical soil that contains high levels of iron and/or aluminum oxides, and is therefore of poor quality.

laterization.   Process by which laterite soils are formed. It can be accelerated by anthropogenic influences, such as deforestation. See *laterite*.

Laurasia.   A supercontinent that broke from the even larger supercontinent Pangea and eventually split and formed into North America (including Greenland), Asia, and Europe. See *Pangea*.

lava.   Magma that reaches the Earth's surface.

law of conservation of energy.   The physical law that energy can neither be created nor destroyed but can only be transformed or transferred to another place.

law of demand.   The tendency for most goods and services that the quantity demanded will increase as the price decreases, all things else being equal.

law of large numbers.   The notion that as the number of observations becomes very large, the incidence of a particular event (as a ratio of all observations) converges to the probability of such an event occurring.

law of supply.   The tendency for most goods and services that the quantity supplied will increase as the price increases, all things else being equal.

LDC.   Acronym for Less Developed Country. See *developing countries*.

leachates.   Soluble compounds dissolved and carried by water percolating through the ground from any anthropogenic sub-surface structure, such as landfills or tailings.

leaching.   The transportation of soluble material (organic and inorganic) from soils or waste sites through the percolation of water. Agricultural run-off may include fertilizers and pesticides that have leached from the soil.

lead.   A widely used metal in batteries and, in some countries, still used as an additive to paints and fossil fuels. Lead readily accumulates in the body and chronic (low-level) exposure may affect the development of children (especially cognitive abilities) and, at high levels of exposure, may even result in death.

lead poisoning.   See *lead*.

leakage.   See *carbon leakage*.

leapfrogging.   In the context of climate change, the notion that developing countries bypass intermediate technologies and immediately adopt the cleanest technologies when they are developed.

least developed countries (LDCs).   Low-income countries which face severe handicaps to growth, such as human resource and infrastructure weakness, natural resource depletion and economic vulnerability. These countries generally operate as a cooperative group within the framework of the United Nations, including international climate negotiations.

least squares.   A method of estimating parameters in a regression model by minimizing the differences between the predicted values from the model and the observed values from the data, multiplied by some weight. If the weights are identical and equal to one, the method of estimation is called ordinary least squares. See *ordinary least squares*.

Le Chatelier principle.   1. The notion that a physical system in equilibrium responds to a shock in a way that minimizes its effects. 2. In economics, the notion that the economic response to changes in prices is greater in the long run when all variables can change than in the short run when some variables may be fixed.

legume.   Plants of the pea family (including beans, lentils, etc.) that have the ability to "fix" nitrogen and enrich the soil because of the nitrogen-fixing bacteria that grow on their roots.

leishmaniasis.   A debilitating and sometimes fatal disease spread by sandflies and caused by protozoa. The incidence of the disease is increased with deforestation and the settlement of equatorial forests that increase human exposure to sandflies that have bitten animals with the disease.

lentic.   Term used for standing or stagnant freshwater.

levee.   An artificially constructed or natural slope along a body of water that acts as a barrier against inundation of surrounding land.

lex non scripta.   Latin term for the common law.

lex scripta.   Latin term for written or statute law.

liability.   A legal rule that makes it possible for individuals, firms, or even governments to be held legally responsible for their actions.

libertarianism.   Term used to describe the notion that individual liberties are paramount and that the state should not interfere or infringe upon these rights. In its extreme, some libertarians argue that taxation contravenes the rights of individuals to spend the money they earn as they see fit.

lichen.   An association of fungi with green algae.

life-cycle analysis.   A scientific accounting of material and the effects of throughput of a natural resource from initial extraction, through production and consumption, and finally to the ultimate emission of residuals to the environment.

life-cycle cost.   The sum of all the fixed and variable costs incurred over the life span of an object. For example, the life-cycle cost of an electricity generation facility includes costs arising from licensing requirements and regulatory approval, planning and construction, operation, and finally decommissioning. See *life-cycle analysis*.

life-cycle model.   A method of valuing risk that models the way a change in external factors causes a change in an individual's probability of dying.

FURTHER READING
Cropper and Sussman (1990).

life expectancy.   The age that half the population is expected to live to or beyond.

lifestyles of health and sustainability (LOHAS).   A market segment in developed countries focused on health and fitness, personal development, the environment, sustainable living, and social justice.

life table.   A mortality table that specifies the average remaining years of life at each age for a defined population.

lifetime.   A broad term for the rate at which the concentration of trace gases declines from a reservoir, such as carbon dioxide in the atmosphere. Often used interchangeably with adjustment time, the period across which a single increase in concentration decays completely. The length of time a greenhouse gas persists in the atmosphere influences the effect of its emissions upon global warming, in addition to the amount of heat that the particular gas traps. See *global warming potential* and *turnover time*.

light green technologies. Processes that may benefit the environment, such as the development of fuel-efficient vehicles, but that were not necessarily developed to reduce environmental degradation. See *dark green technologies*.

light-water reactor. A nuclear reactor that uses ordinary water to moderate the nuclear fission reaction, and also as the coolant to remove heat from the reaction which in the process produces steam for turning a turbine. Ordinary water is an inefficient moderator so the uranium used in light-water reactors must be enriched to a high concentration of the isotope U^{235}.

lignite. A type of coal with properties in-between that of peat and bituminous coal.

likelihood. The probability of an event, outcome or consequence occurring. In the formal terminology of IPCC reports the following terms are used to quantify likelihood: virtually certain (>99%); very likely (>90%); likely (>66%); about as likely as not (33 to 66%); unlikely (<33%); very unlikely (<10%); exceptionally unlikely (<1%). See *Intergovernmental Panel on Climate Change*.

likelihood function. Function defined as
$$L(x, \beta) = f(x_1, \beta) \cdot f(x_2, \beta) \ldots f(x_n, \beta)$$
where $x_1, x_2, \ldots x_n$ are observations of a random variable x and the estimated parameters β are maximum-likelihood estimators that maximize the likelihood function (or a logarithmic transformation of the function). See *maximum likelihood*.

FURTHER READING
Greene (2003).

likelihood ratio test (LR). A widely used statistical test used to evaluate hypotheses. The test compares the values of the log-likelihood function with ($\ln_r$) and without ($\ln_{ur}$) restrictions where

$$LR = -2\{\ln_r - \ln_{ur}\}$$

is distributed as a chi-squared statistic. See *likelihood function.*

likely.   See *likelihood.*

limit.   The value of a function as the argument of the function approaches a particular value.

limited dependent variable.   Econometric models in which the dependent variable can only be observed over a particular range due to either a lack of observations, or the fact that the data only exists for a particular subset.

limited entry.   Term used to describe controls placed on the number of resource users, and possibly their inputs, in the use of natural resources. Limited entry is a common form of regulation in fisheries. See *input controls.*

limited-user open access.   A property-rights regime where the number of resource users is regulated, but no other restrictions exist in terms of how the resource is exploited.

limiting factor.   A factor, element, or resource in the environment that may restrict the growth of individuals or distribution of a species or of communities. In deserts, water is often a limiting factor that prevents their colonization by plants and animals that are not adapted to water-scarce environments.

limits to growth.   See *Club of Rome.*

limnology.   The study of freshwater bodies such as lakes and rivers.

Lindahl equilibrium.   A Pareto-efficient equilibrium involving both private and public goods where individuals agree to the same level of provision of public goods for a set of lump-sum transfers of income. See *Pareto efficiency* and *public good.*

FURTHER READING
Cornes and Sandler (1996).

linear function.   A function where the variables are not raised to any power except one and are not cross-multiplied by any other variable. For example, $y = 3a$, is a linear function but $y = 3a^2$ is not.

Linear growth.   Growth in a variable over time such that it increases by the same fixed amount each time period.

linear probability model.   A regression model in which the dependent variable is either 1 (representing the occurrence of an event) or 0 (representing the absence of an event).

linear programming.   A widely used mathematical programming technique for problems that are linear in the objective function and in the constraints.

FURTHER READING
Simon and Blume (1994).

lipids.   Substances that are soluble in fats but not water.

liquefied natural gas (LNG).   A product of natural gas consisting primarily of methane (CH_4) that is non-toxic, odorless, colorless and non-corrosive, and condenses to a liquid state at close to atmospheric pressure.

liquefied petroleum gas (LPG).   Light hydrocarbons comprising propane or butane, or a combination of the two gases, derived from the distillation of crude oil that can be stored liquefied under pressure.

liquidity.   Term used to describe how quickly an asset can be converted into cash.

liquidity preference.   Term used to describe the demand, or the desire, to hold cash over other assets.

Listeria monocytogenes.   A pathogen found in the feces of some animals that is responsible for widespread food contamination as it can be transferred via water, soil and animal tissues. Although most people can tolerate the pathogen, it can be particularly harmful to pregnant women and can cause miscarriages.

lithology.   The science and study of the structure of minerals and rocks.

lithosphere.   The subsoil layers of the Earth including its core, mantle and crust.

litter fall.   The accumulation of organic debris on the forest floor from falling leaves and other material.

Little Ice Age. A cooling trend in the Earth's temperature, especially in the northern hemisphere, that spanned the period from the sixteenth century up to the early or mid-nineteenth century.

littoral zone. The ecological zone between the low and high water marks of lakes, rivers and the sea.

load shape. A term that refers to the distribution of electricity generated (electricity load) over a specified period of time, which is commonly shown as a chart with the quantity of electricity supplied on the vertical axis and time on the horizontal axis.

load factor. The ratio of the average electricity demand to the maximum peak load that the electrical system can support.

loam. Soil that is a mix of clay, silt and sand.

locally unwanted land use (LULU). A land use that would decrease the utility of local constituents if sited "nearby".

local mean sea level change. Sea level change with respect to a land-based benchmark such as a cliff. See *eustatic sea level change* and *sea level rise*.

local pollution. Pollution from which most or all the degrading effects to human, environmental, and/or ecological well-being occur locally.

local public good. A public good the benefits of which are limited to a small area or a particular population, such as a city park. See *public good*.

FURTHER READING
Cornes and Sandler (1996).

location theory. A set of theories used to explain land use and location of economic activities. German writers, such as Von Thunen and Weber, were the pioneers in location theory. Von Thunen provided a theory as to why farmers produce the crops they do in relation to their markets, and Weber provided a theory about where firms locate their production relative to their raw materials and market.

locavore. A person whose diet consists primarily of locally grown food.

The movement was spawned in the United States after an increased awareness in sustainability issues.

lock-in effect. A term describing the phenomenon whereby technologies remain dominant as a result of large sunk investment costs, complementary technologies and widespread usage.

loess. Silt deposited by wind which originates from glacial plains or desert regions.

logarithm. The power or exponent to which a base must be raised to obtain a particular number. For example, using the base 10, ln $100 = 2$ since $10^2 = 100$.

logarithmic scale. A scale that uses a base equal to 10 such that an increase in the scale by one represents a 10-fold increase in the value of the variable being measured. For example, the Richter scale is logarithmic such that an earthquake of 7.0 generates a 10-fold increase in the intensity of seismic waves compared to an earthquake of 6.0. See *logarithm*.

logistic growth. Natural population growth characterized by a logistic growth function and that is characterized by density dependence. A generalized density-dependent growth function is defined by

$$f(x) = rx\left(1 - \frac{x}{k}\right)^\alpha$$

where $f(x)$ is growth in the population, x is the population, r is an intrinsic growth rate, k is the carrying capacity and α is a parameter. If α is less (greater) than unity the growth function is skewed to the right (left). As x approaches the carrying capacity the population growth approaches zero. See *carrying capacity*.

logistic growth curve. See *logistic growth*.

logit analysis. A method of regression analysis with a limited dependent variable (usually binary) commonly used for discrete choice analysis. The logit model assumes a functional form for the regression in which $y_i^* = \beta x + u_i$, where y_i^* is unobservable, but can be deduced by assuming that an observable dummy, y, equals 1 if $y_i^* > 0$ and 0 otherwise. In logit analysis, u_i is a random variable with a logistical distribution. See *probit analysis*.

logit regression. See *logit analysis*.

log-likelihood.   The logarithm of the value of the likelihood function used to obtain maximum likelihood estimators. See *likelihood function*.

FURTHER READING
Greene (2003).

London Amendment.   A 1990 amendment to the 1987 Montreal Protocol which speeded up the phased out reductions in the production of chlorofluorocarbons (CFCs) that destroy stratospheric ozone. See *Montreal Protocol*.

London Dumping Convention.   First signed in 1972, and now ratified by 75 countries, the Convention on the Prevention of Marine Pollution by Dumping of Waste and Other Matter prohibits the dumping of specified wastes, including radioactive wastes.

longitudinal data.   Data including repeated observations from the subject of study.

longline.   A method of fishing by which lines with baited hooks are held at or below the surface by floats and weights. Longlining is a common method of harvesting large pelagic species such as tuna.

long-range transportation of air pollution (LRAP).   The transportation of pollutants, such as sulfur dioxide, a great distance from the source of emissions.

long run.   Period of time where all factors of production can be varied.

long-run sustainable yield (LRSY).   The culmination of the mean annual increment weighted by the area of all productive and utilizable forest land types in a forest stand. See *mean annual increment*.

lopping.   Cutting of tree tops and branches after felling to increase the amount of slash left near the ground.

Lorenz curve.   A graph which plots the relationship between cumulative income or wealth or some other variable (on the vertical axis) against the cumulative population (on the horizontal axis). If every person has the same income or wealth the curve is represented by a 45° line and the more unequal the distribution the greater the area between the Lorenz curve and the 45° line.

FURTHER READING
Perkins, Radelet and Lindauer (2006).

lotic.   Term used for running fresh water.

Lotka–Volterra model.   A predator–prey model governed by a two dimensional system of differential equations where the population of the two species fluctuates in a perpetual harmonic motion around an equilibrium that is never reached.

FURTHER READING
Grafton and Silve-Echenique (1997).

Love Canal.   A community in upper New York state which was formerly a disposal site for chemicals and plastics. Concerns over seepage of toxic materials from the site into basements and soils, and the potential health hazards from such leakage, led to the area being declared a disaster area in 1980. The incident provided stimulus for CERCLA or the Superfund legislation. See *CERCLA.*

low carbon economy.   An economy that has transformed into being a negligible or very small source of greenhouse gas emissions by virtue of measures that have resulted in the consumption and production of low emissions-intensive goods and services.

low carbon technology.   A technology used to provide goods or services that emits a smaller amount of greenhouse gases than other available technologies. For example, electric powered vehicles.

low hanging fruit.   A term used for actions that are easy to take and readily available for the purpose of achieving an outcome, goal, or target. For example, in the context of improvements in energy efficiency a negative cost measure to reduce electricity consumption is switching lights off in areas when they are not being used.

low income country (LIC).   A World Bank classification for national economies earning less than US$995 GNI per capita (2009 values).

lowest observed adverse effects levels (LOAELs).   A measure derived from animal and epidemiological studies. Exposure to a defined substance above its LOAEL may be associated with potential health problems.

low-intensity logging.   Logging that removes less volume over a given area relative to more conventional logging techniques, such as clear-cutting.

Lucas critique.   Named after Robert Lucas Jr, a Nobel laureate in economics, this states that the expectations, and thus behavior, of economic agents are likely to alter following changes in economic policies. Thus, economic policies should not be undertaken under the assumption that, on average, individuals have naïve expectations.

FURTHER READING
Romer (2006).

LULU.   See *locally unwanted land use*.

LULUCF.   See *land use, land-use change and forestry*.

Lyme disease.   A bacterial disease transmitted by ticks to humans and named after the place it was first discovered on Long Island, New Jersey, in 1982. Increased populations of deer in North America have increased the number of ticks that feed on them and, in consequence, the incidence of the disease in humans.

M

MAB.   See *Man and the Biosphere*.

macroeconomics.   The study of business cycles, economic growth and broad or large-scale economic problems using measures of economic aggregates and gross economic activity. For example, the study of unemployment is a branch of macroeconomics. See *microeconomics*.

"mad cow" disease.   See *bovine spongiform encephalopathy (BSE)*.

magma.   Molten material found beneath the Earth's surface.

Magnuson–Stevenson Act.   A US federal act ratified in 1976 but changed in 1990, and renamed in 1996, that provides guidelines to fishery management councils. See *fishery management plans*.

malaria.   A disease common in most tropical areas that is caused by several different parasites and is spread by the bite of certain types of mosquitoes. Malaria kills and debilitates millions of people, and control measures are often limited to the spraying of insecticides to reduce mosquito populations.

malnutrition.   A generic term for a diet that lacks specific and necessary nutrients, such as protein. Millions of people in developing countries suffer from malnutrition, and it is a contributing factor to morbidity and mortality.

Malthusian.   Term used to describe ideas or models that make pessimistic predictions about future resource availability. Named after the Rev. Thomas Malthus, whose eighteenth-century essay on population predicted that population growth would shortly be checked by food limits and resource shortages. See *cornucopian*.

mammal.   Class of animals (including humans) that have several distinguishing characteristics, including mammary glands used for feeding young.

Man and the Biosphere (MAB).   An interdisciplinary research and training program directed by the United Nations Educational, Scientific and Cultural Organization (UNESCO) to develop biosphere reserves and reconcile the sustainable use of natural resources with ecological integrity.

manganese nodules.   Deposits of iron and manganese found on the sea floor.

mangroves.   Trees found in estuarine environments in subtropical and tropical regions. Mangroves support important ecosystems and provide a "nursery" for some marine species, such as shrimp. In recent years, coastal developments (particularly in Asia) have led to concerns about the destruction of mangroves and the effects of this on the environment.

mantle.   The layer of rocks within the Earth that lies above the core and beneath the Earth's crust.

mare liberum.   Latin term for those parts of the sea or ocean that are freely navigable.

marginal abatement cost (MAC).   The change in the total cost of abatement of pollution or emissions following a marginal change in the level of pollution or emissions abatement. See *abatement*.

marginal benefit.   See *marginal private benefit*.

marginal cost.   See *marginal private cost*.

marginal cost of enforcement (MCE).   The additional cost of providing one more unit of enforcement effort.

marginal cost of exploration.   The cost of finding additional non-renewable resources. It is sometimes used as an indicator of resource scarcity.

marginal-cost pricing.   The pricing of goods such that the price per unit equals the marginal cost of producing that unit.

marginal damages.   The change in damages arising from producing an additional unit of output or emissions. Also, the change in damages from a change in a particular state or policy.

marginal external benefit (MEB).   The benefit obtained by others not party to the consumption or production of the addition unit of a good from which it is derived. See *positive externality*.

marginal external cost (MEC).   The cost obtained by others not party to the consumption or production of the addition unit of a good from which it is derived. See *negative externality*.

marginal extraction cost.   Change in the total cost of extracting a resource following a marginal change in the amount extracted.

marginal physical product (MPP).   The extra output produced from a marginal change in the quantity of a variable input used in a production process.

marginal private benefit.   Change in total private benefits following a marginal change in output.

marginal private cost.   Change in total private costs following a marginal change in output.

marginal product.   See *marginal physical product.*

marginal profit.   The difference between marginal revenue and marginal private cost.

marginal propensity to consume.   The proportion of a marginal increase in income that is consumed, rather than saved.

marginal propensity to save.   The proportion of a marginal increase in income that is saved, rather than consumed.

marginal rate of substitution (MRS).   The rate at which an individual is prepared to substitute one good for another while maintaining the same level of utility. It is the negative of the ratio of the marginal utilities of the two goods.

marginal rate of technical substitution (MRTS).   The rate at which one input can be substituted for another while maintaining the same level of output. It is the negative of the ratio of the marginal products of the two inputs.

marginal revenue.   The change in revenue following a marginal change in sales.

marginal social benefit.   The change in private and external benefits following a change in output.

marginal social cost.   Change in total private costs and the costs imposed on others (due to external costs) following a marginal change in output.

marginal stock effect.   The right-hand-side term in the fundamental equation of renewable resources defined by the ratio of the following partial derivatives

$$\frac{\partial \pi / \partial x}{\partial \pi / \partial h}$$

where x is the level or size of the renewable resource, h is the harvest, B is the current rent from the renewable resource. The marginal stock effect represents the marginal value of the biomass or resource stock relative to the marginal value of the harvest. See *fundamental equation of renewable resources*.

marginal utility.   The change in total utility following a marginal change in the consumption of a good or service.

marginal value.   The change in value attached to a marginal change in the quantity of an asset or resource.

marginal willingness to pay.   The willingness to pay, by a consumer, to acquire one more unit of a good. The first derivative of *total willingness to pay* with respect to quantity.

marine fish farming.   The raising of fish in enclosed pens in ocean waters for commercial purposes.

marine protected area (MPA).   A spatial location where fishing or fishing practices are limited or prohibited. Typically, they are established for conservation purposes but they may also provide spillover benefits to fishers outside MPAs.

market approach.   An approach in which free trade in the market determines the allocation of environmental goods or natural resources. The market approach is thought to achieve either a least cost or economically efficient allocation provided that property rights are well-defined and prices include all externalities.

market barrier.   See *barriers to entry*.

market-based instruments.   Various approaches to resource management that rely on tradeable property rights, where the value of such rights is determined in a market context. Trades in individual transferable quotas is a good example of a market-based instrument in a fishery. In this context, trades will normally occur from high to low marginal cost producers, since low marginal cost producers can pay more for each unit of quota, thus increasing efficiency in harvest. Tradable emission permits is another form of market-based instruments. See *emissions trading scheme.*

market-clearing price.   The price in a market that ensures that the quantity demanded of a good or service exactly equals the quantity supplied.

market economy.   An economy where most of the decisions about what to produce and consume are decided by markets rather than planners. See *command economy.*

market environmentalists.   Term used to describe persons who believe in and promote the use of incentives and market-based instruments to address environmental problems.

market equilibrium.   A situation where the quantity supplied to a market exactly equals the quantity demanded such that there is no tendency for the price to change.

market failure.   The failure of a market to achieve the economically optimal allocation of resources caused by one or more of the following: increasing returns to scale in production that are not exhausted before equilibrium output is reached (e.g., monopolistic production); the presence of externalities in production or consumption; excessive uncertainty within markets; a lack of a sufficient number of producers and consumers (i.e., thin markets); and incomplete property rights. For example, climate change is an example of market failure because the marginal social cost of greenhouse gas emissions has historically not been incorporated into firm and household decision making.

market-oriented reform.   Economic or environmental reform that occurs from the creation or correction of markets by government, rather than direct rules or regulation governing behavior.

market price.   The price at which a unit of a good or service can be purchased in the current market. The market price need not necessarily be a competitive price.

market risk.   The inherent risk in holding equities. It cannot be eliminated by diversification of the portfolio of holdings.

market structure.   The principal characteristics of a market including the barriers to entry, the number of buyers and sellers, the differences among products sold in the market, and other factors.

market stumpage prices.   Prices for timber derived from arm's length transactions between the buyer and seller. Such prices, theoretically, should equal the market price of the wood embodied in the tree less all conversion costs.

Markov transition matrix.   A matrix that defines the probability of going from one state of the world to another, and that can be used to model the dynamics of populations.

MARPOL.   See *London Dumping Convention*.

Marrakesh Agreement, The.   The conclusion of talks under GATT that started in Uruguay in 1986, and ended in Marrakesh in 1994. The agreement eliminated a wide range of tariffs in a number of different industrial sectors and established a framework for liberalized international trade. It also created the World Trade Organization (WTO). The agreement incorporated the need to address environmental concerns in the context of international trade. See *Uruguay Round, The*.

Marshallian demand.   Demand for a good or service that is a function of output prices and income.

Marshallian surplus.   The area under the Marshallian, or uncompensated, demand function between the price of a consumption good and the choke price of that good (i.e., the price that drives consumption to zero). Marshallian surplus can be considered a money metric measure of utility only when the marginal utility of income is unity for all goods. See *consumer surplus*.

marsupials.   Mammals whose young are born live and then develop in a pouch or marsupium, such as a kangaroo.

mass balance.   The balance, or net change, in the mass of a glacier, ice cap, or ice sheet from the accumulation and loss of ice.

mass spectrometer.   An instrument used to measure the relative proportions or masses of different elements and their isotopes.

materials balance principle.   Derived from the first law of thermodynamics, the materials or mass balance principle states that the matter/energy content of a material cannot be destroyed. Therefore, matter extracted from the Earth (say ore) and processed must also be accompanied by waste. The more matter extracted and processed, the more waste is created.

mathematical expectation.   See *expected value*.

matrix.   A systematic array of numbers arranged in rows and columns.

maximin and minimax strategies.   The same strategy, used in models in game theory, described from two different perspectives. Under the *maximin* strategy, all other participants in a game are trying to make a particular participant's payoff as low as possible. A particular participant then chooses the *maximin* strategy, which is the strategy that will yield the highest payoff for him/her, given the others' strategies and states of nature. The strategies pursued by the other participants, which are chosen to minimize the payoff to that particular party, are then collectively described as a *minimax* strategy.

maximum allowable concentration.   The maximum concentration of a pollutant allowed in the workplace.

maximum economic yield.   A sustainable catch or effort level that creates the largest difference between total revenue and the total costs of fishing for a fishery as a whole. A point of maximum economic yield is often associated with a point of maximum profits in the fishery, or a catch or effort level that maximizes resource rents.

FURTHER READING
Grafton, Kompass and Hilborn (2007).

maximum likelihood.   A widely used statistical method of estimation that maximizes the likelihood function from an observed sample for an assumed distribution of the residuals. The maximum likelihood estimators have the greatest probability of drawing the sample data used to derive the estimators. See *likelihood function*.

FURTHER READING
Ramanathan (1992).

maximum permissible dose.   The maximum amount of ionizing radiation an individual should be able to absorb in a given period of time without negative health results.

maximum principle.   Necessary conditions in optimal control theory that must be satisfied to ensure an optimal time path for the control and state variables in a dynamic optimization problem in continuous time. See *optimal control theory.*

FURTHER READING
Chiang (2000).

maximum residue limits.   Limits set by food health and safety authorities that determine the maximum amount of a particular substance (such as a pesticide) allowed on a food (such as grapes).

maximum sustainable yield (MSY).   A point where catch or the harvest of fish is just equal to the growth of the stock of fish (or net additions to the fish stock), which is both sustainable and at a harvest level that generates the largest possible catch. Maximum sustainable yield is generally associated with surplus-production models that show that the sustainable catch at first rises with the stock of fish, reaches a maximum, and then falls with increases in stock of fish to the point of maximum carrying capacity.

maximum value function.   See *optimal value function.*

Mayflower problem.   The notion that the actual exploitation of non-renewable resources is governed by imperfect information and uncertainty such that lower quality resources may be used before higher quality resources. This notion is in contradiction to the theory of David Ricardo, the nineteenth-century British economist. Thus, the "pilgrim fathers" that came to the United States from England on the ship the *Mayflower* settled on poorer soils first as they did not know (nor could they use) better soils that existed to the west.

McKelvey box.   A two-dimensional representation of the likelihood of the existence of mineral reserves (proven, probable, possible and undiscovered) versus the economic value of exploiting the reserves.

MEA.   See *Multilateral Environmental Agreements.*

mean.   See *arithmetic mean.*

mean annual increment (MAI).   Average annual growth in timber volume in a given stand of trees expressed as an increase in volume per unit area.

mean residence time.   See *residence time.*

mean sea level (MSL).   The average sea level recorded across a period of time, in order to remove natural variability from waves and tides. See *sea level rise.*

mean squared error.   Expected value of the square of the difference between the true value of a parameter and its estimated value. It is equal to the variance plus the square of the bias.

mean surface temperature.   See *global surface temperature.*

measurement, reporting and verification (MRV).   A technical term for the process of collecting, analyzing and confirming data pertaining to emissions reductions associated with climate change mitigation.

measure.   A broad term covering practices and processes directed at achieving a specific goal. For example, government funding for solar power research and development is a greenhouse gas mitigation measure. See *policy.*

measurement error.   Errors in the recording of data used in statistical analysis.

median.   The middle value from a sample or population where the observations are arranged from the lowest value to the highest.

mega.   Popularly taken to mean "very large", the term is more precisely used as the prefix signifying a million or 1×10^6. For example, a megabyte is one million bytes and a megaliter is one million litres.

megafauna.   Large animals (such as lions or gazelles) found in an ecosystem. Conservation efforts are often directed to megafauna rather than smaller-sized animals. See *fauna* and *charismatic fauna.*

megalopolis.   See *conurbation.*

melanism.   Term used for the appearance of dark skin, pigment, or hair due to an abnormal increase in the production of melanin.

melanoma.   A malignant form of skin cancer that is linked to exposure to ultraviolet radiation. A decline in stratospheric ozone, in the absence of avoidance measures from sun exposure, will increase the incidence of melanoma.

meltdown.   The accidental melting of nuclear fuel in a reactor due to the breakdown of the cooling system.

membrane.   A pliable and thin tissue consisting of lipids and proteins that separates cells and other structures within plants and animals.

Mendel's laws.   Laws or results named after the plant breeder Gregor Johann Mendel which state that traits are inherited separately from each other and that recessive traits will only appear if there exist two recessive genes.

mercantilism.   Name given to a school of thought that had its origins in the sixteenth and seventeenth centuries in England. Mercantilists supported the notion of the state intervening to promote trade, such as the granting of trade monopolies.

mercury.   A toxic metal found in many different chemical compounds and sometimes used as an amalgam in dental fillings. See *methyl mercury*.

mercury poisoning.   See *Minamata disease*.

mesosphere.   A part of the Earth's atmosphere located around 50–80 kilometers above the Earth's surface.

Mesozoic.   Geological era that spans the period 245 million to 66.4 million years ago.

meta-analysis.   A statistical method of analysis that combines the results of two or more independent studies.

metabolism.   The combined physical and chemical processes within organisms that sustain life.

metagenesis.   Name given for alternatively sexual and asexual reproduction.

metals.   Any of the metallic elements and compounds characterized by a crystalline structure when solid.

meta-modeling.   The creation of a general model from more detailed models that may differ in spatial and temporal scales.

metamorphosis.   A fundamental change in the form of an organism during its life, such as in the life cycle of a moth from a caterpillar to its adult form.

meta-population.   The population of a species as defined over its entire spatial distribution. A widely dispersed meta-population reduces the probability of extinction, as environmental shocks to one region or area that may lead to the extirpation of a species still allows the species to continue in another locale.

meteor.   A rocky object that becomes visible upon entry to the Earth's atmosphere.

meteorite.   A meteor that reaches the Earth's surface.

meteorology.   The study of weather and the atmosphere.

meter.   Unit of length equal to 3.28 feet.

methane (CH_4).   An important greenhouse gas that is 21 times more potent than carbon dioxide. The primary component of natural gas. Major sources of methane emissions are coal mines, natural gas fields, livestock production, and decaying organic matter in landfills and bogs. Large deposits of methane locked in arctic permafrost are steadily being released as global temperatures rise. See *climate feedback* and *global warming potential*.

Methanol (CH_3OH).   An alcohol that is used as an alternative or additive to standard gasoline. The use of methanol has some environmental benefits such as lower carbon monoxide emissions, especially when produced from organic materials. See *biofuel*.

methyl bromide.   A widely used pesticide that has a high ozone-depletion potential.

methyl mercury.   A toxic form of mercury created by the breakdown of mercury in the environment. Several important cases of mercury poisoning have occurred in people who had consumed foods (fish and shellfish) with high levels of methyl mercury. See *Minamata disease*.

methyl tert butyl ether (MTBE).   A gasoline additive used in the United States by refineries to assist them in complying with state and federal regulations for "cleaner" gasoline. MTBE is an oxygenate that enables gasoline to burn more cleanly but it is highly water soluble, such that gasoline leaks and spills can contaminate water sources. Concerns over water pollution have led the State of California to phase out and commit to discontinuing its use from January 1, 2003.

metric system.   Also called the SI, it is the set of standard international units of measure. Further details are available in Appendix 3.

microclimate.   A variation in the climate limited to a small spatial area, and often due to the physical landscape.

microeconomics.   The study of economic behavior and problems at a household, individual, or firm level. For example, evaluating the benefits of pollution permits to control sufur dioxide emissions is a part of microeconomics.

micrometer.   A unit of measure equal to 1/1000 of a millimeter.

micronutrient.   A nutrient required in very small quantities.

micro-organism.   Any organism that can only be seen with a microscope.

migratory species.   A species that moves over considerable distances in its life cycle. Many fish and bird species are migratory. See *sedentary species*.

Milankovitch Theory.   A hypothesis developed by the mathematician Milutin Milankovitch in the early twentieth century to explain the Earth's long-term climate cycles. Milankovitch cycles involve periodic changes in: the Earth's orbit, its angle to the sun and the precession of the equinoxes. These factors collectively affect the amount of solar radiation reaching the Earth, although individually they may reinforce or negate the others. The subsequent cycles in global temperature have been used to explain the existence and characteristics of ice ages. See *ice age*.

FURTHER READING
Ruddiman (2007).

mile.   Unit of length equal to 1.61 kilometers.

Millennium Development Goals (MDGs).   A set of eight goals adopted in 2000 to alleviate conditions related to extreme poverty. The general themes are: poverty, hunger, education, women, child mortality, maternal health, HIV/AIDS, environmental sustainability and development partnerships. All 192 UN member states agreed to achieve the goals by the year 2015.

Minamata disease.   A disease named after a fishing village in Japan where its people were first diagnosed with a debilitating (and sometimes fatal) nervous disorder, which was subsequently found to be caused by the consumption of fish containing high levels of methyl mercury.

mineral.   Any non-living substance that is extracted from the surface or sub-surface of the Earth by mining or quarrying.

mineralogical threshold.   A hypothesized threshold at which the cost of extracting certain metallic elements increases sharply due to a change in how the metals are chemically bound into various ores and rock bodies.

mineral reserve.   The amount of a mineral that is known to exist with a reasonable degree of certainty, and which can also be profitably extracted.

minimum environmental stock.   The minimum level of environmental goods and services required to support sustained economic activity.

FURTHER READING
Barbier (1993).

minimum-tillage farming.   An agricultural technique that minimizes the disturbance of the soil surface to reduce erosion.

minimum utilization rate.   A standard sometimes imposed on forest companies when harvesting timber where they have cutting rights but do not own the land. It specifies the maximum height of the tree stumps, and other rules, to encourage the most efficient utilization of the trees.

minimum utilization standard.   See *minimum utilization rate.*

minimum value function.   See *optimal value function.*

minimum viable population.   The population size (or stock in fisheries) below which death and emigration exceed birth and immigration.

Populations that fall below the minimum viable population size will become locally extinct.

Miocene.   Geological epoch of time from 23.7 to 5.3 million years ago.

mist belt.   A contiguous area with persistent precipitation that condenses directly on vegetation. See also *occult precipitation.*

mitigation.   The act of reducing the impact or prevalence of an undesirable circumstance or entity. For example, the mitigation of greenhouse gas emissions by anthropogenic measures such as fuel switching, changes in land use, use of more fuel-efficient technologies and carbon sequestration. See *climate change mitigation.*

mitigation potential.   The amount of mitigation that could be realized in the future under current circumstances. In the context of climate change it is unrealized greenhouse gas mitigation by households, businesses, or economy-wide resulting from limited utilization of existing technological and economic factors.

mitigative capacity.   A broad term for the ability of a country to reduce greenhouse gas emissions given its endowment of skills, institutions and resources.

mixed economy.   An economy where the decisions about what goods and services are produced, how many, and at what prices are made through markets as well as interventions by government and/or state enterprises.

mixed forest.   A forest made up of both deciduous and coniferous species.

mixed-use development.   A plan that involves different uses of the same site rather than dedication to a single purpose (e.g., forest lands that support wildlife habitat and recreation, not just timber production).

mobile source.   A source of pollution that can be non-stationary, such as an automobile. See *stationary source.*

mode.   Most frequent observation in a sample or population.

modeling.   A method of inquiry by which a simplified abstraction is used to represent and predict phenomena. Economic models often consist of equations, but a model may also be written in sentences or represented by graphs, flow charts, or other means.

modes of climate variability. Broad term for natural patterns of variability in the climate system that occur on hemispheric or global scales, and across seasons or multiple years. These cycles occur as a result of dynamic interactions between the atmosphere, land and ocean. For example, the El Niño Southern Oscillation.

modus operandi. Latin term meaning "the way things are done".

modus vivendi. Latin term for a working compromise between parties.

moisture seasonality. Seasonal patterns in the availability of moisture over the year.

molecule. At least two atoms of one or more elements chemically combined. For example, carbon dioxide is a molecule that consists of one carbon atom and two oxygen atoms.

mollusks. A common and widespread group of marine and freshwater animals that are also found on land. Mollusks are capable of producing a hard shell from secretions of calcium carbonate, and have a tongue-like organ. Mollusks include scallops and mussels, as well as snails.

monetary policy. Policy of the central bank regarding the control of the money supply and its effects on output, inflation and the exchange rate.

monitoring. Data collected by a third-party or regulatory authority concerning adherence to environmental regulations. In the context of climate change, commonly refers to quantitative examination of project-based and national mitigation activities occurring within a formal reporting framework.

monoculture. Single-species cropping.

monopolistic competition. A market structure where there are many buyers and sellers, ease of entry and exit, and in which each firm produces a slightly different product.

monopoly. A market structure where there is only one seller that can influence the price of the good it sells by changing the quantity it chooses to produce.

monopsony. A market structure where there is only one buyer in the market that can influence the price it pays for a good or service by changing the quantity it chooses to purchase.

monotypic genus. A genus of plants or animals containing only one extant species. Humans are a monotypic genus.

monsoon. A term for the seasonal reversal of wind and associated rainfall that occurs in tropical and subtropical regions. The cause of the monsoon is the large annual variation in temperature between large land masses and the ocean. The strongest monsoons occur in southern and eastern Asia. A summer monsoon is generally associated with heavy rainfall on land.

montane. A sub-alpine environment characterized by a cold winter and coniferous trees.

Monte Carlo simulation. Data-simulation approach to testing models and hypotheses involving multiple runs of randomly generated parameter values.

montmorillonite. A clay derived from the weathering of rocks that is noted for swelling in wet periods and contracting in dry periods.

Montreal Process. An initiative started by the Government of Canada, which hosted a meeting in Montreal in September 1993 to develop a series of criteria and indicators for sustainable forest management. The final meetings were held in Santiago, Chile, in 1995 and produced two documents: the Santiago Declaration and "Seven Criteria and Associated Quantitative and Qualitative Indicators". See *Helsinki Process*.

Montreal Protocol on Substances that Deplete the Ozone Layer. An agreement first signed in 1987 in Montreal, Canada, and often called the Montreal Protocol, under which signatories agreed to reduce their production of ozone-damaging chlorofluorocarbons (CFCs) by half by the year 2000, and which continued the process begun in 1985 with the Vienna Convention for the Protection of the Ozone Layer. The Protocol was subsequently strengthened in 1990, and again in 1992, with the London and Copenhagen Amendments, under which all CFC production would stop by 2000.

moral hazard. A problem of asymmetric information whereby the actions of one party to a transaction are unobservable. For example, if people were able fully to insure their valuables (with no deductibles) they might be less inclined to undertake actions that would reduce the probability of theft than if they were obliged to cover some of the costs of loss. See *asymmetric information*.

morbidity. The incidence of disease. A morbidity rate is usually defined as the number of reported cases of a disease in a population per 100000 people in a given period of time.

more-is-better property. The assumption that larger quantities of any good generate higher levels of utility from the consumption of that good. Also known as local non-satiation.

mortality rate. The death rate of a population over a given period of time. Human mortality rates are defined as the number of deaths per 100000 people per year.

most-favored nation status. A status accorded to countries by the US Congress in order to promote trade between the United States and that country. Pressure groups have lobbied to remove some countries from the most-favored list so as to pressure certain states to accept democracy, human rights and environmental causes.

Mount Pinatubo. A volcano in the Philippines that erupted in 1991 and which pushed a large amount of debris into the stratosphere. This material reduced the solar radiation reaching the Earth's surface and led to a cooling of the northern hemisphere in 1991 and 1992.

Mount Tambora. Site of a major volcanic eruption in 1815, in what is today known as Indonesia, that had a major short-term impact on the Earth's weather and resulted in significant cooling.

moving average (MA) process. A stochastic process by which the current value of a variable is a function of past errors. An MA process of order n is defined by

$$x_t = \alpha + e_t + a_1 e_{t-1} + a_2 e_{t-2} + \ldots + a_n e_{t-n}$$

where x_t is the variable in time t and e_t is an unobservable and uncorrelated random error with a zero mean and a constant variance.

MTBE. See *methyl tert butyl ether*.

mud flats.   Low-lying areas along the shoreline that are exposed during periods of low tide and typically found in estuaries.

Mullerian mimicry.   Biological mimicry whereby poisonous or unpalatable species mimic each other in appearance, thus reinforcing the trait and accentuating their ability to avoid predation. See *Batesian mimicry.*

multicollinearity.   In statistical or econometric estimates, a situation where explanatory variables are highly correlated or interconnected. In such cases, it becomes difficult to disentangle the separate effects of each of the explanatory variables on the variable to be explained.

multi-factor productivity (MFP).   See *total factor productivity*.

Multilateral Environmental Agreements.   Agreements worked out between different countries over environmental issues ranging from regional to international problems. See *Convention on International Trade Endangered Species* and the *Montreal Protocol.*

multilateralism.   Term used to describe a co-ordinated and cooperative approach to international relations to promote trade and peace, and to address global and regional environmental problems among more than two different countries.

multi-media pollution.   Pollution that occurs across several vectors that may include air, water and soils.

multinational corporations (MNCs).   Companies that have plants and operations in several different countries. Some multinationals are so large that their sales dwarf the gross domestic products of many smaller countries.

multiple chemical sensitivity (MCS).   Term used to describe a complaint where patients become ill from contact with very small amounts of chemicals that normally would not be a cause for concern.

multiple-purpose trees (MPTs).   Trees, especially those used in agroforestry endeavors, that provide more than one important resource including fodder, fuelwood, timber, windbreaks and, in the case of leguminous MPTs, nitrogen for the soil.

multiple regression. Analysis where a single dependent variable is estimated from at least two independent or explanatory variables. See *regression analysis*.

multiple use. Term used to describe the joint use of habitats or ecosystems. In forestry, multiple use may involve some tree harvesting combined with the maintenance of corridors to preserve ecological integrity.

Multiple-use management. The management of a single resource (e.g., a forest) for multiple and jurisdictionally overlapping uses (e.g., camping, hunting and timbering).

multiplier. An economic term that describes the change in income following a change in expenditure.

multistage sampling. Sampling where the population is first divided into groups that are then subdivided, followed by a random sampling of the subdivisions, from which a final random sample of observations is made from the picked subdivisions.

municipal solid waste (MSW). Domestic and business waste materials collected by local authorities, including recyclable materials, organic matter and plastics. Although municipal solid waste was formally disposed of mainly in city landfills, recycling facilities have become more commonplace. See *landfill*.

municipal tipping fee. A payment charged by a municipality for the right to dump refuse at a site.

mutagen. Any agent capable of causing genetic mutation, including various chemicals and short-wave radiation.

mutation. A significantly different, but inheritable, change in the genome of an organism.

mutatis mutandis. Latin term meaning "with the necessary changes".

mutualism. A mutually beneficial relationship between at least two different species. For example, mutualism exists between flowering plants and bees because the flowers provide nectar and the bees help pollinate the flowers.

myopic behavior.   Behavior, especially economic behavior, that does not take into account associated future costs and benefits of actions or events and focuses upon a short-term horizon.

N

NAFTA. See *North American Free Trade Agreement*.

Nash equilibrium. A concept in game theory that describes a game outcome where an individual participant has no reason to change his or her own strategy, given the strategy chosen by all the other participants.

natality rate. See *birth rate*.

national action plan (NAP). A country's stated objectives for pollution control, and the policies designed to meet those goals. In the context of climate change, this is the name given to plans formally presented to the UNFCCC outlining national measures to reduce greenhouse gas emissions.

national ambient air quality standards (NAAQS). US federal standards that set the levels of legally permitted concentrations of defined pollutants in a defined volume of air.

national debt. The cumulative borrowing by a government. Debts in excess of 60 percent of the gross domestic product are considered to be a potential economic problem.

national emission standards for hazardous air pollutants (NESHAP). US standards for point sources of emissions of potentially hazardous pollutants.

National Environmental Policy Act (NEPA). Legislation passed in the United States in 1970 that created both the Environmental Protection Agency (EPA) and the requirement for federal agencies to evaluate the potential impact from agency actions or proposed legislation on the environment through an Environmental Impact Statement (EIS), followed by an Environmental Impact Assessment (EIA), if there is found to be an impact. The formal preparation and incorporation of environmental assessments into decision making has subsequently been adopted by a number of countries.

FURTHER READING
Callan and Thomas (2007).

national forests. Forest-lands held by the government for the benefit of the country as a whole. In the United States, national forest-lands refer to those lands managed by both the US Forest Service and the Bureau of Land Management, all of which are owned by the federal government.

national greenhouse gas inventory (NGGI).   A national register of sources and sinks of greenhouse gas emissions. This accounting measure enables verification of emission reduction commitments within the UNFCCC framework.

national income.   The total income of all residents in an economy, less the costs of capital depreciation.

national income and product accounts (NIPA).   A set of accounts that measure economic activity in an economy, including estimates of national income, output and expenditures.

nationalization.   The appropriation of companies or industries by the state, with compensation to the previous owners, so as to create state-owned enterprises.

nationally appropriate mitigation actions (NAMAs).   Policies and measures adopted by developing countries to reduce their greenhouse gas emissions. Originating from the Bali Road Map, this expression is used in UNFCCC formal documents to recognize the differentiated nature of responsibilities attributed to developed and developing countries. Whilst the former group are expected to observe binding emissions reductions, poorer countries are expected to only pursue measures within their existing capacity that do not compromise economic development. The latter, bottom-up approach also encompasses significant financial assistance from wealthy countries.

National Oceanographic and Atmospheric Administration (NOAA).   US agency charged with the federal management of oceans and the marine environment.

national pollutant discharge elimination system (NPDES).   A US system designed to control the discharge of effluents from designated industrial polluters.

national primary drinking water regulations (NPDWR).   US regulations designed to ensure domestic drinking water meets designated minimum health standards.

national priorities list (NPL).   A US list of hazardous waste sites that have the highest priority, in terms of control, so as to prevent negative environmental effects.

native rights.   The human and property rights of native or aboriginal people to use their traditional land and preserve their traditional way of life.

native species.   A species that has been found in a particular region for a very long period of time.

natural attenuation.   Naturally occurring processes that can reduce the mass or toxicity of contaminants in polluted soil or water.

natural capital.   A collective term for all aspects of the environment, including natural resources.

natural gas.   Gas formed by hydrocarbons, which consists mainly of methane, ethane and propane. Natural gas deposits are found in porous rock formations underground.

natural increase.   The positive difference between the birth rate and death rate in a population.

naturalism.   1. A reverence for natural processes and objects. 2. The belief that scientific inquiry can explain processes.

natural monopoly.   Term used to describe industries that exhibit increasing returns to scale such that one producer or firm is able to produce at lower cost per unit than if the same quantity were produced by several different firms.

natural radioactivity.   The background radiation level given off by the decay of naturally occurring isotopes.

natural rate of unemployment.   See *non-accelerating inflation rate of unemployment*.

natural resource economics.   A sub-discipline of economics concerned with the management and use of renewable and non-renewable resources.

natural resources.   Resources in the environment that include both renewable (such as fisheries) and non-renewable resources (such as oil and gas deposits).

natural selection. Process described by Charles Darwin where traits that are best suited to the existing environment tend to favor the reproductive success of individuals with such traits. See *Darwinian evolution.*

FURTHER READING
Darwin (1859).

natural subsidy. The availability of natural energy resources in a form that is easily accessed and used by the human economy. Energy resources, such as East Texas petroleum, that require little labor and capital to be extracted and processed are said to have a natural subsidy relative to other more costly energy resources, such as Alaskan petroleum.

FURTHER READING
Daly (1996).

nature-based tourism. Tourism that has, as its main attraction, access to nature. Nature tourism differs from ecotourism in that nature tourism does not necessarily meet criteria designed to ensure low environmental impacts.

naturally occurring radioactive elements. Term for radioactive elements that are naturally present in the environment, including uranium and thorium.

Nature Conservancy. A non-governmental organization that focuses its efforts on purchasing private land so as to conserve endangered ecosystems and areas of particular beauty or natural significance.

nautical mile. Unit of distance equal to 1.151 miles, or 1.852 kilometers.

NDVI. See *Normalized Difference Vegetation Index.*

necessary conditions. Conditions that must be satisfied at the optimum, or the solution to an optimization problem.

negative externality. An action which imposes costs on others but the person(s) undertaking the action do not account for these "external" costs in their decision making. For example, a firm that pollutes a river without consideration of the effects on downstream users is causing a negative externality. See *externality* and *positive externality.*

negative feedback. See *feedback.*

negligence. Legal term used to describe the behavior of individuals, firms or governments where due diligence has not been undertaken regarding decisions or actions.

nektonic. Term used to describe any sea creature that can move using its own power.

neoclassical economics. The most popular mode of analysis of economics at present. It is based on the premises of rational behavior by economic agents, optimization by individuals and firms who do the best they can given their constraints, and equilibrium in the sense that systems will not move from a state of rest without a shock.

neoclassical growth theory. Models used to explain economic growth using economic aggregates, paying particular attention to capital/output and capital/labor ratios. Neoclassical growth models imply that an increase in the savings rate, population growth or changes in the level of technology do not affect the steady-state per capita growth rate in output which always equals zero in the long run.

FURTHER READING
Barro and Sala-i-Martin (2004).

neoteny. Term for juvenile or immature characteristics that are retained by an organism in its adult form.

NEPA. See *National Environmental Policy Act.*

neritic. Marine zone usually defined as less than 200 meters in depth.

net domestic product (NDP). The gross domestic product less an allowance for depreciation of the capital stock.

net energy. The work done from an energy source less the work needed to find, extract, process and use the energy source.

net national product (NNP). The gross national product less an allowance for capital depreciation.

net national income (NNI). The net national product plus subsidies but less indirect taxes (such as value-added taxes). See *net national product.*

net present value (NPV).   The sum of the discounted net benefits (benefits minus costs) in each time period, over the lifetime of an investment. See *discount rate*, *present value*, and *cost–benefit analysis.*

net primary productivity (NPP).   The total energy per time period stored by photosynthesis in an environment less the total amount of energy used by the plants in respiration. NPP provides a measure of the biological productivity of an ecosystem.

net production.   The energy of an organism devoted to growth and reproduction.

net production efficiency.   Proportion of total energy consumed by an organism that is devoted to reproduction and growth.

net social surplus.   The net value to society from the production (extraction) and consumption of a good or service measured as the total benefits from consumption minus the total costs of production (extraction).

netting.   A sub-program of emission credit trading set up in the United States to allow some flexibility in meeting air pollution control targets. It allows a firm to modify its plant without going through an involved new source review. Netting allows a firm to increase its emissions for a given pollutant from one point provided that there is a corresponding decrease in emissions at another point so that net emissions do not increase overall for the firm.

FURTHER READING
Tietenberg and Lewis (2009).

neurotoxin.   Any substance that affects the functioning of the nervous system.

neutral tax.   A tax that, in theory, does not distort economic behavior. A lump-sum tax that is paid irrespective of the actions of individuals or firms is considered to be the closest to a neutral tax.

neutron.   An atomic particle that contains no electric charge. This is the part of the atom that is dislodged in nuclear fission. See *nuclear fission* and *nuclear fusion.*

newly industrialized country (NIC).   A group of developing countries experiencing high rates of economic growth. The growing global economic status of these countries permeates into international political structures such as the G20, and is associated domestically with rising living standards. NICs include: India, China, Brazil, Mexico, Malaysia, Thailand, Turkey and South Africa.

new source performance standards. US-based technological standard of pollution control to which new plants must conform.

newton. A measure of force equal to a mass of 1 kilogram that has an acceleration of 1 meter per second per second.

NGOs. See *non-governmental organizations*.

niche. The function or part played by an organism within an ecosystem.

niche requirements. The biological needs of a particular species within the environment, such as the availability of food, habitat, as well as the constraints placed upon it by physical characteristics, such as temperature, water salinity, and availability of other resources.

NIMBY. Acronym for Not in My Back Yard. The term is used to describe individual and community efforts to prevent waste disposal plants and other undesirable facilities near their own communities or places of employment. See *NIMTOO* and *locally unwanted land use*.

NIMTOO. An acronym for Not In My Term Of Office, used to describe public officials' unwillingness to support controversial developments because of localized public concern in the area where the development will take place. See *NIMBY* and *locally unwanted land use*.

nitrates. A natural salt formed by micro-organisms in the soil that can be used by plants as a source of nitrogen. Excessive use of nitrogenous fertilizers has contaminated groundwater sources because nitrates can leach from the soil into aquifers.

nitrites. A salt formed by micro-organisms in the conversion of nitrogen into nitrates.

nitrogen. The principal constituent of the Earth's atmosphere, it is also an essential element required by plants and animals.

nitrogen cycle.   The biochemical processes through which atmospheric nitrogen is fixed and assimilated in the biosphere via soils and lakes, then returned to the atmosphere through denitrification.

nitrogen dioxide (NO$_2$).   A very common atmospheric pollutant that is formed by automobiles when nitric oxide combines with oxygen in the combustion process. Nitrogen dioxide can produce tropospheric (surface) ozone in the presence of sunlight.

nitrogen fixing.   The act of removing nitrogen from the Earth's atmosphere or from mineral compounds in the soil or fresh water by bacteria and algae and converting them into compounds available for use by other biological organisms.

nitrogen-fixing bacteria.   Bacteria found in leguminous plants that convert atmospheric nitrogen and nitrogen found in soil and freshwater into compounds that can be used by plants and other organisms.

nitrogen oxides (NO$_x$).   A collective term for the pollutants nitric oxide, nitrous oxide and nitrogen dioxide. NO$_x$ is partly responsible for photochemical smog and, in the stratosphere, can contribute to depletion of ozone.

nitrous oxide (N$_2$O).   A gas used as an aerosol propellant and which is generated as a byproduct in the manufacture of fertilizers and the production of fossil fuels. Nitrous oxide is a potent greenhouse gas with a global warming potential of 310. See *global warming potential*.

NOAA.   See *National Oceanographic and Atmospheric Administration*.

NOAA panel.   The distinguished panel of experts convened by the National Oceanographic and Atmospheric Administration (NOAA) in 1992 to examine the contingent valuation methods (CVM) of non-market valuation. Its findings provide a set of guidelines on the approaches that should be followed in CVM studies.

Noah's ark problem.   The problem that not all endangered species, or species at risk, may be saved. The problem poses difficult choices as it may be necessary to decide which species should be given priority in terms of preservation.

noise and number index.   An index for measuring noise pollution at or near airports, based on noise levels and the number of aircraft landing and taking off at a given point in time.

nominal fishing effort.   A simple measure of the total inputs applied to a fishery, usually measured as days fished or trawling hours. Nominal fishing effort is typically used in measures of catch per unit of effort

nominal value.   Value, in terms of currency, before considering the rate of inflation. See *real value*.

non-accelerating inflation rate of unemployment (NAIRU).   The supposed unemployment rate below which a reduction in unemployment would result in an acceleration in the inflation rate. See *consumer price index*.

FURTHER READING
Romer (2006).

non-Annex I countries.   Mainly developing countries that are not included in Annex I of the United Nations Framework Convention on Climate Change, and which were not obligated to control their levels of GHG emissions to 1990 levels by the year 2000. See *United Nations Framework Convention on Climate Change*.

non-Annex B countries.   Mainly developing countries not included in Annex B of the Kyoto Protocol, and thus were not obligated to control their level of emissions during the Protocol's 2008–12 commitment period. See *Kyoto Protocol*.

non-attainment area.   A term in the United States for an air quality control region that fails to meet the designated air quality standards.

non-biodegradable materials.   Materials that will not naturally decompose in the environment within a reasonable timeframe.

non-compliance.   Term that indicates a firm or individual is not currently meeting a defined environmental regulation or standard.

non-consumptive benefits.   Benefits from resources or the environment that do not arise from any direct extraction or harvesting. See *non-consumptive use*.

non-consumptive use.   Use of the environment or natural resources, such as photography, that does not detract from the benefits these provide.

non-cooperative game theory.   The branch of game theory that involves models of behavior where individual participants are assumed to choose actions that will only maximize their own well-being.

non-diversionary water uses.   Those uses, such as fishing, boating, or the provision of habitat, that require in-stream water flows.

non-exclusive.   An asset or resource which others cannot be prevented from using. See *exclusivity*.

non-extractive use value.   The consumptive value of a good *in situ*, in nature.

non-fuel minerals.   Mineral resources excluding those used for the production of energy (e.g., petroleum, coal and uranium) but including aluminum, copper, and other metals and ores.

non-governmental organizations (NGOs).   Registered not-for-profit organizations that may be funded by individual members, industry and/ or a country, but which are not directly controlled by the government. NGOs play an important role in providing linkages and technical assistance to developing countries and are influential in providing information and influencing policies on the environment in many countries.

non-jointness in inputs.   A production relationship where different outputs can be produced using separate production functions.

non-linearity.   Any relation or function where the dependent variable does not vary proportionally with changes in other variables of the model. For example, non-linearity may be a feature of the complex and dynamic nature of the climate system, and therefore a tipping point may exist whereby increases in temperature reach a level that could initiate catastrophic climate change. See *climate threshold*.

non-market costs.   Costs that cannot be directly valued through market-based measures. For example, destruction of a rare forest incurs market costs through loss of tourist revenues, and also non-market costs from loss of the existence value placed upon the forest by society. See *existence value*.

non-market valuation.   The attempt to place a monetary value on goods, especially public or environmental goods, that are not traded directly in the market. See *contingent valuation* and *hedonic analysis*.

non-material values.   Those human values, such as spiritual beliefs, which are not associated with material goods or money.

non-pecuniary rewards.   Benefits that are not financial in nature, such as goods or improved social status.

non-point source.   Source of pollution where emissions and discharges cannot be traced to a specific point or location. Agricultural run-off is a common source of non-point pollution. See *point source.*

non-price rationing.   Restricting the quantities available for consumption by means other than price, such as rationing.

non-renewable resource.   A natural resource that has a finite stock and cannot be renewed.

non-rivalness.   See *rivalrous.*

non-stationary.   See *stationary state.*

non-tariff barriers.   Measures or procedures, such as quotas, which are placed by countries on the importation of foreign goods, are not tariff-based, and that effectively restrict imports. Examples might include product safety or labeling requirements designed to either exclude foreign products or impose higher costs on them due to safety or environmental concerns.

non-timber forest products.   Natural products, other than timber, that can be harvested from a forest stand, usually without significant damage to standing trees. Examples include latex, syrup, vines, fibers, fruits and game.

non-timber resource values.   Values that emanate from forests but do not require the harvesting of forest stands. Non-timber resource values include values that come from the existence or enjoyment of biological diversity, recreation, fisheries, wildlife, water resources and water quality, minerals, and cultural and historical resources.

non-timber values.   Value of the forest separate from the commercial or timber value of the trees. For example, forests provide amenity values for recreational purposes that are non-timber values.

non-use benefits.   Benefits from resources or the environment that do not involve its direct use (consumptive and non-consumptive) such as the value placed on the existence of blue whales. See *non-use value*.

non-use value.   Also intrinsic value, existence value, or preservation value. The total willingness to pay to preserve, maintain, or possess a natural resource or environmental good in excess of the willingness to pay to use that good *in situ*. Non-use value = total economic value – use value.

FURTHER READING
Bergstrom and Randall (2010).

non-zero sum game.   A game or a set of transactions whereby all players or persons participating have the potential to be better off. See *zero sum game*.

no observed adverse effects levels (NOAELs).   Measures obtained from animal studies or epidemiological studies that are used as guidelines for determining maximum levels of exposure and risks associated with different exposure levels.

no-regrets policy.   Policy that generates net benefits whether or not the problem that it is designed to address happens. Often used in the context of climate change. For example, reducing greenhouse gas emissions by implementing measures to increase energy efficiency reduces energy expenditure whether or not climate change is a problem.

normal distribution.   A distribution where the mean, mode and median are identical, so that is symmetrical around the mean. If the frequencies of a normally distributed variable are drawn on the vertical axis and the values are given on the horizontal axis, the resulting curve or distribution would resemble the shape of a bell where the top of the bell is the value of the mean, mode and median. Many characteristics of a sufficiently large enough population (such as height of individuals) can be characterized by a normal distribution.

normal good.   A good where the demand varies directly with income. See *inferior good*.

Normalized Difference Vegetation Index (NDVI).   An index of ground cover used to characterize vegetative canopy, ground cover and biomass. NDVI has been used extensively to describe remotely sensed data, especially from AVHRR sources. See *AVHRR*.

normal profit.   The difference between total revenue and total cost, including all opportunity costs. See *opportunity cost* and *profit*.

normative analysis.   See *normative economics*.

normative economics.   A term used to describe economic analysis which focuses upon "what should be", such as recommendations for economic policy, and that involves some value judgements.

norms.   Commonly accepted forms of behavior within a society. Changing norms from "consumption and disposal" to "consumption, recycling, and reuse", can improve overall environmental quality.

North Atlantic Oscillation.   See *annular modes*.

North American Free Trade Agreement (NAFTA).   An agreement between the United States, Canada and Mexico that entered into force on 1 January 1994 to promote free trade among the three countries. The agreement includes various environmental clauses and commits the signatories to promoting sustainable development.

Northern Annular Mode.   See *annular modes*.

"No Surprises" clause.   Under the Endangered Species Act in the United States, a landowner who is correctly implementing a Habitat Conservation Plan (HCP) is guaranteed to be excluded from any additional burden imposed by changes in regulations or restrictions on the use of their property. See *Habitat Conservation Plan*.

not-for-profit organization (NPO).   A broad term for organizations that use any profits from their activities to further their goals. In the field of the environment, this term is often used to describe issue-based non-governmental organizations that rely on supporter and government donations.

NOx.   See *oxides of nitrogen*.

nuclear energy.   Electrical power generation from nuclear fission. Although nuclear energy creates dangerous radioactive waste products, it directly involves few greenhouse gas emissions and is therefore a relatively clean source of energy in the context of climate change. As a result, nuclear energy is often considered a short-term solution for mitigating global warming.

nuclear fission.   The splitting of uranium atoms by neutrons to produce lighter elements with at least two neutrons. Fission is the basis of the nuclear power industry.

nuclear fusion.   The fusing of two lighter elements (deuterium and tritium) to form hydrogen and a free neutron.

nuclear waste.   Radioactive material generated from nuclear tests and the operation of nuclear power stations. Low-level nuclear wastes can be disposed of relatively easily but highly radioactive material requires very costly methods of storage that, because of the long half-life of some waste, must last for thousands of years. See *radioactive waste.*

nuclear winter.   A hypothesis that a major nuclear conflict would spread so much particulate matter in the atmosphere that surface temperatures in higher latitudes would fall such that summer would be as cold as a normal winter.

null hypothesis.   The hypothesis to be tested in a statistical test. For example, if we are testing whether the means of two populations are equal, then the null hypothesis, denoted by H_0, is that the means are equal and the alternative hypothesis, denoted by H_1, is that the means are not equal. The hypothesis test allows us to either fail to reject or reject the null hypothesis. If we reject the null hypothesis then we accept the alternative hypothesis. See *hypothesis* and *alternative hypothesis.*

nutrient concentrations.   The availability of organic and inorganic material from both external and internal sources that can be utilized by plants and animals within an ecosystem.

nutrient cycle.   A cycle in which minerals and nutrients are transported via soils, water, air and organisms.

O

objective function. A function that is maximized or minimized in a model.

observer data. Data obtained by direct observation, usually for scientific purposes.

Occam's Razor. A principle of scientific thought that holds simplicity and parsimony in explanation superior to complexity, all else equal.

occult precipitation. A mist-like form of precipitation in which condensation forms directly on vegetation.

ocean acidification. The increasing acidity of the Earth's oceans as a result of higher atmospheric concentration of carbon dioxide. As part of the carbon cycle, carbon dioxide gas dissolves in the oceans. This process initiates chemical reactions that raise acidity levels, thus dissolving naturally occurring calcium carbonate essential to the body structure of organisms such as coral and mollusks. Scientists believe that current rates of ocean acidification are unprecedented and pose an existential threat to marine ecosystems.

ocean anoxia. See *anoxic event*.

ocean circulation. The large-scale movement of water within the Earth's oceans that is driven by wind and the thermohaline circulation. The interaction between the circulation of wind and ocean currents performs the continuous process of energy transfer underpinning the climate system. See *thermohaline circulation*.

ocean conveyor belt. See *ocean circulation*.

ocean heat uptake efficiency. A measure of the rate at which the Earth's oceans absorb surface temperature rise, an important parameter for determining climate sensitivity.

oceanography. The science and study of oceans, including both biotic and abiotic systems.

oceans. The contiguous bodies of salt water that encompass about 70 percent of the Earth's surface and provide a home to an abundance of marine life. Oceans also moderate temperatures along coastal areas.

ODA.   See *Official Development Assistance.*

OECD.   See *Organisation for Economic Co-operation and Economic Development.*

Official Development Assistance (ODA).   Flows of finance or resources to developing countries from official state agencies on concessional terms and directed towards economic development.

offsets.   Emissions reductions initiated to offset emissions in another location or country, thus resulting in zero net emissions across the two activities. See *carbon offsets scheme.*

offset provider.   A company managing or owning a carbon offset project, usually for profit.

off-shore oil drilling.   The extraction of petroleum and petroleum-related resources from the ocean floor by drilling.

off-site waste management.   The treatment of waste at a location different from that where it was generated.

offtake.   Harvest.

Ogallala.   An alluvial fan system in the United States that stretches from South Dakota to Texas in the south and includes parts of Wyoming, Colorado, Nebraska, Kansas and Oklahoma. The system is a very large and widely used aquifer where the withdrawal rate greatly exceeds the rate of regeneration.

oil sands/shale.   Petroleum deposits in which oil is bound into absorbent sands and sedimentary rock. Extraction is generally an expensive and a highly mechanical process that causes serious environmental degradation.

oil spills.   The accidental emission of petroleum or petroleum-related products into the environment.

old growth.   Term used to describe forest of a climax community that may contain live trees of various ages from seedlings to very old and dead trees, but which is dominated by mature trees. Depending upon the environment, an old-growth forest may take hundreds of years to develop and often has high non-timber and timber values.

Oligocene. Geological epoch 36.6 to 23.7 million years ago.

oligopoly. A market structure with a few sellers, each of whom can influence the market price.

oligotrophic. A body of fresh water that is low in nutrients and productive capacity.

OLS. See *ordinary least squares.*

omitted-variable bias. A bias that alters coefficient values and other statistical measures as a result of omitting an otherwise key explanatory variable from the estimation.

omnivorous. Animal that can feed on both plants and other animals, such as a bear.

one-shot game. Models used in game theory where the participants do not interact with one another again after they have completed the game. See *repeated games.*

one-tailed test. A statistical test that only uses one critical region when testing a null hypothesis. For example, $H_o : b_1 = 0$ is a one-tailed test if the alternative hypothesis is $b_1 < 0$. See *two-tailed test.*

on-farm efficiency. The relationship between the amount of water delivered to a farm and the amount of water actually used by crops.

open access. Term used to describe the lack of property rights over resources whereby no controls exist over the number of users or how much of the resource they may extract. Open access almost always leads to biological and economic overexploitation of natural resources.

open-access externalities. The imposition of costs upon others by individual actions, created by the open-access nature of the resource. See *externality* and *prisoner's dilemma.*

open-access resources. See *open access.*

open-cast mining.   Also known as strip mining, the removal of overlying surface material to access an underground mineral deposit.

open economy.   An economy where no restrictions exist in terms of imports and exports, investment and currency transfers. See *closed economy.*

open forest.   Forest that has tree crown cover of between 10 and 30 percent. This would also include savannas.

open loop.   A control system whereby the control variable does not respond to changes in the variables of the system.

open-loop recycling.   The reuse of residual material products in other productive activities. Examples include the use of ground-up glass in road surface materials and shredded rubber in flooring products. See *closed-loop recycling.*

open system.   A term used to define a physical system where there is an exchange of energy and mass or matter with the environment. Most natural environments are open systems. See *closed system.*

operable forest.   A forest, or portion of a forest, in which timber can be profitably harvested under current market conditions.

opportunity cost.   The cost of an economic activity in terms of the other goods and services, in the next best alternative, that must be forgone to be able to undertake the activity. For example, an opportunity cost associated with the installation of a pollution control device may be the forgone production that would otherwise have occurred if the firm had, instead, invested in additional capital equipment.

optimal control theory.   A set of mathematical methods and a theory developed to model dynamic systems that is closely related to the calculus of variations. The theory is used to solve problems where an objective function is maximized over time subject to set of constraints. A basic result of the theory is Pontryagin's maximum principle, which is concerned with the necessary conditions for optimality.

FURTHER READING
Chiang (2000) and Takayama (1985).

optimal rotation.   The age at which a stand of replanted trees is harvested that yields the highest expected value of the forest. The optimal rotation depends on whether or not the land will be replanted after harvesting, and whether or not non-timber values are accounted for. See *Faustmann rotation* and *Hartman rotation.*

optimal value function. The function formed by substituting the optimized choice variables of an optimization problem into the original objective function.

FURTHER READING
Simon and Blume (1994).

optimal yield. A term that is not well-defined, but is frequently used to refer to a level of harvest from a renewable resource that will maximize overall benefits to society.

optimization. Term used for the maximization or minimization of an objective function subject to a set of constraints. See *objective function.*

optimum yield. See *maximum sustainable yield.*

option price. The maximum amount an individual would pay to preserve the option to use a resource in the future even if the individual is unsure whether she will demand that resource in the future. The option price is equal to the expected use value plus the option value.

option value. A value, beyond an expected use value, derived from the willingness to pay for the option of using a resource at a future date.

FURTHER READING
Wiesbrod (1964).

orbital forcing. The external force applied to the Earth's climate system as a result of variations in its orbit. See *Milankovitch Theory.*

order. A classification of organisms by family. For example, the order of Primates includes the family of Hominidae which includes humans. See *family.*

order of a matrix. The number of rows and columns in a matrix. A matrix of order (2×3) has two rows and three columns.

order of integration. The number of times a non-stationary variable needs to be differenced in order to make the differenced variable stationary. The order of integration is written as I(n) where n represents the differences required to ensure stationarity. See *differencing* and *stationarity*.

ordinal utility.   A preference structure in which individuals are assumed to be able to rank different combinations of goods and services. Ordinal utility forms the basis for demand theory in economics.

ordinary least squares (OLS).   A statistical method of analysis for estimating parameters in a relationship between a dependent variable and one or more independent variables. OLS estimates the parameters by minimizing the sum of the squared deviations between the estimated relationship and the actual observations. See *regression analysis.*

Ordovician period.   Geological period of time from 505 to 438 million years ago.

organic compounds.   Material or substances containing carbon bound with hydrogen.

organic farming.   A set of agricultural practices in which no artificial fertilizers, hormones, pesticides, or herbicides are used. Organic farming is becoming increasing popular in developed countries as consumers are becoming more concerned over the potential health hazards of chemical residues in food.

organic fertilizer.   A mixture derived from organic constituents made up of various compounds that aid in plant growth.

organic matter.   Commonly defined as organic material found in soils that, in general, improves fertility and helps to retain moisture.

organic soil.   Soil containing greater than 30 percent organic matter.

organism.   A single animal or plant that is capable of functioning as an individual unit.

Organisation for Co-operation and Economic Development (OECD).   Formed in 1961, it is the so-called "Rich Man's Club" of countries. Its mandate is to foster economic growth, financial stability and high employment in member countries.

Organization of Petroleum Exporting Countries (OPEC). Formed in 1960, it consists of major petroleum exporting countries. Current members include: Algeria, Angola, Ecuador, Iran, Iraq, Kuwait, Libya, Nigeria, Qatar, Saudi Arabia, the United Arab Emirates and Venezuela. The oil price shocks of the 1970s were, in large measure, due to the collusion of OPEC members.

organochlorides. A large number of organic chemical compounds containing chlorine including PVCs, PCBs, CFCs and even DDT. Organochlorides are very persistent in the environment and accumulate in fatty tissue, and some are known carcinogens.

organochlorines. See *chlorinated hydrocarbons*.

orogenesis. Tectonic movements that build mountains. Orogenesis has been used by some scientists to explain the timing and existence of ice ages.

orthogenesis. The notion that evolution's progression over time leads to more complex organisms.

orthogonal. A statistical term that indicates independence of two variables.

otolith. Bone in the ear of fish that is frequently used for determining the age of fish.

Our Common Future. The 1987 final report of the World Commission on Environment and Development that is commonly known as the Brundtland Commission after its Chair, Gro-Harlem Brundtland. The report was a major catalyst for governments to begin implementing policies to encourage sustainable development.

FURTHER READING
World Commission on Environment and Development (1987).

outfall. A point source of water pollution.

outlier. An observation in a sample that differs markedly from other observations.

output controls. Controls on the individual output of resource users. Output controls, in contrast to input controls, attempt to change the incentives faced by resource users by giving them a long-term interest in the resource. See *Individual Transferable Quotas* and *input controls*.

output orientated measure of technical efficiency.   A measure of actual output relative to maximum possible output given a set of inputs.

outputs.   The end result of a production process whereby inputs, or factors of production, are combined to produce a good or asset. See *inputs*.

overcapacity.   Capacity in excess of that required for a desirable level of output. Overcapacity is a common problem in many fisheries due to a lack of well-defined property rights over the resource. See *capacity*.

overcapitalization.   Term used for a situation when capital utilization exceeds one such that too much capital is being used relative to the ideal level.

overexploitation.   Biological or economic over-use of a natural resource. A common reason for overexploitation is a lack of well-defined property rights.

FURTHER READING
Devlin and Grafton (1998).

overfishing.   Fishing at a level that exceeds what is desirable from either a biological or economic perspective.

overhead.   See *fixed costs*.

overland flow.   The flow of water outside of a water channel.

overlapping generations (OLG).   A class of economic models in which the consumption of every generation, all equally long-lived, occurs simultaneously with at least one other generation.

overparameterized.   A statistical model where the number of parameters to be estimated exceeds the number of observations.

overshoot.   1. A description for the surpassing of an environment's carrying capacity, which results in a population crash. 2. A term used in reference to climate change mitigation strategies that "overshoot" the targeted stabilization level for greenhouse gas concentration, before arriving at the target from above.

overstorey.   The layer of vegetation in a forest formed by the canopy of the dominant trees.

overvaluation.   Placing too high an economic value upon a certain activity or resource without taking into account all of the associated economic costs.

oxidation.   The physical process whereby a molecule, atom, or ion loses electrons.

oxides of nitrogen.   See *nitrogen oxides.*

oxygen.   A gaseous element that, after nitrogen, is the largest component of the Earth's atmosphere. Oxygen is a component of all organic compounds and is present in most organic processes.

oxygen cycle.   The biological cycling of oxygen through the atmosphere and the biosphere. The primary mechanisms of the cycle take place through the uptake and release of oxygen through green plants.

ozonation.   A method of treating water to destroy micro-organisms.

ozone (O_3).   A gas that, in the stratosphere, helps protect the Earth's surface from ultraviolet radiation from space. Near the surface, or the troposphere, ozone can be created from various sources including vehicle emissions and can be damaging to humans, plants and animals.

ozone depletion potential (ODP).   A value which represents the potential effect of a chemical on stratospheric ozone depletion. The value of 1 ODP is assigned to CFC-11 while some halons have an ODP as high as 10. Substitutes for CFCs, such as HCFCs, can have an ODP of less than 0.05.

ozone hole.   An area centered around Antarctica where, in the spring, a significant reduction in stratospheric ozone occurs due to chemical reactions between chlorofluorocarbons (CFCs), halons and ozone. This allows a greater amount of ultraviolet radiation to pass through the atmosphere, increasing the incidence of skin cancer over the affected areas. Although CFC emissions have been dramatically reduced, largely from the success of the Montreal Protocol, past emissions will remain in the atmosphere for a lengthy period.

FURTHER READING
Ahrens (2007).

ozone layer.   See *stratospheric ozone*.

P

Paasche index. A commonly used index that employs weights from the current period. For example, a Paasche price index may be defined as $p_1 q_1/p_0 q_1$ where q_1 is the quantity consumed in the current period, p_0 is the price of the good in the base period and p_1 is the price of the good in the current period. See *Laspeyres index*.

PAC. Acronym for pollution abatement and control.

Paleocene. Geological epoch of time from 66.4 to 57.8 million years ago.

paleoclimate record. The historical record of past climatic conditions.

paleoclimatology. The study of past climates using proxy records constructed from physical samples, such as ice cores and pollen sediment. Knowledge of earlier climate dynamics, when prevailing conditions were often very different to today, is an important tool for understanding the complexities of the modern climate system and predicting future climate change. See *ice-core samples*.

paleoecology. The study of past ecological systems using the fossil record and associated sedimentary strata.

Paleozoic. Geological era from almost 570 million years to 245 million years ago.

pampas. Grassland region that predominates in Argentina.

panchayat forests. Forests in the Himalayas which are managed by statutory village councils. In Nepal, panchayat forests are often administered jointly by several villages and are managed in the interest of the village communities.

FURTHER READING
Arnold and Campbell (1986).

pandemic. An epidemic that affects various locations simultaneously. The 1919 "Spanish Flu" pandemic simultaneously appeared at various places in the world and was responsible for up to 20 million deaths.

panel data.   Data for which observations are available over time, and for individual units of observation. See *cross-sectional data* and *time-series data*.

Pangaea.   The name given to a supercontinent that existed from 280 to 180 million years ago.

panspermia.   The hypothesis that life in the universe can be "seeded" from planet to planet via the transmission of micro-organisms in meteors or comets.

parallel evolution.   Evolution of geographically separated, but related, species that results in very similar organisms in terms of appearance or behavior.

parameter.   Fixed values used in a model or optimization problem. Models should be analyzed to determine the sensitivity of the results to values of the parameters.

parameterization.   In the context of climate change modeling, this term refers to using simplified parameters to represent an overly complex or small-scale process to facilitate a workable model.

parasites.   Organisms that live on or in other creatures. The relationship sustains the parasite but often weakens and sometimes kills the host. Many diseases are caused by parasites, including malaria, and have been extremely important in determining evolution in both plants and animals.

Pareto criterion.   A social welfare criterion that finds a change in the state of the world to be welfare improving if, for every individual, the expected utility that results from the change is at least equal to the status quo level of utility, and strictly greater for at least one individual. See *Pareto improvement*.

Pareto efficiency.   A situation where it is not possible to change the allocation of goods and services to make any individual better off without making someone else worse off.

Pareto improvement.   A change in the state of the world in which the expected utility of every individual in the new state is at least equal to the status quo level of utility, and strictly greater for at least one individual. See *Pareto criterion*.

Pareto optimality.   An allocation of resources or goods in which there exists no other feasible allocation that could improve the utility of one individual without diminishing the utility of at least one other individual.

parthogenesis.   Development of offspring from unfertilized eggs.

partial correlation.   The correlation between two variables accounting for the effects of another variable.

partial equilibrium analysis.   Economic analysis which only considers the effects of changes in one sector or market, or part of an economy, without considering the possible feedbacks in the rest of the economy.

participation rate.   The proportion of individuals in a defined population who are active in the paid workforce.

particulate loading.   The combined weight of particles of a given substance found in a given volume of air or water.

particulate matter.   Fine microscopic matter suspended in air or water. Common particulates in the air include smoke and dust. See *suspended solids*.

parts per billion (ppb).   Parts of a constituent substance present per billion parts of a material.

parts per million (ppm).   Parts of a constituent substance present per million parts of a material. Often used in the context of climate change to describe concentrations of greenhouse gas. See *greenhouse gas concentration*.

passive adaptive management.   A method of environmental management where managers choose a model and management strategy based on the best available evidence, and then assume that the model and strategies are correct.

passive gear.   Fishing gear that is stationary (such as a gill net). See *active gear.*

passive solar design.   The efficient use of solar energy through simple building design principles and materials, rather than mechanical conversion processes. For example, incorporating large south-facing windows in the Northern Hemisphere to capture light and heat.

passive use value.   See *non-use value*.

pasteurization.   A method of heat-treating food and drinks to destroy pathogens.

pathogenic organisms.   Organisms, usually micro-organisms, that can cause sickness or death, especially in humans.

pathogens.   Micro-organisms, including bacteria and viruses, that can be harmful to human health.

patterns of climate variability.   See *modes of climate variability*.

pay-as-you-earn (PAYE).   A common method of collecting income taxes where the estimated amount owed at the end of the financial year is automatically deducted out of the pay packet of wage and salary earners.

payback period.   The period of time that it takes for an investor to repay or cover the initial cost of the investment. The shorter the payback period, everything else equal, the more lucrative is the investment.

payment for ecosystem services (PES).   A practice whereby an economic incentive is offered in exchange for land management linked to a particular ecological service. The transaction typically consists of a voluntary action which results in a payment from the beneficiary to the provider for a well-defined ecological service, contingent on service delivery.

PCBs.   See *polychlorinated biphenyls*.

peaking power plants.   Power plants used to generate electricity for relatively short periods of time during periods of peak demand. These are usually low capital cost, gas-fired plants that incur high operating costs. See *base-load power plants*.

peaking scenario.   A scenario for global greenhouse gas concentrations where mitigation policies result in emissions reaching a maximum level and then declining over time, rather than stabilizing at the maximum.

peak-load pricing.   Pricing that differs depending upon the demand for the good, typically varying directly with the amount demanded. For example, the cost per unit of electricity could vary with the time of the day, with the price increasing at times of peak demand, such as from 17:00–21:00.

peak nitrogen.   A term for the moment when the amount of nitrogen available for extraction from natural gas begins to decline. Fertilizer manufacture principally involves nitrogen obtained from natural gas. Thus, natural gas stocks affect the volume of future fertilizer production, an important factor in agricultural productivity.

peak oil.   The point in time where the rate of global crude oil extraction reaches a maximum and afterwards terminally recedes. Many experts predict that the occurrence of the peak is imminent or has already occurred; however, its timing is highly contested.

peak phosphorous.   The stage at which the rate of phosphorous extraction, principally from mining rocks, declines. Similar to nitrogen, phosphorous is a key determinant of soil productivity for crops. Scientists believe that peak phosphorous has already occurred, increasing the challenge of feeding the world's growing population.

peak water.   A modification of the concept of peak oil, such that the rate of water production peaks and then declines terminally. There are three forms of water peaking, namely: peak renewable water beyond which renewable flows are consumed at a faster rate than they are replenished; peak nonrenewable water beyond which groundwater is consumed or contaminated faster than it is recharged; and peak ecological water beyond which the environmental degradation of consumption exceeds its economic benefits.

peat.   Partially decayed organic material, usually from moss, that accumulates in swampy environments. When dried it is highly combustible, due to a lack of exposure to oxygen during formation, and is consequently used as a fuel in many parts of the world. However, the burning of peat releases large amounts of carbon dioxide into the atmosphere. See *greenhouse effect*.

peat fires.   Large smoldering fires occurring in underground peat deposits, particularly Indonesia. It is estimated that a major incidence of Indonesian peat fires in 1997 emitted the equivalent of between 13 to 40 percent of global greenhouse gas emissions that year. The primary cause of peat fires in Indonesia is land-clearing by loggers and farmers.

peatland.   A major natural source of carbon dioxide and methane gas, two of the prominent greenhouse gases. The carbon material in peatland readily oxidizes to form carbon dioxide in the presence of air. Organic decomposition in active peatland, such as bogs, causes methane emissions. There is a concern that methane from peatland trapped beneath frozen permafrost in sub-Arctic regions will be released as global temperatures rise, causing a sudden upsurge in atmospheric greenhouse gas concentration. See *climate feedback* and *permafrost*.

pecuniary externality.   An externality the effects of which are fully accounted for in the price of goods and services such that no distortion exists to the efficient allocation of inputs or outputs. See *externality* and *technological externality*.

pedogenesis.   The formation of soils through natural processes.

pedosphere.   The portion of the Earth consisting of the three layers of soil above the lithosphere.

pelagic species.   Fish species commonly found near the surface, such as herring or tuna.

pellagra.   A debilitating disease, associated with poverty, that arises from a deficiency of the B-group vitamin, niacin.

per capita emissions.   A measure of national greenhouse gas emissions per head of population.

per capita income.   The total income earned in a region or country in a 12-month period divided by the total number of people in the population.

perennial.   Plant whose life cycle goes beyond two years.

perfect competition.   A hypothetical benchmark in economics where there exists an identical product; many buyers and sellers, each of whom cannot influence the market price; market participants who all have perfect information about relevant markets; and no barriers to entry and exit of firms.

perfluorocarbons.   Greenhouse gases (GHGs) that can have a global warming potential up to 9000 times greater than carbon dioxide. Perfluorocarbons are a byproduct of aluminum smelting and have been used as a substitute for chlorofluorocarbons (CFCs) in semiconductor manufacture.

performance-based standard.   A standard for pollution control that specifies the amount of permissible discharge, but does not dictate the type of technology required to achieve the standard.

performance bonds.   See *environmental bonds*.

periodic.   A repetitive cycle or change in a variable.

periodic harvesting.   Spasmodic harvesting or exploitation of a renewable resource. Periodic harvesting can allow the resource to recover or increase in size so that exploitation can occur at high levels of abundance, thus reducing harvesting costs.

permaculture.   A broad term referring to sustainable agriculture systems or techniques integrating natural cycles and enhancement of the surrounding environment.

permafrost.   Ground that has been frozen for a period of at least two years. See *peatland*.

permanence.   An essential requirement of a carbon sequestration activity, such as an avoided deforestation project within the framework of the REDD mechanism. As such projects are designed to offset emissions elsewhere, if the reductions are not permanent then the net result is a rise in greenhouse gas emissions. In order to ensure this condition is upheld, carbon credit producing activities are subject to ongoing monitoring.

permanent income hypothesis.   The notion that people's consumption is governed by their permanent income, and thus individuals tend to smooth out temporary increases or declines in consumption by saving or borrowing.

permeability.   Measure of how well a substance permits liquids or gases to pass through it.

Permian.   Geological time period lasting from 286 to 245 million years and which ended with a mass extinction of species.

permit.   A legal document authorizing an activity. See *emission permit* and *pollution permit*.

persistence.   Term used to describe substances or processes that linger in the environment. See *persistent organic pollutants*.

persistent organic pollutants (POPs). A variety of pollutants which persist for a long time in the environment and can bio-accumulate, such as DDT or PCBs.

persistent organochlorines (POs). See *chlorinated hydrocarbons*.

PERT. Acronym for pilot emission reduction trading. See *pollution permits*.

perverse incentives. Incentives that lead to undesirable behaviors. For example, in a fishery with no property rights a perverse incentive exists to catch before someone else any fish that can be profitably harvested, regardless of the long-term consequences for the resource.

pest. A living organism that is harmful to human interests, such as crops, forests and fishing. Some examples include: the cane toad in Australia and the pine beetle in North America.

pesticide resistance. Genetic development, through mutations in pests, that leads to the use of pesticides becoming less effective.

pesticides. Agents, usually chemical, that deter or kill animals (especially insects) that might damage crops or other biological resources.

petrol. See *gasoline*.

petroleum. Also referred to as liquid crude oil. A naturally occurring flammable liquid consisting of a mixture of hydrocarbons that is the base ingredient of conventional gasoline. Petroleum was formed millions of years ago by the deposition of partially decayed organic matter on the ocean floor during anoxic events. See *fossil fuels*.

pH. Abbreviation for potential of hydrogen. A logarithmic measure of the concentration of hydrogen ions in a solution that is used to determine whether a liquid is acidic, neutral or base (alkaline). A pH measure less than 7 is acidic, 7 is neutral and greater than 7 is basic (alkaline). See *logarithmic scale*.

phase diagram. Diagrams which are used to convey the dynamic properties of differential and difference equations.

FURTHER READING
Shone (2002) and Simon and Blume (1994).

phenology. Study of the effect on the behavior of animals of periodic changes in the weather and climate.

phenotype. The observable appearance and characteristics of an organism, which are a function of its genetic make-up and the environment.

pheromone. Substances secreted by animals and used to influence the behavior of members of their own species, often in terms of attracting mates. Pheromones are widely used in integrated pest management as an alternative to pesticides.

Phillips curve. An apparent inverse relationship between the rate of increase in wages and the unemployment rate that was first observed by Alban Phillips in a paper published in 1958. The implication of the so-called Phillips curve is that there exists a trade-off between the rate of increase in the price level and unemployment.

phosphates. Sources of phosphorus that occur naturally but which were also found in detergents in the past. Excessive phosphate run-off from fertilizer applications can be a major water pollutant and can accelerate eutrophication of lakes and rivers.

phosphorus. A non-metallic solid element and one of the essential biological nutrients, the availability of which is a determinant of biological productivity. Phosphorus is a structural component of nucleic acids, phospholipids, and bone in living organisms.

phosphorus cycle. The movement of phosphorus through soil and water in the environment. See *phosphorus*.

photochemical. Term used for chemical reactions that arise in the presence of sunlight. See *photochemical smog*.

photochemical smog. Air pollution is neither smoke nor fog, but is formed by the interaction of hydrocarbons and nitrogen oxides in the presence of sunlight. In large cities with many automobiles, and where air circulation is limited, photochemical smog can be a major problem and health hazard.

photoperiodism. The seasonal response of plants and animals to changes in the hours of sunlight.

photosynthesis. A fundamental process of life on Earth whereby plants absorb solar radiation and convert it into chemical energy with chlorophyll and, in the process, take in carbon dioxide and water, store carbohydrates and release oxygen. See *carbon sequestration*.

photovoltaic effect.   The physical process fundamental to the working of solar panels. Light from the sun contacts with a material, usually silicon based, generating energy from electron movement that can then be used as electricity. See *solar panel*.

phylogeny.   The evolutionary history of a species, family, order, or class.

phylum.   (Plural: phyla.) Classification of organisms by classes. See *class*.

physical capital.   A broad term used in economics for non-labor inputs to production, such as machinery and infrastructure.

physical water scarcity.   A lack of water caused primarily by environmental factors, such as low precipitation and high evaporation rates. However, physical scarcity may also occur as a result of non-environmental factors, such as upstream diversion or dams, and soil degradation from unsustainable farming practices. See *drought*, *economic water scarcity*, and *water scarcity*.

physics.   The study of the general principles that govern the physical world.

physiocrats.   A school of thought in eighteenth-century France that promoted free trade and believed that the basis of economic wealth was agriculture.

phytoplankton.   Plankton capable of photosynthesis. See *plankton* and *zooplankton*.

Pigouvian subsidy.   See *pollution subsidy*.

Pigouvian tax.   Tax named after the economist A.C. Pigou. A Pigouvian tax (commonly a per unit charge) is a tax applied to emissions or pollution and is designed to internalize an environmental externality.

FURTHER READING
Anderson (1994).

pioneer community.   First group of plants and animals to first arrive or colonize a site following a major disturbance, such as a forest fire or volcanic eruption.

piosphere.   Degraded soils and vegetation surrounding a watering hole or well, usually found in the arid and semi-arid regions of Africa.

placebo effect.   An apparent beneficial effect from patients in a clinical trial who receive a non-beneficial substance, but which they believe might be a treatment.

plankton.   The aggregate term for a large group of floating, drifting, and sometimes mobile plants and animals found in the water column. See *phytoplankton* and *zooplankton*.

planned economy.   An economy where the state or central government, rather than market forces, makes most of the production and pricing decisions. The Soviet Union was a planned economy.

planned adaptation.   See *anticipatory adaptation.*

planetary albedo.   The aggregate ability of the Earth surface and atmosphere to reflect solar radiation. Loss of ice and snow cover reduces planetary albedo, thereby amplifying global warming.

plastics.   Materials commonly formed from petroleum and which have a wide variety of uses. Their production, use and disposal can cause significant environmental problems.

plate tectonics.   The study of planetary changes in the Earth's crust based on the understanding that the continents "float" on molten rock beneath them.

play.   A geographic area containing oil or gas deposits, generally with similar geological characteristics.

Pleistocene. Geological epoch from around 1.6 million years to 10 000 years before the present, and which coincided with several ice ages.

Pliocene.   A geological epoch period from 5.3 million to 1.6 million years ago.

plutonium.   A metal formed by nuclear fission of the isotope uranium 238, having a half-life of over 24 000 years.

PM$_{10}$.   Particulate matter of a diameter less than 10 micrometers.

PM$_{2.5}$.   Particulate matter of a diameter less than 2.5 micrometers.

point source.   An identifiable point or location of emissions and discharges, such as a factory stack. See *non-point source.*

Poisson distribution.   Named after the French mathematician, Siméon Poisson, it is widely used for predicting the number occurrences of a random event over time and spatially. The cumulative distribution function is defined by
$$\Pr(X = x) = (e^{-D} * D^x)/x!$$
where D is the mean of the distribution and $x!$ is x factorial.

policy.   An action or plan adopted by a government, group, or individual that is intended to achieve a specific outcome. For example, greenhouse gas reduction policy. See *measure.*

policy decentralization/centralization.   The process of making policy where decision making either takes place within one centralized location within an organization (which can either be a geographical location or a hierarchical level within the organization in the case of policy centralization), or is dispersed throughout an organization (either geographically or among different management levels, or across different decision makers in the case of policy decentralization).

policy failures.   Government policies or laws that induce environmental problems, such as subsidies that encourage exploitation of vulnerable resources.

political economy.   Term formerly used for the discipline of economics and which today is used to describe the relationship between economic outcomes and policies.

pollen analysis.   A dating technique that involves collecting pollen from geological features such as river sediment, and determining age from the type and abundance of pollen.

pollutant standards index (PSI).   A US index of air quality that combines readings of concentrations of particulate matter, carbon monoxide, ground-level ozone, nitrogen oxide and sulfur dioxide. An index of greater than 100 is considered unhealthy, while an index in excess of 300 is viewed as hazardous.

polluter-pays principle.   The idea that those persons responsible for the immediate cause of pollution should also be responsible for paying the costs to society of pollution. One method of implementing the principle is to impose Pigouvian taxes on polluters. See *Pigouvian tax.*

pollution.   Any substance that, at the concentrations found in the environment, can have a deleterious effect on the health of individuals, species, habitats and ecosystems.

pollution abatement.   The reduction in the level of pollution by the use of end-of-pipe treatment and cleaner technologies, or by the reduction in the output of the product that results in pollution.

pollution allowances.   See *pollution permits.*

pollution charge.   See *Pigouvian tax.*

pollution credits.   See *pollution permits.*

pollution havens.   Countries with low environmental standards, or a lack of regulations, that could provide a low-cost place for high-polluting industries to operate.

pollution permits.   A method of pollution control where firms are assigned a right to emit a given pollutant in a set period of time, and are allowed to trade this right with other firms in the program. Pollution permits with heterogeneous firms allow firms with high costs of pollution abatement to meet their pollution targets or obligations by purchasing pollution permits from low abatement cost firms. Trading of pollution permits has the potential to reduce the overall cost of achieving reductions in the total level of pollution from an industry. See *emissions trading scheme.*

pollution subsidy (per unit).   A per-unit payment for each unit of pollution reduced from a defined level.

polychlorinated biphenyls (PCBs).   Organochlorines that have been widely used as an electrical insulator, and in such products as brake linings and paints. PCBs accumulate in fatty tissue and are a known carcinogen in animals, as well as being linked to various health complaints in humans.

polychlorinated dibenzodioxins (PCDDs).   Commonly called chlorinated dioxins, these are a type of organochlorine. PCDDs bioaccumulate in the environment and are highly toxic.

polymer.   A compound formed by a repeated sequence of molecules, such as plastics.

polymorphism.   The existence of more than two distinct physical forms of a species, that is not a result of sexual dimorphism. See *dimorphism.*

polynomial equation.   An equation of the form
$$a_0 + a_1x + a_2x^2 + \ldots + a_nx^n = 0$$
where $a_0 \ldots a_n$ are real numbers and a_n does not equal 0. The values of x that solve a polynomial equation are called its roots.

polynomial function.   A function of the form
$$P(x) = a_0 + a_1x + a_2x^2 + \ldots + a_nx^n$$
where $a_0 \ldots a_n$ are real numbers. See *polynomial equation.*

polythene.   See *polyethylene.*

polyethylene.   A type of plastic that can be recycled and, if disposed of by combustion, generates carbon dioxide and water.

polyvinylchloride (PVC).   Polyvinylchloride is a widely used plastic that is made from the highly toxic vinylchloride monomer. PVC does not break down easily and, when burned, can produce toxic compounds.

Pongidae.   Family of species that includes chimpanzees and gorillas.

Pontryagin principle.   See *maximum principle.*

pooled cross-sectional data.   Data that combines cross-sectional observations on a sample of a population over time.

population.   The total number of individual members of a species in a defined geographical area at a particular point in time.

population density.   Population per unit of area.

population dynamics.   Fluctuations in the population size of a species over time.

population ecology.   The study of the interaction of species, genera, or other taxa at the level of populations.

population explosion.   1. The rapid increase in the human population on Earth. 2. Any massive increase in a population of a species.

population growth.   Net increase in the number of individuals in a defined group over a defined period of time.

population pyramid.   Hierarchical ranking of the population of different organisms found in a specific area (grouped by either species or common characteristics), under which those organisms found in a specific rank or row will prey or consume those organisms found in the rank immediately below. These rankings are distinguished by the fact that the biomass and range of the organisms found in the different rows decreases with increasing rank.

Porter hypothesis.   The hypothesis that environmental regulations can stimulate innovation and increase productivity by providing incentives for individuals and firms to develop new production processes.

portfolio choice.   Term used for the theory and practice of choosing the appropriate set of investments under the assumption that, for a given expected return, investors prefer less risk. See *risk* and *capital asset pricing model.*

positive analysis.   See *positive economics.*

positive economics.   A term used to describe economic analysis that views the world as it is (rather than as it should be) and, supposedly, is free of value judgements. See *normative economics.*

positive externality.   An action which generates benefits for others. These benefits are not taken into account in the decision making of the persons responsible for the action. See *externality* and *negative externality.*

positive feedback.   See *feedback.*

posterior distribution.   Distribution obtained after an event or following the receipt of new information. See *Bayes theorem.*

post hoc.   Latin term meaning "after this" or "following this" in a temporal sense.

post-material values.   Those human values associated with a society in which material goods are considered unimportant and the primary focus is placed upon human equality, spirituality, and sustainability.

post-SRES emissions projections.   Projections for greenhouse gas emissions without future mitigation that have been constructed after the release of the widely used SRES scenarios in 2000. Many recent estimates have significantly upgraded emissions growth predictions, rendering climate change mitigation a more urgent concern. See *Special Report on Emissions Scenarios*.

FURTHER READING
Garnaut et al. (2008).

potential energy.   The energy which exists in an object by virtue of its position, this being the product of its height above the ground, its mass and its acceleration due to gravity.

potential GDP.   The total amount of goods and services in an economy that can be produced in a year if all factors of production are fully utilized.

potentially responsible parties (PRPs).   Term used in determining liability for US hazardous waste sites. It includes all those persons and firms that may, at one time, have been involved with either the operation or transportation or treatment or disposal of substances at the site.

potential Pareto improvement (PPI).   A change, perhaps hypothetical, in the state of the world in which aggregate benefits of the change outweigh the costs. It requires that the potential beneficiaries of the change be able to compensate potential losers so that the expected utility of each individual could, potentially, be at least equal to the status quo levels, and greater than status quo levels for at least one individual.

poverty.   A relative and absolute term for the least well-off in a society. Poverty is a denial of choices and opportunities and the inability to participate effectively in the basic aspects of society. Among other impacts, it has important environmental consequences, as short-term day-to-day survival may prevent the poor from taking a long-term perspective or making investments to improve the quality of the environment.

FURTHER READING
Perkins, Radelet and Lindauer (2006).

power.   In terms of electricity generation and consumption, power is the rate at which energy is converted. The standard measurement of power is watts (W).

power plants.   Facilities used to produce electricity. Many base-load power plants, such as coal-fired electricity plants, produce greenhouse gas emissions. As a result, significant research and funding have been invested in renewable technologies, leading to the increasing usage of energy sources such as solar and wind. See *base-load power plants*.

ppb.   See *parts per billion*.

ppm.   See *parts per million*.

prairie.   Grasslands found in semi-arid areas of North America, particularly in the Midwest United States and the provinces of Alberta, Saskatchewan and Manitoba in Canada.

Precambrian.   A period encompassing the time when the Earth was formed 4550 million years ago to 570 million years before the present.

precautionary principle.   A principle designed to lead to the sustainable use of the environment and natural resources. It states that actions should be taken to avoid damages in the future even when there exists a lack of full scientific certainty over the potential consequences in the future. The precautionary principle provides a justification for acting today to address possible problems associated with climate change due to anthropogenic emissions of greenhouses gases (GHGs).

FURTHER READING
Perrings (1991).

precipitation.   The rain, snow, hail and water vapor that reaches the Earth's surface.

precision.   A measure of the variability of an estimator. Precision can usually be increased (variability reduced) the larger is the sample size.

predation.   The act or process by which predators harvest their prey.

predator.   An animal that hunts other animals, its prey, for food.

predator–prey model.   A model that explicitly considers the interactions between predators and prey in predicting their population dynamics. See *Lotka–Volterra model.*

FURTHER READING
Grafton and Silve-Echenique (1997).

predetermined variable.   Lagged endogenous variables in a simultaneous equation model.

preferences.   The underlying wants and needs of individuals.

preferential use doctrine.   A bureaucratic means of water allocation in which end uses are given hierarchical rankings that determine the priority of access to water.

pre-industrial.   Adjective used to describe conditions prior to the Industrial Revolution that began in the eighteenth century. Often used for reference to greenhouse gas concentrations or temperatures prior to the initiation of major anthropogenic climate change.

present value.   The value of returns in the future in terms of current dollars. See *current value (dollars).*

present value of benefits (PVB).   The discounted numeric value of future or past benefits.

present value of costs (PVC).   The discounted numeric value of future or past costs.

preservation.   The practice of environmental management that seeks to ensure that the environment is left, as much as possible, undisturbed and free from anthropogenic influences. See *conservation.*

preservation costs.   The costs associated with preserving or maintaining a natural area or wildlife population.

pressure (atmospheric).   Commonly measured in millibars and frequently used to predict changes in the weather.

pretreatment standards.   Environmental standards on the quality of effluent entering treatment facilities. It may include, for example, restrictions on how much of a certain compound can enter a municipal sewer system.

prey.   Animals that are hunted by and fed on by predators.

price ceiling.   Regulations that impose a maximum price that can be charged for a good or service.

price discrimination.   The selling of the same good at different prices for reasons other than differences in costs.

price elasticity.   The percentage change in the quantity demanded (if demand price elasticity) or supplied (if supply price elasticity) following a 1 percent change in the price. A good is price elastic if a 1 percent change in the price results in a more than 1 percent change in the amount demanded or supplied, and is inelastic if a 1 percent price change leads to a less than 1 percent change in the amount demanded or supplied.

price floor.   See *price support*.

price signal.   A change in the price of a good or service that signals to consumers or producers the need to alter their consumption or production behavior, respectively. For example, an increase in electricity prices arising from greater demand by consumers signals to suppliers the need to invest in more capacity.

price support.   The payment by the government or a state enterprise of a minimum price for a good irrespective of the market price. Price supports most commonly exist for agricultural products.

primary effects.   The direct impacts of different policy actions or interventions on the users of a resource stemming from changes in the resource created by the policy actions (e.g., improved water quality in a reservoir may lead to increased tourism). See *secondary effects*.

primary energy.   A term for the primary sources of all energy that comprises three categories of energy: fossil fuels, renewable energy and nuclear power. See *energy conversion chain*.

primary forest.   A mature forest that has developed with little direct intervention or human activity.

primary fuels.   Sources of energy that can be used in, more or less, the same state as they occur naturally, such as natural gas, coal and oil (and its derivatives).

primary pollutant.   Polluting compounds and materials that are directly discharged into the environment.

primary producer.   1. In economics, an agent or firm that grows or harvests a resource and converts it into an economic commodity. 2. Green plants that convert sunlight into organic materials.

primary sewage treatment.   The first stage, and most basic type, of wastewater treatment in which physical operations are used to remove contaminants. The most common method is to use a grid chamber and sedimentation tank to remove sediments. See *secondary sewage treatment*.

primates.   An order of mammals that includes humans, the great apes, lemurs and monkeys.

principal components analysis.   A statistical approach to help overcome problems of multicollinearity that tries to extract much of the variation present in the independent variables in a regression. See *multicollinearity*.

FURTHER READING
Greene (2003).

prions.   Proteins that have been linked to some diseases, such as "mad cow disease". See *bovine spongiform encephalopathy*.

prior appropriation doctrine.   The notion that the first user of a resource, such as a river or a stream, has the uninterrupted right to the resource equal to the amount initially withdrawn or used by the individual.

prisoner's dilemma.   In game theory, a model in which two prisoners (agents) are given the opportunity to confess to a crime or deny a crime with the understanding that the outcome of this decision depends on the decision of the other prisoner. If both deny the crime, both face a lenient penalty, if one confesses and the other denies then the one who confesses faces the most lenient penalty while the one who denies faces the most severe penalty, and if both confess both face intermediate penalties. The prisoner's dilemma demonstrates why individuals may choose to engage in non-cooperative behavior, even when cooperative behavior has a higher combined payoff, and is used to describe more general games and real-world situations where the same incentives hold.

private good. A good where the owner can exclude others from using it, its use being rivalrous in the sense that it precludes others from using it simultaneously. An example of a private good is an automobile.

private property rights. See *private rights.*

private rights. A property-rights regime where ownership of goods and resources rests with individuals or firms.

FURTHER READING:
Grafton, Squires and Fox (2000).

privatization. The divestiture of ownership or control of assets, or the right to use resources, by the state to individuals or groups of individuals. Privatization has been undertaken to increase revenues for the state by the sale of state assets, and also to increase economic efficiency.

probability. The chance or likelihood of an event occurring, defined by $Pr(x)$ where x is the event. If the event will occur with certainty, then $Pr(x) = 1$ and if the event will never occur, then $Pr(x) = 0$.

probability density function (PDF). See *probability distribution function.*

probability distribution function. Sometimes called the probability density function, it is the function that assigns a probability of occurrence for each possible value of a discrete random variable. Thus for a random variable X, the probability function can be used to obtain $Pr(X = x)$ where x is any value that the random variable can obtain.

probit analysis. A method of regression analysis with a limited dependent variable (usually binary) that is commonly used for discrete choice analysis. The probit model assumes a functional form for the regression in which $y_i^* = \beta x + u_i$, where y_i^* is unobservable, but can be deduced by assuming that an observable dummy, y, equals 1 if $y_i^* > 0$ and 0 otherwise. In the probit analysis, u_i is a random variable with a cumulative normal distribution.

process and production methods (PPMs). Term used to describe how products are produced, which may have important trade implications.

The term generally refers to non-product-related PPMs where there is very little difference in final products, despite differences in the production process. Trade law under the World Trade Organization (WTO) prohibits discrimination of products on the basis of non-product-related PPMs. Discrimination on product-related PPMs, where there exist significant differences in the final products, such as for genetically modified potatoes versus organically grown potatoes, is permitted under trade law.

FURTHER READING
Gallagher (2008).

producer surplus.   Returns over and above those necessary to compensate a firm to supply a given quantity of a good or service.

production costs.   The direct costs associated with the production or manufacture of a good, asset, or output. See *variable costs*.

production frontier.   A production function that shows maximum output for any given inputs, typically used in reference to either a deterministic (without randomness) or stochastic (with random effects) production frontier.

production function.   A function where the dependent variable is output, and the independent variables are inputs in the production process.

production possibilities frontier (PPF).   Also known as the production possibilities curve, this represents a mapping of the maximum quantities of the different possible combinations of goods and services that can be produced in an economy in a given period of time, using the current resource base and technology.

productivity.   A measure of total output divided by a measure of the inputs to produce the output. A common productivity measure is labor productivity which divides the value of output by the number of worker hours or days taken to produce it.

productivity–diversity relationships.   The relationship between the number of species in an ecosystem and the rate of growth of one or more resources in that system that is subject to biological competition.

FURTHER READING
Abrams (1995).

productivity index. A measure of productivity usually indexed to a base period or a particular observation or numéraire.

product substitution. The degree to which one product will replace or be replaced in consumption by another product following a change in its price.

profit. The difference between total revenue and total cost. See *normal profit*.

profit decomposition. A decomposition of changes in relative profit performance into differences in output prices and input price adjusted, for example, in the case of a fishery, for their importance in the catch (outputs), fishing effort (variable inputs) and fixed inputs, such as vessel size.

profit function. A function in terms of output and input prices that defines the maximum level of profit for a firm at any given set of prices.

profit maximization. A common economic assumption about firm behavior that supposes firms optimize by maximizing their profits (that is, revenues less all costs of production).

Programme for the Endorsement of Certification Schemes (PECS). A global non-profit organization engaged in certifying sustainable forest management, particularly for small forest owners.

progressive tax. A tax which takes proportionately (as well as absolutely) more from people with higher incomes than with lower incomes. Most income taxes are, to some extent, progressive in that the highest marginal rates of taxation are imposed at higher levels of income.

projection. A forecast of a future value or circumstance based on a wide range of information, including current and past conditions, as well as informed assumptions. See *emissions projection*.

prokaryote. Organisms that do not have nuclei in their cells. See *eukaryote*.

property rights. The *de jure* and de facto rights of individuals or groups of individuals to use or benefit from resources, goods and services.

FURTHER READING
Devlin and Grafton (1998).

property-rights regime.   A way of describing or grouping property rights with similar characteristics. A commonly observed property regime is private rights.

prospect theory.   Psychological theory that states, among other things, that individuals value losses from a given state or status quo much more than they do gains of equivalent magnitude.

prospective case study.   A case study of a site, system or process that is initiated prior to some externally introduced change. This enables data collection that can be used to assess the consequences of this change. For example, water quality sampling of a river prior to upstream construction of a timber mill. See *retrospective case study*.

protandry.   The change in gender of an organism during its life from a male to a female.

protective/recovery measures.   Measures to protect a species, as required under the United States Endangered Species Act. Protective measures include restrictions on activities that may harm a species, while recovery measures give the United States Federal Government the authority to improve the condition of a listed species.

proteoid.   Plants characterized by thickened leaves.

Proterozoic era.   A geological era from 2500 to 570 million years ago.

protozoa.   Single-celled organisms that are responsible for a variety of human diseases, including leishmaniasis.

proved recoverable reserves.   The quantity of a resource's proved reserves that is recoverable using existing technology under current and expected future economic conditions.

proxy.   An entity that is used in place of another for a specific purpose. For example, estimates of climatic conditions prior to the advent of modern instruments are approximated by proxy measurements obtained from the atmospheric content of ice-core samples.

psychrosphere.   Part of the sea or ocean where the temperature is low and varies little over the seasons.

public bad. A good that generates disutility or harm to users and is both non-exclusive (it is not possible to prevent others from being harmed by the good) and non-rival (the use of the good by any one user does not diminish the harm incurred by others).

public choice theory. Economic theory concerned with how the government and the state actually make decisions. It considers political self-interest to be an explanatory factor.

public debt. See *national debt*.

public good. A good that generates benefits to users and is both non-exclusive (it is not possible to prevent others from using it) and non-rival (the use by any other user does not diminish any benefits from using the good by others). An example of a public good is a lighthouse.

public investment. The aggregate expenditures by a government on public goods.

public water system. A term for infrastructure such as pipes and dams that are used to provide water for human consumption at a community, city, or national level.

pulp. A product formed from wood chips that is used for making paper. Some of the chemical processes to derive pulp can generate considerable water and air pollution.

pulp and paper mill. A factory that converts wood chips into pulp and paper. Pulp and paper mills can be major point sources of both air and water pollution.

pulpwood. Wood that, upon harvesting, is directly transformed into pulp and paper.

pulse fishing. See *periodic harvesting*.

punctuated equilibrium. The notion that the evolution of life is characterized by long periods of relative stability punctuated by comparatively short periods of major evolutionary change.

purchasing power parity (PPP). A method of comparing the value of currencies across countries using a suitable basket of goods and services.

For example, if the basket of goods costs $A400.00 in Australia and $US200.00 in the United States then the purchasing power parity is $A2 for $US1. The PPP may differ from the actual exchange rate.

pure compensation. Term used to describe population growth models where declines in the population are compensated for by increases in the population growth rate.

pure competition. See *perfect competition.*

pure existence value. The value ascribed to a good solely for its existence and divorced from any possible *in situ* use value it might have.

pure rate of time preference. The rate at which future utility is discounted. It is a measure of future value relative to the present. A higher rate of time preference coincides with a higher discount rate. See *discount rate.*

purse-seiner. A fishing vessel that employs a net that encircles schools of fish and is then closed from below, trapping the fish. This method of fishing is employed to catch tuna, herring and other pelagic species, as well as salmonids.

PVC. See *polyvinylchloride.*

pyrolysis. See *biochar.*

pyrotoxins. Toxins produced by fire.

Q

QED. See *quod erat demonstrandum*.

quadrats/quadrant sampling. Sampling or measurements done within square-sided plots (e.g., looking at the density of different plant species by the amount of area covered by the plants).

quality of life. A broad and holistic measure of the well-being of individuals within a society that includes consideration of health and health care, education, crime, environmental quality, and so on. See *genuine progress index*.

quality of title. A property-right characteristic that refers to the *de jure* rights of the holder.

quantified emissions limitations and reductions objectives (QELROs). A term used in the text of formal UNFCCC documents, such as the Kyoto Protocol, to refer to emissions targets. See *binding emissions targets*.

quasi-option value. The welfare gain that comes from delaying a decision to use a natural resource when there are uncertainties about the payoff of alternative uses of the resource, and when at least one of the uses is irreversible.

FURTHER READING
Arrow and Fisher (1974).

quasi-rent. Earnings or returns to a fixed factor of production that is temporarily fixed in supply. Quasi-rent is the surplus over and above the variable costs of production but may still be less than the fixed costs and, thus, positive quasi-rents do not necessarily imply positive profits.

Quaternary period. Geological period of time from 1.6 million years ago to the present.

quid pro quo. Latin term for a mutual agreement to exchange. The term generally refers to the implicit understanding that a favor granted from one person to another should be reciprocated.

quod erat demonstrandum (QED). Latin term meaning "that which was to be proved". The term is commonly used at the conclusion of a proof.

quota.   A regulated amount of a good that can be produced, imported or exported by a firm or country. In the case of fisheries, a regulated amount of catch. A total quota in a fishery refers to the total allowable catch allowed in a given season while an individual quota refers to the maximum permissible harvest by an individual or vessel.

R

race to fish. Term used in so-called derby fisheries where the total harvest is controlled, but the individual harvests of fishers are not. Consequently, fishers compete or "race to fish" among themselves to harvest as much as they profitably can before the fishing season closes. Such behaviour is typical in input controlled fisheries.

race to the bottom. The simultaneous dismantling of regulation so as to gain a competitive advantage. For instance, a country may relax pollution laws to attract business from those countries with stricter regulations.

rad. Unit of measure of absorbed ionizing radiation.

radiation. A broad term for energy that is transmitted in the form of particles or waves. Examples include: infrared and ultraviolet radiation.

radiation absorption. The accumulation of radioactive energy within animal tissues and plant matter from exposure to radioactive sources.

radiation budget. An accounting procedure that tracks the amount of radiation that reaches and leaves the Earth's surface. The budget consists of sub-budgets that account for reflection and absorption by the Earth's atmosphere. Changes in the proportion of greenhouse gases in the atmosphere alter the Earth's radiation budget, thus causing global warming. See *greenhouse effect*.

radiation feedback. See *radiative forcing*.

radiation reflection. See *albedo*.

radiation sickness. Sickness from exposure to radiation involving nausea and diarrhea and, in the long run, increased rate of cancers.

radiative forcing. Also known as climate forcing, this refers to the perturbation to the energy balance of the Earth's atmosphere following a change in the system, such as an increased concentration of a greenhouse gas. This would be described as positive radiative forcing, as it involves an increase in mean surface temperatures.

FURTHER READING
Archer and Rahmstorf (2010).

radical. A chemical compound in which there is a free electron, thus making the compound highly reactive.

radioactive waste. Waste material from nuclear fission. High-level radioactive waste, such as spent fuel rods in nuclear reactors, present a particular environmental hazard. Various proposals have been made for the long-term storage of high-level radioactive waste in various countries, but have been met with widespread concern regarding the potential leakage of the material.

radiocarbon dating. The process of determining the age of organic material based upon the decay of the unstable carbon-14 isotope. This form of dating provides accurate information for substances up to approximately 60 000 years old.

radiometric age determination. See *radiocarbon dating*.

radionuclides. Isotopes of different atoms, from both natural and man-made sources, that decay and release ionizing radiation that can be hazardous to human health.

radon. Naturally occurring gas produced from radium, which can seep into the basements of buildings and, thus, create a health hazard.

radon-222. A radioactive gas that can pose a severe risk to human health, and that has been identified as a hazardous pollutant.

rainfall deficit. A period of below average rainfall or drought.

rainforest. Mainly evergreen forests where there is usually an abundance of precipitation throughout the year. Tropical rainforests are very important ecosystems that include many of the world's terrestrial species. Rainforests also exist in subtropical areas, such as southern Queensland in Australia, and in temperate areas, such as the United States, the Canadian Pacific west coast and the west coast of the South Island of New Zealand. The Amazon Rainforest sequesters a large proportion of anthropogenic carbon dioxide emissions annually.

FURTHER READING
Birmingham, Dick and Moritz (2005).

rain shadow. Dry or arid area on the leeward side of mountains.

Ramsey growth model. Model of economic growth developed by the English economist Frank Ramsey, which includes the specification of consumer behavior.

FURTHER READING
Barro and Sala-i-Martin (2004).

random effects. Statistical term that refers to a model where group effects on individual units of observations arise through a group specific disturbance term. See *fixed effects.*

FURTHER READING
Greene (2003).

random parameters model. Econometric models in which the values of the parameters do not stay fixed, but change due to one of several reasons including a change in regimes or processes, or because the parameters depend upon other variables, or because the parameters are simply random in nature.

random sampling. A sampling technique where each observation is drawn at random from a population. Each draw gives an observation that is independent of previous draws such that no one observation is more likely to be selected than another.

random shocks. Unexpected and non-deterministic events, the occurrence of which cannot be predicted in advance.

random utility model (RUM). A model of consumer choice in which the consumer is assumed to choose among goods or activities in order to maximize their "random utility", and which consists of a deterministic core plus a good-specific random term.

FURTHER READING
Alberini and Kahn (2006).

random variable. A variable that can take on a certain value with a defined probability. A discrete random variable is a variable that assumes only a particular finite or countable set of values (such as the number of boats in a fishery). A continuous random variable can assume any value within a certain range.

random walk. The movement or motion of a variable over time that moves in discrete jumps from one point in time to another at a fixed probability. A variable that follows a random walk may have an apparent tendency to move in a particular direction, especially if there exists an absorbing barrier that may prevent it moving past a particular point.

range. The difference between the highest and lowest value in a sample of a population.

range management. The regulation of grazing and other activities on open land.

rapid climate change. See *abrupt climate change*.

rapid rural appraisal (RRA). A multidisciplinary assessment method in which teams of researchers gather and present information concerning site-specific farming conditions, techniques, and outcomes. RRA is intended to be an informal tool to evaluate farming impacts on the environment, and also a means for fostering co-operation between local people and researchers.

raster. The representation of spatial or geographic data in a grid.

rate of return. The net returns from an enterprise as a ratio of the total amount invested.

rational agents. Individuals or other economic agents who act in a manner that will maximize their own well-being.

rational expectations. The hypothesis that market participants use all the available information to predict the future and that, on average, they are correct, barring non-systematic or random errors. Rational expectations led to a questioning of the ability of governments to systematically trick or fool consumers or firms through their policies. See *adaptive expectations* and *Lucas critique*.

FURTHER READING
Sheffrin (1996).

Rawlsian justice. The notion expounded by John Rawls that decisions should be taken so as to consider the least fortunate or well-off. A Rawlsian social welfare function maximizes the minimum level of utility and has been applied in studies on intergenerational equity.

FURTHER READING
Rawls (1971).

reaction curves.   Also known as reaction functions. The mathematical expression of a participant's set of preferred actions within a game that may, or may not, be based on the actions of other participants.

real GDP.   The value of goods and services produced in an economy in a year measured in terms of real prices, or relative to a base-year index.

real interest rate.   The current interest rate less the current rate of inflation.

real numbers.   All numbers consisting of integers (. . . −1, 0, 1, 2 . . .), ratios of integers and functions of integers.

real option pricing.   The method under which the opportunities for a firm to invest in real assets (new technology, plants, or equipment) are valued.

real value.   The inflation-free value of a good or service, adjusted to represent the exchange value of that good or service for a representative basket of market goods.

real wage.   A measure of earnings by workers divided by an index of the price level.

rebound effect.   An increase in the demand for energy or water attributable to the lower per unit cost of energy that arises from the adoption of energy or water efficient technologies or goods.

receptor locations.   Specific sites that incur the environmental impacts of pollution.

recession.   Term used to describe a small decline or slowdown in the level of economic activity. In some countries, three successive quarter (three-monthly) declines in the gross domestic product is officially defined as a recession.

recharge rate.   Rate at which water enters an aquifer.

reconstruction.   The use of indicators, such as measurements of atmospheric composition from ice-core samples, to determine the dynamics of past climate.

reciprocity.   The mutual assistance provided between individuals and between groups of individuals. Reciprocity helps explain the success of human beings, it being an important component of the collaboration and co-ordination that exists between individuals in a society.

RECLAIM.   See *Regional Clear Air Incentives Market.*

reclamation.   The process of restoring an environment to a usable state. Reclamation commonly occurs after mining.

recoverable reserves.   That portion of the mineral body or deposit still unmined but which can be economically extracted.

recovery plans.   Plans or strategies designed to restore an ecosystem or environment to a desirable state, such as after a mining operation.

recreational fisheries.   See *sport fisheries.*

recreational site choice models.   Models of consumer choice that reflect the consumer's decision to visit one or more multiple recreational destinations. Recreational site choice models provide a means to estimate the non-market values of outdoor recreation sites, as well as environmental amenities at these sites.

recreation benefits.   The benefits from an aspect of the environment associated with recreation. Recreation benefits (such as walking in the forest) are often non-market values but can be estimated through a variety of techniques such as the contingent valuation methods and travel cost methods.

recruitment.   The weight or number of a species that, upon reaching a certain level of maturity, is considered to be part of the biomass or resource stock.

recruitment overfishing.   Fishing that reduces the number of reproductive fish in a population that, in turn, reduces future recruitment and the future population.

recyclables.   Any material that, after its use, can be used as an input in a production process, such as glass and steel.

recycling.   Process of collecting and reprocessing used materials. Curbside or blue-box recycling programs are widespread in many rich countries.

recycling ratio.   The amount of a good recovered through the recycling process expressed as a percentage of the amount entering the process.

REDD.   See *Reducing Emissions from Deforestation and Forest Degradation.*

red data species.   Rare and endangered species listed by the International Union for the Conservation of Nature. See *International Union for the Conservation of Nature.*

red queen hypothesis.   The notion that species need to continuously evolve to survive.

red tide.   An algal bloom that occurs in coastal waters and which can produce toxins that can be fatal to marine species and humans who eat contaminated animals, especially shellfish. The phenomenon can occur naturally but is also enhanced by the run-off of phosphates into the sea. See *algal bloom.*

reduced form.   A simultaneous equation model rewritten such that the endogenous variables are a function of the undetermined parameters, the disturbance term and lagged endogenous variables.

redox (oxidation-reduction) reaction.   A chemical reaction where electrons are transferred between two substances or molecules (i.e., where reduction and oxidation occur in a complementary fashion).

Reducing Emissions from Deforestation and Forest Degradation (REDD).   A formal policy structure providing incentives for reducing greenhouse gas emissions from deforestation and forest degradation in developing countries. REDD is a carbon offset scheme that reduces the cost of mitigating developed country emissions through finance of avoided emissions in developing countries. A typical REDD project would involve a government or private entity within a developing country preserving forest that would otherwise be logged, thus producing a carbon credit that can be purchased to offset emissions in a developed country. In order to ensure that such projects achieve real and permanent emissions reductions, the whole process is subject to substantial measurement, reporting, and verification procedures. In international climate policy, REDD exists as a complement to binding national emissions reductions, rather than a substitute. The related concept of REDD+ also incorporates: sustainable management, conservation, and reforestation. See *carbon offset scheme*.

reductio ad absurdum.   Latin term meaning "a reduction to the absurd". For example, if one glass of red wine at dinner helps reduce the risk of heart disease then the belief that drinking a cask of red wine every day eliminates the chance of a heart attack altogether is *reductio ad absurdum*.

reduction.   The physical process whereby a molecule, atom, or ion gains electrons.

reference case.   See *baseline*.

reference concentration (RfC).   Term used by the US Environmental Protection Agency that estimates the level below which daily exposure is unlikely to lead to deleterious health effects in the human population.

referendum.   A question format in contingent valuation surveys in which the respondent is asked to respond "yes" or "no" to a hypothetical trade-off between some amount of an environmental good or service and something else of value (especially money). The referendum format is considered by many to be superior to other question formats, but may require strong assumptions about the functional form of utility in order to derive from it willingness to pay measures.

FURTHER READING
McConnell (1990).

reflectivity.   See *albedo*.

reforestation.   The restocking or regeneration of trees and forest after the land has previously been harvested for trees or converted to another use, such as agriculture. Reforestation can be both a natural process (from seeds) or an anthropogenic process (from planting seedlings). As trees sequester large amounts of carbon dioxide, reforestation is one of many actions contributing to climate change mitigation. See *carbon sequestration*.

refund system.   See *deposit-refund system*.

refuse-derived fuel.   Energy derived from the combustion of waste material.

reg.   An arid landscape characterized by large areas of gravel.

regeneration.   The regrowth of a tree crop through natural or artificial means. See *reforestation*.

Regional Clean Air Incentives Market (RECLAIM).   An air pollution permit system in southern California that controls emissions of nitrogen oxides and sulfur dioxides.

regression.   The tendency of offspring of parents who are above or below the mean in a particular characteristic (such as height) to revert or regress to a value closer to the mean of the population.

regression analysis.   A statistical analysis whereby a relationship between a dependent variable against one or more independent variables is estimated using data. For example, the relationship between the rate of savings (dependent variable) and the level of income (independent variable) can be estimated in a savings–income model. See *ordinary least squares (OLS)*.

regression to the mean.   See *regression*.

regressive tax.   Tax that imposes a proportionately greater burden on low-income households than on high-income households. Some green taxes, such as a carbon tax, may be regressive.

relative abundance.   The frequency with which different organisms or resources are found in different ecosystems.

relative humidity.   The proportion of water in the air relative to the maximum that the air is capable of holding at the current temperature.

relative risk.   The ratio of the incidence of a complaint, malady or illness to those exposed to a risk factor, relative to the incidence among those not exposed to the risk factor.

relative sea level.   See *local mean sea level change*.

remedial actions.   Environmental clean-up activities designed to return a degraded site or resource to its original condition.

remediation.   Process and practices associated with the restoration of habitats and ecosystems to achieve desired outcomes. Remediation is commonly practiced at waste sites and old industrial sites so that the land can be used for other purposes.

remote sensing.   A method of obtaining physical data from a distance. Many satellites use remote sensing to provide data on various phenomena, including weather.

removal unit (RMU).   Tradeable emissions credits generated through projects that reduce greenhouse gas emissions from land-use change, such as afforestation.

renewable energy.   Energy derived from a source that is undepleted through use or withdrawal, such as wind, solar, geothermal and tidal. A major focus of climate change mitigation is investment, and research and development in renewable energy technologies. See *climate change mitigation* and *renewable technology*.

renewable resource.   A natural resource, like a fishery, that has the potential to be self-perpetuating.

renewable technology.   Machinery, equipment or tools that utilize or facilitate the use of renewable energy sources. For example, solar panels and wind turbines. Often used more broadly in reference to technologies contributing to greenhouse gas mitigation.

rent.   See *economic rent*.

rent dissipation.   Term given for the over-use of scarce resources in the harvesting of a resource or the production of a good. Rent dissipation is common in open-access fisheries where the "race to fish" encourages the over-use of inputs and, thus, an overall reduction in returns from the resource.

rent gradient.   Term for the decline in the rent paid for land the further its distance from a central core.

rent seeking.   Term used to refer to activities and lobbying by individuals or firms to acquire scarce property rights that can be assigned by a central authority. The term was first used with respect to the activities of importers who wished to have access to import licenses or quotas.

repeated games.   Games in which the participants interact more than once and which allow for the possibility of participants modifying their strategy in response to observing the choices of others.

replacement rate.   The fertility rate that is necessary to maintain a population of fixed size such that is neither growing nor shrinking in size.

replicate.   The process by which some organisms duplicate themselves. Bacteria reproduce by replication.

reproduce.   The process by which organisms produce offspring.

reproducible capital.   Capital, usually human-made, that can be reproduced exactly; for example, machinery and buildings.

reproductive success.   The proportion of offspring that are alive after a given period of time.

required compensation locus.   Government policies to manage risk often impose costs on certain individuals, including changes in net income and direct utility costs. The required compensation locus (RCL) may be an infinite number of pairs of state-dependent compensations that would exactly, and at the least cost, return the individual to his or her original, status quo, expected utility.

FURTHER READING
Hazilla and Kopp (1990).

res communes.   Latin term for community property rights. See *community rights.*

research and development (R&D).   The systematic process of advancing human knowledge and available technologies. For example, research and development for climate change mitigation could investigate potential efficiency improvements for solar power generation.

reserves-to-use ratio.   The volume of reserves of a non-renewable resource divided by the current consumption of the resource. The lower the ratio, the sooner the resource will, theoretically, be depleted.

reservoir.   1. A large or additional supply of a resource, such as a water reservoir. 2. A carbon reservoir.

residence time.   The period of time that molecules remain in a physical region or system. See *lifetime*.

residual.   1. The amount of materials, energy or radioactivity introduced into the process of production or consumption that remains unused, and is returned to the environment. 2. A statistical term for the difference between the estimated value of a variable and its observed value.

residual claimant.   The party entitled to whatever profits or returns are left after all factors of production, such as labor, have been paid.

resilience.   A term used to describe the ability of natural populations to return to their previous state following a shock. Usually it is measured by the speed or time it takes to return to its former state.

FURTHER READING:
Holling (1973).

res nullius.   Latin term that refers to natural resources which have free or open access and are owned and managed by no one. Many fisheries in the high seas outside the jurisdiction of coastal states may be described as res nullius. See *commons, tragedy of*.

resource allocation.   The distribution of goods and assets among individuals.

resource base.   The natural resources, including land and water, available to sustain a community or society.

resource recovery maximization.   The attempt to recycle and reutilize as much of the waste generated by industrial activity as possible before discarding it into refuse facilities.

resource rent.   Loosely defined as the net return after all the economic costs (including opportunity costs) of production have been deducted from total revenues.

resource rent tax.   A tax imposed on the extraction of non-renewable resources based upon the cash flow from the operations.

FURTHER READING
Garnaut and Ross (1975).

respiration.   The process by which organisms convert food into energy and, in the process, release carbon dioxide and water.

response accuracy.   Error that arises in a survey where the respondent inadvertently gives an incorrect or inaccurate answer.

response bias.   The bias that arises when the information supplied by a respondent differs from reality. Various techniques can be used, including asking the same question in different ways, to reduce response bias.

response rate.   The ratio of people who respond to a survey to the number of people asked to participate.

response time.   A term for the length of time taken for the Earth's climate system to respond to internal or external forcing and reach a new equilibrium state.

res privatae.   Latin term for private property rights. See *private rights.*

res publicae.   Latin term for state property rights. See *state rights.*

restoid.   Vegetation having a low, bushy form similar to that of the genus *Restio* found in Africa.

restoration costs.   Expenditures associated with helping to restore an environment to a previous state.

retroactive liability.   A legal rule that makes it possible for individuals, firms or even governments to be held legally responsible for their actions in the past even if those actions were, at the time, legal. See *liability.*

retrofitting.   The conversion of existing equipment or materials to meet pollution or emission standards. Retrofitting is often an inexpensive option for mitigating pollution because it does not prevent the primary activity causing the pollution. See *scrubbers.*

retrospective case study.   A case study of a site, system or process that is initiated after some externally introduced change. Such studies are usually initiated as result of observed and often negative changes to environmental variables. For example, poor water quality levels that are suspected to be the result of pollution from an upstream timber mill. See *prospective case study.*

retrovirus. Viruses, such as the virus responsible for AIDS, that can mutate quickly and use the cells of their host to make DNA copies of their RNA.

return flows. The water returned to a water system via seepage and runoff.

returns to scale. The proportional change in output from a given proportional change in all inputs. Constant returns to scale implies that a doubling of *all* inputs would exactly double output. Decreasing (increasing) returns to scale implies that a doubling of all inputs results in less (more) than a doubling of output.

revealed preference. An approach developed by the Nobel laureate in economics, Paul Samuelson, that is used to identify the underlying preferences, and thus demands of individuals, based upon the choices each reveals in their consumption. Thus, if a bundle of goods A is bought when another bundle B is available and affordable, then bundle A is revealed to be preferred to bundle B.

reverse osmosis. A cost-effective method of filtering water by which untreated water is forced at pressure through a membrane that traps compounds, parasites and other undesirable matter.

reversibility. See *irreversibility*.

Rhine Convention. Name given to a series of conventions on the use of the Rhine river. The convention signed in Bonn in 1976 regulated the discharge of specified chemicals into the river by France, West Germany, Switzerland, The Netherlands and Luxembourg.

Rhine spill. A 1986 discharge of pesticides into the Rhine river that originated from a chemical factory in Basle, Switzerland.

rhizomatous perennials. Herbaceous plants characterized by the presence of a central enlarged root which produces the stem and leaves.

ribonucleic acid (RNA). An organic code used in the transfer of genetic information.

Ricardian rent. Also known as differential rent, this is the economic surplus that arises from the diminishing returns from using successive units of a factor of production, given that other factors of production are fixed in supply. Those units of the variable factor that generate a value of their marginal product that exceeds their cost obtain a Ricardian rent.

Richter scale. A logarithmic scale for measuring the intensity of earthquakes and tremors from seismic waves. The scale ranges from 1 to 9 where a unit increase in the scale corresponds to a 10-fold increase in the intensity of the seismic waves. See *logarithmic scale.*

Ricker curve. See *stock-recruitment model.*

rickets. A disease that affects bone formation, arising from a lack of vitamin D due to insufficient sunlight or from an inability to absorb calcium. The disease is a symptom of poverty and was widespread in northern countries when children were employed inside factories or mines.

rill erosion. Erosion caused by small rivulets and streams.

Rio plus 10. The follow-up conference to the United Nations on Environment and Development that was held in Rio de Janeiro in 1992.

Rio plus 20. The 2012 United Nations conference concerning sustainable development. The conference follows on from the Rio plus 10 in 2002, and the initial Rio Summit.

Rio Summit. See *United Nations Conference on Environment and Development.*

riparian. A term meaning near or adjacent to streams, rivers, lakes or wetlands.

riparian rights. The property rights associated with the access to, and use of, streams, lakes and rivers.

riparian water doctrine. A system of property rights in which access to water (from a natural water course) requires ownership, or other tenure, of land adjacent to that water.

rising sea level. See *sea level rise.*

risk. Term for a set of circumstance where probabilities exist for all potential events or outcomes. See *uncertainty.*

risk assessment.   Evaluation of the risks posed to health and/or the environment from a given activity.

risk averse.   Term used to describe an individual who prefers a certain outcome with a given return than an uncertain outcome with the same expected return.

risk–benefit analysis.   A type of risk assessment whereby the risks or costs associated with an activity are compared to the potential benefits. For example, breast screening through mammograms imposes some health risks but the benefits in early detection of cancers may, for older women, outweigh the possible health costs.

risk loving.   Term used to describe an individual who will prefer an uncertain outcome with a given expected return, over a certain outcome with the same return.

risk management.   An approach to decision making that explicitly considers the risks associated with decisions and investments.

risk neutral.   Term used to describe an individual who is indifferent concerning a certain outcome with a given return and an uncertain outcome with the same expected return.

risk premium.   The extra return required to induce an individual to hold a riskier asset (returns that exhibit greater variability).

rivalrous.   Term used to describe a resource or asset where if someone increases his or her use, less is available to others. All common-pool resources, such as a fishery, are rivalrous in use. See *common-pool resource*.

river basin.   The geographic area that contains all the drainage waters of a specific river.

river blindness.   See *schistosomiasis*.

river corridor.   The land bordering a river channel that is often important in terms of wildlife habitat and species diversity.

RNA.   See *ribonucleic acid*.

roadless areas.   1. Areas in which no roads have been built. 2. A term used in the United States to designate National Forest land that falls under special management rules.

robustness. See *resilience*.

Roentgen. A unit measure of absorbed radiation from X- or gamma rays that was named after Wilhelm Roentgen.

roots. 1. The part of plants that enables them to uptake nutrients and water from soils. 2. The values that solve a polynomial equation. See *polynomial equation*.

rotation. The process of cutting and replanting a stand of timber.

rotation age. The age at which a forest is scheduled to be harvested. Rotation age can refer to the age of biological maturity, biologically maximal productivity, or the economically optimal age for harvest. See *optimal rotation*.

Rotterdam Convention. The Rotterdam Convention on the Prior Informed Consent Procedures for Certain Hazardous Chemicals and Pesticides in International Trade provides a set of procedures to ensure that importing countries are adequately informed about potentially hazardous materials that are internationally traded.

roundwood. Merchantable sections of tree stems free from branches, and sometimes free from bark, that may include logs, posts, pilings and bolts.

royalty. A payment made to the owner of a natural resource for exploiting the resource, and that is based on the value of the resource extracted.

Roy's identity. A mathematical result used to obtain the demands for a good from an indirect utility function, and that is the negative of the ratio of the partial derivative of the indirect utility function with respect to the good's price to the partial derivative of the indirect utility function with respect to the consumer's income.

r-squared (R^2). See *coefficient of determination*.

r-strategists. Animals that ensure their reproductive success by producing large numbers of offspring, such as insects.

rule of capture.   The norm, and often the law, that allows a person who captured or harvested a resource to enjoy its benefits. Rule of capture applies in many renewable resources, such as fisheries.

runaway climate change.   The theoretical state where human activities cause sufficient disruption to the climate system that, by virtue of feedback loops, no amount of greenhouse gas mitigation could halt damaging and rapid climate change. See *climate feedback*.

run-off.   The water that is transported across the land and deposited into water bodies such as rivers and lakes. Run-off from urban centers may also include a variety of contaminants, including oils and particulates.

run-of-river dams.   Hydroelectricity projects which have either no reservoir or only a small storage capacity. These dams utilize the natural river flow to rotate turbines, rather than a highly managed flow from a large reservoir. Run-of-river dams generally cause less harm to the environment as compared to large-scale reservoir dams. See *hydropower*.

runs test.   Test used to detect serial correlation that is often used in time-series analysis. The test compares the sequences of positive and negative residuals with their expected number.

S

saddle-point. A point on a multivariable function that is neither a maximum nor a minimum, but is where the value of the function does not change with small changes in the values of the choice variables.

"Safe Harbor" policy. An agreement in the United States under which non-federal landowners can restore, enhance, or maintain habitats for endangered species with the assurance that additional restrictions will not be placed upon them if endangered species either appear on the land or populations increase.

safe minimum standard. An approach to resource management that tries to reduce the probability of potentially high costs or losses by ensuring the level of exploitation is such that the population does not fall below a safe minimum standard. See *precautionary principle.*

Sahara Desert. The desert region in north Africa that stretches from the Atlantic ocean to the Red Sea.

Sahel. An arid area in western Africa, south of the Sahara Desert, which encompasses (from west to east) Senegal, Mauretania, Mali, Burkina Faso and Chad.

sales tax. Tax levied as a percentage of the retail sales price of a good or service.

salinity. The proportion of dissolved salts in water. Water with salinity levels in excess of 500 parts per million (ppm) is unsuitable for domestic use and salinity levels in excess of 2000 ppm makes the water unsuitable for irrigation purposes. Excess salinity in soils can be a major agricultural problem, especially in poorly drained soils. A major cause of excess salinity in soils is irrigation that can contribute to the rise in the ground-water table and lead to excessive evaporation of water on the surface, such that salt levels increase over time.

salinization. The accumulation of mineral salts in soil over time from the evaporation of water. A problem commonly found in irrigated soils in more arid conditions.

salmonids. Collective term for salmon species, including trout.

salt marshes.   Marshy areas influenced by ocean waters.

saltwater intrusion.   The movement of seawater into underground fresh-water aquifers. This phenomenon may be caused by pumping of water from the aquifers, reduced runoff, or sea level rise.

salvage value.   The value of an asset at the end of a particular period of time, usually its estimated working life.

sample.   A selection of members or elements from a larger population. A great deal of statistical analysis involves tests on a sample to infer the characteristics of the population.

sample bias.   Bias or differences that arise between the mean of a sample and the population from which it is drawn because the sample is not representative of the population.

sample selection bias.   The error that arises from making inferences about a population from a sample or samples that are not representative of the population.

sampling error.   The difference between the estimator obtained from a sample and the estimator obtained from a population.

Samuelson conditions for public goods.   A condition that ensures a Pareto-efficient outcome in the provision of a pure public good that requires the sum of the marginal rates of substitution for the private and public goods (absolute value of the slope of the consumers' indifference curves) for all individuals to equal the marginal rate of transformation (absolute value of the slope of the production possibilities frontier).

FURTHER READING
Cornes and Sandler (1996).

sanitary landfill.   A municipal or industrial waste or dump that meets defined standards to avoid leaching and contamination of soils and groundwater.

Sankey diagram.   An energy flow diagram originally used by the Irish engineer M.H.P.R. Sankey in the nineteenth century, and commonly used today to provide a snapshot of the magnitude of energy flows (measured in joules) relating to one or more energy conversion chains. See *energy conversion chain*.

saprophyte. Organisms that live off the dead, decaying or non-living parts of other organisms.

satellite accounts. Financial accounts, especially in systems of national accounts, that consider economic and physical stocks and flows (e.g., the value of oil reserves) that relate to traditional accounts (e.g., gross domestic product), but are not considered part of the system of core accounts.

satisficing. Term coined by the economist Herbert Simon, denoting that individuals and firms seek to achieve certain levels or goals rather than the highest possible values. For example, firms may wish to achieve a certain profit level that is less than the maximum possible, after which they may seek to achieve other goals.

savanna. A biome characterized by little rainfall and grasslands with scattered trees, that is found in tropical and subtropical regions.

sawntimber. Wood that has been harvested and sawn either into lumber or into large pieces of timber with square sides.

Say's law. A law, attributed to the nineteenth-century economist Jean-Baptiste Say, commonly defined as the idea that supply creates its own demand.

scale efficiency. Production at the level of output, and with an amount of plant or capital equipment, that would maximize profits if the firm were economically efficient. See *allocative efficiency, economic efficiency*, and *technical efficiency*.

scarcity. A measure of the relative or absolute abundance of a good or resource. For example, a widely used scarcity measure is the ratio of proven reserves to current annual use of a non-renewable resource – a measure that may actually increase over time due to the discovery of new reserves or a decline in consumption due to rising prices.

scarcity pricing. A dynamic method of pricing where the supply is a temporarily or permanently fixed supply and the current price is set in a way that accounts for the effect of current consumption on the future supply. The method has been proposed for urban water pricing whereby the volumetric price rises as the amount of water available in storages declines.

scarcity rent. See *resource rent*.

scenario. A hypothesized set of plausible conditions formulated from assumptions about fundamental variables and the links between them. In the context of climate change, constructed scenarios for future emissions and climate are important analytical tools used by scientists and policy makers. See *Special Report on Emissions Scenarios.*

Schaefer–Gordon model. See *Gordon–Schaefer model.*

Schaefer model. A surplus production model of a fishery where the growth in the biomass is defined by a logistic growth function less the harvest of fish. See *logistic growth* and *surplus production models.*

schistosomiasis. A sometimes fatal group of water-borne diseases caused by parasitic worms. The disease is contracted from larvae that can penetrate the skin of persons in water and is common in Africa, Asia and South America. A common form of the disease, bilharziasis, is endemic in most parts of Africa and has increased with the building of dams and the creation of large bodies of water, which have increased the population of snails that serve as hosts in a stage of the life-cycle of the parasites.

scientific method. An approach to understanding and explaining the world around us that requires all ideas to be questioned and tested. The method requires that ideas, theories or hypotheses be formed in a dynamic interaction of observations, tests and experiments.

Scitovsky compensation criterion. One social state is preferred to another when, using the Kaldor criterion, it always leads to that state being preferred and when the Kaldor criterion does not lead to contradictory results. This is equivalent to the Pareto criterion if there are no restrictions on reaching any social state from any other social state.

sclerophylous. The condition by which evergreen shrubs and trees survive prolonged droughts by producing tough, thick-skinned leaves.

scrubbers. 1. Attachments to industrial vents that chemically filter emissions and prevent harmful pollutants from being released into the atmosphere. A common application is the reduction of sulfur dioxide emissions from coal combustion using limestone, or lime-rich solutions. 2. Proposed geoengineering technology that processes air away from emissions sources, in order to remove carbon dioxide and mitigate climate change.

scurvy. A disease due to a dietary deficiency of vitamin C. Its symptoms include painful joints and membranes, and skin eruptions.

sea ice. Ice that is formed from frozen seawater. See *Arctic shrinkage*.

sea level. A broad term for measurement of the level of the Earth's oceans. See *local mean sea level change* and *mean sea level*.

sea level rise. Term used for the absolute rise of global sea level across the Earth's oceans. Also used in reference to localized increases such as those resulting from tidal movements. Sea levels have risen and fallen across the paleoclimate record in sync with the formation and melting of ice during glacial cycles, as well as changes in ocean temperature. In a modern context, rising sea levels are one of the most readily apparent and damaging impacts of anthropogenic global warming. Scientists believe that current rates may result in sea levels rising up to 2 meters by 2100, rendering many coastal cities, communities, and even countries, such as the Maldives, underwater. Many planned climate change adaptation measures aim to deal with this problem, including the construction of barriers and reduction in the erosion of coastal dunes through revegetation. See *glacier retreat* and *thermal expansion*.

sea wall. See *levee*.

sea surface temperature (SST). A temperature measurement obtained from the top few meters of the ocean. Sea *global surface temperature*.

seasonal allocation. A yield or allocation of water to holders of entitlements for a given cropping season.

secondary effects. The indirect impacts created by changes in the use of a resource on those activities that do not directly involve the resource itself. For example, improved water quality in a reservoir may lead to improved tourism which in turn may increase the demand for seasonal help. The increased demand for help would be a secondary effect. See *primary effects*.

secondary forest. A regenerated forest that has grown after harvesting, or following a major disturbance, such as a forest fire.

secondary pollutants. Pollutants that are formed from subsequent chemical reactions once primary pollutants are discharged into the environment (such as the production of ozone by the interaction between nitrogen oxides and volatile organic compounds in the atmosphere).

secondary sewage treatment.   A second step in waste treatment in which micro-organisms consume most of the solids remaining after primary treatment. A common method of secondary treatment is to use a trickling filter that sprays wastewater over rock beds coated with micro-organisms. See *primary sewage treatment*.

second best.   See *theory of the second best.*

second fundamental theorem of welfare economics.   An economic theorem that states that any Pareto-efficient outcome can be obtained from a perfectly competitive market, under a certain set of conditions and with appropriate lump sum transfers or initial wealth distribution between individuals. See *Pareto efficiency*.

second growth.   Forest that has regenerated following a harvest, or a natural disaster, such as a forest fire, and which has yet to reach maturity.

second law efficiency.   A measurement of the energy efficiency of an apparatus, which is expressed as the ratio of the amount of input energy to the minimum amount of energy theoretically needed to perform a required task. This measurement corresponds to the commonly used definition of energy efficiency used in climate change policy. See *energy efficiency* and *first law efficiency*.

second law of thermodynamics.   Commonly called the entropy law. It states that the amount of unavailable energy (entropy) in a closed system (where there is no exchange of matter) continuously and irreversibly increases over time. As a result, in a closed system, low entropy forms of energy decrease over time and reduce the available energy for work. See *entropy* and *law of conservation of energy*.

FURTHER READING
Georgescu-Roegen (1973).

second-order conditions.   Conditions for an optimum that are obtained from the second-order partial derivatives of an optimization problem.

FURTHER READING
Simon and Blume (1994).

sector. A subdivision of the economy based on a particular form of economic activity, such as the agricultural sector.

sedentary species. Species that remain in one locale or region throughout most of their life. For example, oysters are sedentary species.

sedimentation. The process by which suspended, water-borne materials settle out and become deposited on surface substrates. Sedimentation usual occurs when "in-stream" water velocities fall below a minimum speed necessary to maintain particle suspension.

sediment discharge. The deposition of fine particles or sediment within a streambed and into reservoirs. Increased upstream erosion can lead to increased sedimentation leading to shallower river channels and reducing the holding capacity of reservoirs.

seed trees. Trees that are intentionally spared from harvest in order to provide seed stock for forest regeneration on a site.

seemingly unrelated regression (SUR). A method of regression developed by Arnold Zellner where a system of equations is estimated. SUR accounts for cross-correlation in error terms across equations, and thus provides for more efficient estimators with lower standard errors.

SEGS. An acronym for Solar Electricity Generating Systems.

seismic risk. The potential for damage to structures from earthquakes.

selective logging. Timber harvest in which only selected trees are removed from a stand. Trees may be selected based on species, age, form, or other criteria. See *highgrading*.

selfish gene. Term popularized by Richard Dawkins that explains why individuals may cooperate and act in an altruistic way. The selfish gene refers to genetically inheritable altruistic traits in closely related individuals. If such behavior increases reproductive success, then altruism is a "selfish gene" in that altruism conveys no personal advantage to those who display it, but nevertheless the gene for altruism will, over time, become more widespread within a species. See *altruism*.

FURTHER READING
Dawkins (2006).

Sellafield.   Site in Cumbria, north-west England, of a nuclear power plant and a nuclear reprocessing facility. The Sellafield facilities have generated a lot of controversy due to the disposal of low-level radioactive waste into the Irish Sea and because of higher than normal radioactivity near the plant. Some believe that the increased radiation is an explanatory factor in the cluster of childhood leukemia in the local area.

semi-renewable resources.   Resources that require significant inputs of material and/or energy in order to maintain their initial state such as soil quality, ecological systems, and the assimilative capacity of the environment.

senescence.   The dieback of vegetation, often associated with ripening or age.

sensitivity.   The extent to which an environment or species responds to changes or shocks.

sensitivity analysis.   A method of analysis whereby different results are generated and compared by varying the assumptions and/or parameters used in the model.

separability.   An economic concept used to describe whether individual variables in a function (e.g., utility function or production function) can be aggregated without loss of information about the predicted or dependent variable (e.g., level of utility or output). Separability requires a special set of conditions; for example, a utility function is separable into n groups or aggregates of variables only if the marginal rate of substitution between any pair of variables in any group is independent of the values of variables in any other of the n groups.

septage.   Untreated sewage.

sequence.   A list of numbers from which it is possible to deduce past and future numbers in the series. For example, ... $-2, 0, 2, 4$ ... is a sequence as it represents a series of numbers where the next number is always 2 greater than the previous number.

sequential games.   A type of game in which players do not move simultaneously, but move in sequence. Often, the choices made by players in previous moves are known to players in later moves.

sequestration.   See *carbon sequestration.*

seral stages.   Distinct plant and animal communities that are in a transitional stage before the climax stage.

seral succession.   The sequential development of a community, especially plants.

sere.   The entire sequence of different plant and animal communities that follow one another in a given area through the process of succession.

serial correlation.   See *autocorrelation.*

series.   A sum of a list of numbers. See *geometric series.*

set.   A grouping of objects called elements.

severance tax.   Tax imposed upon the extraction of resources based upon the volume removed.

Seveso.   A town in northern Italy that suffered from an industrial accident in July 1976 that led to the release of the chemical TCDD that, in turn, contributed to birth deformities and a debilitating skin disease. See *TCDD* and *chloracne.*

sewage.   Human and animal excreta that may also contain other contaminants.

sexual selection.   The selection of mates with specific characteristics. For example, peahens select peacocks with large and colorful plumage.

shadow price.   The price of an asset or resource that reflects its marginal value in use. The shadow price is derived from an optimization problem and will not, in general, equal the market price of the asset or resource.

shale.   Soft, fine-grained rock that is generally a mixture of mud and clay.

shallow ecologists.   Term coined by "deep ecologists" for persons who may ascribe value to nature based on anthropogenic measures. See *deep ecology.*

Shannon–Wiener diversity index.   A widely used index of species diversity within a community. It is defined as the negative of the sum of the relative abundance of each species multiplied by the natural log of its relative abundance. For example, with two species and a relative abundance of 0.5 and 0.5 for each, the diversity index would be $(0.5 \times \log_e 0.5 + 0.5 \times \log_e 0.5) = 0.69$.

shared stocks.   Renewable resources that are shared between two or more jurisdictions. For example, Canada and the United States share common fishery resources that go back and forth between the two countries' exclusive economic zones.

shellfish.   Generic term for mollusks (such as mussels and oysters) and crustaceans (such as lobsters and crabs).

shelter belt.   A line of trees or other vegetation planted to protect an area downwind from the prevailing winds.

Shephard's lemma.   A mathematical result in which the partial derivative of a cost or expenditure function with respect to input prices can be used to obtain input demands.

shifting cultivation.   Also called slash-and-burn agriculture or swidden agriculture. Agricultural practices where land is cleared of trees and grass by burning, or other means. When the fertility of the soil declines to an unacceptable level, the farmers and/or communities move to another area and repeat the practice.

short run.   Economic term that describes the period of time in which only some of the inputs or factors of production can be varied. See *long run*.

sibling species.   Species that strongly resemble each other, but are incapable of breeding.

sic.   Latin term meaning "thus".

sick building syndrome.   Health problems caused by people working or living in buildings with poor air quality. These may be caused by volatile organic compounds, particulate matter or other pollutants.

side payments.   In game theory, transfers between participants outside of the prescribed payoffs within the game. In cooperative games, such transfers allow participants to share the gains from cooperative behavior.

Silent Spring. Title of a best-selling book by Rachel Carson which became synonymous with the potential environmental consequences of business-as-usual.

FURTHER READING
Carson (2002).

silt. Fine soil particles.

Silurian period. Geological period of time from 438 to 408 million years ago.

silviculture. The management of trees and forests from planting to pruning to harvesting so as to achieve defined objectives.

simplex method. A widely used algorithm used to solve linear programming problems.

Simpson diversity index. An index of biological diversity in an environment defined as
$$D = 1 - 3p_i^2$$
where p_i is the proportion of individuals from species i.

simulation. Term for calculating the values of variables in a model over time under defined assumptions and parameter values.

simultaneity bias. The bias that exists when a system of equations is estimated, but the simultaneity between variables in the system is not taken into account. Two-stage least squares estimation is an approach that can help overcome the problem.

simultaneous equation model. A set of equations where some of the endogenous variables are determined simultaneously.

sine qua non. Latin term meaning "an indispensable condition".

single-period rotation. See *Fisher rotation*.

sink. See *carbon sink*.

site attributes. A list of quantifiable descriptors of the amenities and disamenities of a recreational destination. Site attributes are the primary explanatory variables in models of recreational site choice.

SI units.   Système International d'Unités; or more commonly known as the metric system. See Appendix 3.

size limits.   Restrictions imposed on harvesters of a resource that prevent them from capturing individuals below a certain size. Size limits are often imposed to protect juveniles from exploitation and to ensure the sustainability of the resource.

skewness.   The degree to which a probability distribution is not symmetric. Positive (negative) skewness exists when the distribution has a long tail to the right (left).

SLAPP.   An acronym for Strategic Lawsuit Against Public Participation and that is used to describe legal actions filed against activists designed to preempt public protest by proponents of potentially controversial developments and/or actions.

slash.   Material left over after the harvesting of trees or removal of vegetation for other purposes.

slash-and-burn agriculture.   See *shifting cultivation*.

Slutsky equation.   An equation commonly used in economics that shows the change in the demand for a good following a price change in another good in terms of a substitution effect (change in consumption from the price change alone but holding the individual to the same level of utility) and an income effect (change in consumption due to the change in real income because of the price change).

Small is Beautiful.   Best-selling book by E.F. Schumacher that argued in favor of small and traditional technologies and practices, and against large-scale industry as a means to promote economic development and welfare.

FURTHER READING
Schumacher (2010).

smart grid.   An electricity grid that uses digital technology to provide a two-way interface between producer and consumer. For example, in peak periods customers could opt to have certain appliances remotely controlled to reduce overall demand on the system.

smart meter.   A component of a smart grid that provides consumers with up-to-date and detailed information about their energy consumption, thereby encouraging efficiency of use.

smog.   See *photochemical smog.*

smog trading.   See *pollution permits.*

social accounting.   A method of accounting that includes allowances for changes in environmental and other socioeconomic variables in calculating national accounts (such as costs incurred from a reduction in environmental quality).

social accounting matrix (SAM).   An accounting matrix in which income flows and capital stocks are tracked across all social institutions, including households.

social capital.   The institutions and norms of behavior that facilitate social and economic cooperation. Social capital is important in explaining economic performance and environmental quality.

social cost.   The private and external costs associated with an economic activity. For example, the social cost associated with a firm's production of a good involving greenhouse gas emissions would include the costs arising from environmental damage caused by the emissions, in addition to the firm's costs.

social discount rate.   The discount rate at which society chooses to discount future costs and benefits. This social discount rate would normally be lower than a private or individual discount rate. See *discount rate.*

social ecology.   A school of thought that suggests that many societies' institutions are exploitative and hierarchical. According to this view, exploitation on the human dimension is reflected in exploitation of nature, which is the detrimental to the environment.

socially optimal use.   The allocation of resources that will yield the most benefit for society as a whole, taking into account all costs and benefits.

social optimality.   See *socially optimal use.*

social security.   State-supported welfare transfers to poor, disabled and retired members of a society.

sociopolitical sustainability. A concept related to sustainable development, where future generations should experience equal or greater access to social resources. See also *environmental sustainability* and *economic sustainability*.

social welfare. 1. A measure of the collective well-being of a society. 2. Social security.

social welfare function. A function that supposedly represents the aggregate preferences of individuals in a society. Social welfare functions are used to consider the society payoffs under different sets of policies.

social welfare rule. A rule, which may be arbitrary, that determines how individual welfare is aggregated across individuals and time so as to find a measure of social welfare.

soft energy. Term coined by Amory Lovins to describe energy derived from small producers relying on renewable resources, such as solar power, wind power, and hydropower. See *hard energy*.

soft water. Water that has a low proportion of alkaline compounds and has a low salinity. See *hard water*.

softwoods. Usually defined as cone-bearing tree species such as pines or firs.

soil. A mix of organic material, micro-organisms and weathered rocks on which plants grow and obtain nutrients. Soil quality is affected by cultivation and requires careful management to avoid soil erosion and loss of fertility.

soil carbon. A broad term referring to carbon that is present in soil as a result of interaction with either living or decaying vegetation. The carbon content of soil can be increased by techniques such as application of biochar. Higher carbon content enhances soil fertility, which in turn increases the amount of atmospheric carbon dioxide sequestered by trees and plants through photosynthesis. See *biochar* and *carbon sequestration*.

soil compaction. The loss of spaces within the existing soil structure that reduces the productivity of soil and may lead to increased run-off.

soil conservation. Methods and practices designed to avoid or reduce soil erosion and preserve soil fertility. Practices may include terracing on slopes, planting of shelter belts and no-till agriculture.

soil degradation.   The diminution in the productive capacity and fertility of the top soil.

soil erosion.   See *erosion*.

soil fertility.   The inherent productive capacity of the soil.

solar cell.   A device capable of producing an electric current from sunlight. See *photovoltaic effect.*

solar constant.   Average rate at which the Earth's surface receives radiant energy from the sun.

solar panel.   An assembly of solar cells working in conjunction to produce electricity. Solar panels may range in size from only a few cells powering small devices such as torches, to larger panels of many square meters for household or industrial use. As technology has improved, numerous sites have been developed where very large numbers of panels are arranged together to generate power for a central electricity grid.

solar power/energy.   Energy derived from sunlight may be converted into an electrical form through the use of photovoltaic panels or concentrated solar power technology. Solar energy may also be used directly to heat living spaces and or water. As solar power is a renewable resource with negligible greenhouse gas emissions, its development and deployment plays an important role in climate change mitigation. See *photovoltaic effect.*

solar radiation.   Radiation from the sun in the form of electromagnetic waves that includes infrared, visible and ultraviolet radiation.

solar thermal energy.   A term for technologies that directly transfer heat generated by solar energy, such as rooftop water heaters. In recent times, a high temperature technology known as Concentrated Solar Power has become increasingly used in commercial operations. See *Concentrated Solar Power.*

solid load.   The material in a water body that is solid material.

solid waste.   Waste material, primarily of municipal origin, that is in solid form. The proper disposal of solid waste is an important environmental issue.

Solow growth model.   See *neoclassical growth theory*.

Solow–Hartwick rule.   See *Hartwick's rule*.

solvent.   A fluid that is able to dissolve other substances and bring them into solution.

Southern Annular Mode.   See *annular modes*.

Southern Oscillation.   See *annular modes*.

South, the.   A collective term for the developing countries of the southern hemisphere.

sovereign debt.   Debt accrued by the governments of independent countries. Unlike individual or corporate debt, governments cannot be compelled to repay a loan. A creditor's leverage is restricted to a borrower's concern for reputation loss and subsequent difficulty with future loans.

sovereign wealth fund.   A fund created by a national or sub-national government to provide long-run benefits to its residents. Such a fund could be used to be managed to smooth consumption over time by enlarging the fund in prosperous times and drawing from the fund in recessions.

space mirror.   A proposed geoengineering solution to global warming whereby large mirrors are placed into orbit to reflect a small proportion of solar radiation away from the Earth. Also referred to as a sunshade. See *geoengineering*.

Spaceship Earth.   Term popularized in the 1960s, originally coined by Kenneth Boulding, that emphasizes that the Earth is a closed system in terms of materials (but not energy).

spacing.   The selective removal of trees in a stand of an early age in order to promote growth, wood quality, or other forest management objectives.

spatial data.   Data that includes observations from at least two different locations and where differences across locations are of interest.

spatial distribution of forest disturbance. The pattern of activities that disrupt tree growth, including fire, disease and harvesting, and which can be characterized by differences in intensity, dispersion and size.

spawning biomass. The total weight of all individuals in a population that is sexually reproductive.

Special Report on Emissions Scenarios (SRES). A group of scenarios projecting future unmitigated greenhouse gas emissions that was developed in the lead-up to the IPCC Third Assessment Report in 2001. Although these scenarios have been widely used in climate and economic modeling, they are now commonly considered as outdated and under-representing emissions growth. The scenarios offered multiple projections based upon alternative parameterizations of future globalization, economic growth and attitudes towards the environment. See *post-SRES emissions projections*.

speciation. The process by which new species are formed.

species. The major subdivision of a genus composed of organisms that are genetically and morphologically similar, and are capable of successfully breeding with other organisms of the same species.

species composition. The percentage of a forest stand that is comprised of each species, based upon number of stems, basal area, or gross volume.

species diversity. See *biological diversity*.

species–energy hypothesis. The hypothesis that habitats with greater primary production will support a greater diversity of species.

species extinction. See *extinction*.

species hot spots. Area of unusually high species diversity.

species prospecting. See *biodiversity prospecting*.

specific gravity. Density of a substance relative to the density of water.

spectra. The distribution of different wavelengths of extraterrestial solar radiation. The wavelengths associated with various colors become attenuated to varying degrees as they pass through the atmosphere and obstacles such as clouds, water and vegetation.

speculation.   Using differences in the prices of goods or assets across time to generate a return; for example, buy at a low price today to sell at a high price sometime in the future.

sphagnum.   A family of mosses living in bogs or marshy areas. The living organism from which most peat originates. See *peat*.

spill-overs.   See *externality*.

spore.   A non-active form of a micro-organism that is capable of withstanding extremes in its environment.

sport fisheries.   A defined system of recreational fishing encompassing anglers and fish populations harvested for non-commercial, non-subsistence, recreational purposes.

spot market.   Market for commodities where delivery from the buyer to seller is immediate.

spot price.   Current price in a spot market. The price of crude oil is frequently given as a spot price rather than the price paid by buyers in a long-term supply contract.

spring flush.   The rapid melting of snow and ice in the spring in mountain environments and high latitudes. Where the land is contaminated with pollutants and the precipitation is acidic, the spring flush can be very damaging to the environment.

square matrix.   Matrix where the number of columns equals the number of rows.

square mile.   Unit of area equal to 2.58 square kilometers.

stability.   Term used to describe the behavior of a variable that converges or returns to a particular point.

stabilization level.   A term describing the global target for greenhouse gas concentration believed to be optimal from a scientific and economic perspective. See *2° Celsius*.

stabilization policy. Government policy designed to reduce fluctuations in national outcome. For example, in a fast-growing economy the government may wish to increase interest rates to reduce the possibility of growth that may be lead to inflation. By contrast, in a slow-growing economy, public works may be initiated to provide income for the unemployed.

stagflation. High inflation coupled with high unemployment.

stand. A group of trees within an area in proximity to one another and relatively homogeneous in terms of age, structure, composition and physical environment. A stand is one of the basic units of forest management.

standard. A rule or regulation set to achieve a desired outcome. In environmental regulations, standards often exist in terms of maximum permissible discharges or concentrations.

standard deviation. A measure of the spread of the distribution of a random variable. The standard deviation of a random variable is the square root of its variance, which summarizes the expected difference from the mean or central tendency of a random variable.

standard error. Standard deviation (or square root of the variance) of the distribution of a statistic from a sample.

standing trees. Living, intact trees (in contrast to trees felled by timbering, clearing, fire or natural disturbances).

starting point bias. Term used to describe the potential bias that may exist in surveys designed to elicit individual valuations of an aspect of the environment due to the initial bid supplied in the survey instrument. See *contingent valuation*.

stasis. A period of rest or equilibrium where there is relatively little change in an ecosystem. In high latitudes, winter is a period of stasis relative to summer.

state-determined stock. A stock that is endogenous or determined within a model.

stated preferences. The relative economic desirability of different goods (or combinations of goods) as expressed by an individual.

state planning.   Economic plans, sometimes over a five-year period, through which the state attempts to achieve economic and social objectives via administrative fiat rather than markets.

state rights.   A property-rights regime where the rights are held by a centralized authority. For example, in many countries state rights exist over national parks.

FURTHER READING
Grafton (2000).

state variable.   A variable in an optimal control problem that can only be affected indirectly via control variables and which represents the "state of nature". For example, in an optimal control problem of a fishery the size of the fish population is the state variable while the control variable would be the level of harvest. See *control variable* and *optimal control theory*.

static efficiency.   See *economic efficiency*.

static reserve life.   The number of years before an economically extractable (mineral) reserve is exhausted at current rates of extraction.

FURTHER READING
Tietenberg and Lewis (2009).

static sink life.   A measure of resource extraction that accounts for the fact that metals extracted from the Earth's crust end up on its surface.

FURTHER READING
Holmberg, Robert and Erikson (1996).

stationarity.   A variable the mean and variance of which are time invariant. Most time-series economic data are non-stationary such that the expected value of the variable depends on time. In regression analysis, using time-series data where the variables are non-stationary, spurious correlations among variables may arise. To correct for this potential problem, procedures have been developed to derive stationary variables from non-stationary variables. See *differencing*.

stationary point.   Also known as a critical point in mathematics, this is where a function is neither increasing nor decreasing such that the first derivative is zero.

stationary resource. A natural resource that is not mobile, such as a forest.

stationary source. A source of pollution that is fixed in location, such as a factory. See *mobile source.*

stationary state. A point where the value of a function is unchanging with respect to time.

statistical efficiency. Term used to compare statistical models where the estimate of a parameter with the lowest variance is the most efficient. See *variance.*

statistical inference. The procedures by which observed data are used to draw conclusions about the population from which the data came, or about the unknown process that generated the data.

statistical life. A unit of statistical measure corresponding to the average lifespan for an average individual. Statistical lives are used in aggregate measures such as income/statistical life or value/statistical life.

status quo. Latin term for the current state of affairs.

steady state. A point at which a dynamic system is in a state of equilibrium.

steady-state economy. Term popularized by H.E. Daly (1973) that refers to an economy that experiences no change in its stocks of capital and wealth, no change in its population, and a minimum flow of throughputs so as to minimize environmental impacts.

steady state equilibrium. A resting point where all variables are growing at the same rate. If the growth rate of all variables is zero, the equilibrium point is stationary.

stem volume. The volume of the principal or main woody stem of a plant.

steppe. A biome of grasslands characterized by the feature of low rainfall and extreme ranges in temperature.

Stern Review. A report authored by Lord Nicholas Stern in 2007 analyzing the economics of climate change mitigation. Stern broke with the economic consensus of delayed and minimal mitigation by presenting a case for strong and urgent global action. Many critics argued that the discount rate used in this analysis was too low and, hence, Stern's findings were not robust. However, the Review stimulated an academic discourse challenging the traditional economic approach to mitigation, in addition to laying the intellectual foundation for the ambitious targets of current international climate policy. See *climate change mitigation*, *discount rate*, and *Dismal Theorem*.

FURTHER READING
Stern (2008).

stochastic models.   Models the outcome of which are determined, in part, by stochastic processes. See *stochastic process*.

stochastic process.   A sequence designating the value of a random variable or a group of random variables through time. A serially independent stochastic process is one where the realization or exact value of a random variable at a point in time is independent of the value of the same random variable in previous periods. Serially correlated stochastic processes imply that the value of a random variable at a point in time in part depends on its past values.

stochastic production frontier.   The maximum amount of output produced for any given set of inputs, allowing for random variation.

stochastic production frontier analysis.   A combination of a stochastic production frontier and an inefficiency model, forming an assessment of the variables that determine output and the firm-specific variables that result in differences in efficiency.

stock.   The quantity of a resource at a given point of time. For example, the biomass of a fishery is a stock, but the harvest of fish is a flow. See *flow*.

stock assessment.   A study of the population of a species in a given fishery to determine the likely stock-recruitment relationship and to estimate its biomass.

stock pollutant.   A pollutant, such as a PCB, that accumulates in the environment. See *polychlorinated biphenyls*.

stock-recruitment model.   Model that relates the stock or biomass of a population, or the spawning biomass, with the number of recruits. Stock-recruitment models are used in fisheries and they usually suppose that the smaller the stock, the greater is the rate of recruitment.

FURTHER READING
Grafton et al. (2010).

stocking rate.   Rate at which a given area of land is used in terms of the number of animals grazed, or the number of trees growing per hectare.

storable resource.   A natural resource that can be stored or accumulated for future use, such as water.

storm surge.   A temporary increase in sea level due to high winds or storm events. Storm surges may cause flooding in beachside or riparian areas in coastal regions with altitudes close to sea level. See *sea level rise.*

straddling stocks.   Resources that are found in the jurisdiction of one state but can also be exploited outside of its jurisdiction. For example, most of Canada's fishery resources lie within its 200-nautical-mile exclusive economic zone, but some of its fisheries (nose and tail of the Grand Banks) lie outside its direct jurisdiction, or that of any other country.

strange attractor.   A set of points to which a chaotic dynamic system converges and which are highly sensitive to the initial conditions. See *chaos.*

strategic bias.   In contingent valuation, the potential that a respondent's stated willingness to pay (or accept) may be different from her true willingness to pay (or accept) if she believes that her answer will influence the provision of the good or service in question. See *contingent valuation.*

strategic form.   Also called a normal form. A graphic description of any game involving two different participants using a table made up of different cells where one participant's strategies are displayed in horizontal rows and the other in vertical rows of the table. The resulting payoffs to each participant from a combination of the different strategies are then found within the cells formed by the intersection of the two strategies.

strategic interaction.   The strategies or behavior pursued by a participant that benefits others, but that also serves to improve the participant's own well-being.

strategic petroleum reserve (SPR).   Reserves of petroleum of about 725 million barrels of crude oil owned by the US government and stored in case of a major disruption to oil supplies. See *buffer stocks*.

strategic plan.   A land allocation plan that provides both land-use objectives and implementation strategies.

strategic rationality.   Choosing a strategy that is consistent and believable to the other parties.

stratified random sampling.   Random sampling of observations in defined strata or classifications. The approach is widely used to provide adequate representation of strata (such as income levels) of interest.

stratosphere.   A region in the Earth's atmosphere from 10 to 50 kilometers above the Earth's surface (as low as 8 kilometers in polar regions and as high as 15 kilometers in tropical regions) and which contains the ozone layer that helps shield the Earth's surface from ultra-violet radiation. See *stratospheric ozone*.

stratospheric ozone.   Ozone found in the stratosphere and which protects life on the Earth's surface from ultraviolet radiation from the sun. See *ozone hole*.

stream flow.   The amount of water passing through a stream channel and typically expressed in volumetric units over a specific period of time.

stream load.   The amount of material transported by a stream or river in a given period of time.

strict liability.   A legal term that holds individuals or firms liable for their actions, even if their actions were not negligent. See *liability*.

strip-cropping.   Growing crops in raised strips that follow the contour of the land so as to reduce erosion.

strip cut.   Pattern of harvesting wood in long narrow strips.

strip mining.   Open-pit mining characterized by the removal of the overlying soil and rock.

strong sustainability.   The principle that there should be no net depletion of natural resources from economic activity. This contrasts with weak sustainability, which allows for the conversion of one type of capital stock into another type, so long as the aggregate value of the net capital stock does not decrease.

FURTHER READING
Neumayer (2003).

structural change.   A term used to describe deep economic, social and cultural transformation. For example, climate change mitigation requires structural changes in energy production and consumption.

structural diversity.   Variations in the physical landscape (such as a hill or a lake) that increase biological diversity.

structural equations.   Equations in a simultaneous equation model that include equilibrium conditions (such as supply equals demand) and behavioral equations (such as a supply equation and a demand equation).

student's t-test.   See *t-test*.

stumpage price.   See *stumpage value*.

stumpage rate.   The charge that is paid to the owner for the right to harvest a given volume of wood from a forest.

stumpage value.   The value placed upon a tree on the stump. Generally, stumpage is the market value of a log less transportation, milling, and harvesting costs. See also *market stumpage prices*.

subalpine.   The habitat zone found just below the alpine zone (differenti-ated by the limit of tree growth).

subcanopy.   Layers of forest vegetation occurring beneath the forest canopy.

subduction.   A physical process where one entity moves under another, such as tectonic plates or heterogeneous masses of water. See *thermohaline circulation*.

subeconomic resource.   A natural resource for which the marginal cost of extraction or harvest exceeds the marginal benefit of extraction.

subgame perfect.   A concept in game theory in which a particular strategy holds not only for the equilibrium outcome, but also for any of the other possible outcomes.

FURTHER READING
Binmore (2007).

sub judice.   Latin term for an issue or case before the courts.

sublittoral zone.   The environment found below the littoral zone and marked by a transition from aquatic plants to blue-green algae below which light does not effectively penetrate. See *littoral zone*.

sub-Saharan Africa.   African countries located south of the Sahara desert.

subsidence.   Fall or depression in the Earth's surface. Too rapid a rate of groundwater withdrawal can be a major cause of local subsidence.

subsidy.   Any payment from a central authority to producers or firms that reduces their costs of production. Subsidies frequently exist in agriculture in developed countries where many governments provide price and income supports to farmers.

subsistence.   The ability to survive or subsist but with no economic surplus or savings should circumstances worsen. Persons living at the subsistence level have few degrees of freedom in terms of their decisions and do not have the resources to ensure their own survival or the sustainability of the environment and the resources from which they obtain their livelihood.

subsistence agriculture.   See *subsistence.*

sub-species.   Different groups of organisms with different physical characteristics or behavior, but which are capable of interbreeding.

substitute.   A good or commodity that is consumed or produced in place of another good. For example, some people drink tea as a substitute for coffee.

substitution effect.   The change in consumption of a good following a price change due to the substitution to other goods, but holding the consumer's utility unchanged. See *income effect.*

subsurface.   Materials that are found beneath but close to the Earth's surface, such as water or shallow mineral deposits and rocks.

subtidal kelp.   Kelp found below the intertidal zone.

subtractable.   See *rivalrous*.

subtropics.   The climatic region between the tropic and temperate regions. Although exact locations vary by continent, they are usually found between 25 and 35 degrees of latitude in both hemispheres.

suburbanization.   The conversion and incorporation of small communities and towns into suburbs of larger urban areas. See *urbanization*.

succession.   The orderly and predictable development of distinct stages within a given area, and marked by different species and processes over time.

succulent.   A plant with thick leaves and stems suited for the retention of moisture.

sufficient conditions.   Conditions that, if satisfied, ensure a particular outcome. Sufficient conditions are particularly useful in the solution of optimization problems. See *necessary conditions*.

suffrutex.   A shrub often found in the understorey of (African) forests.

sulfur.   A non-metallic element and commonly occurring solid element, the compounds of which are one of the principal constituents of acid rain and atmospheric pollution, but it is also a necessary part of many organic processes.

sulfur dioxide (SO_2).   A gas formed when sulfur is burnt in the presence of oxygen, and a major anthropogenic cause of acid rain. An important source of sulfur dioxide is coal-fired electric utilities that burn sulfurous coal to produce electricity.

sulfur dioxide trading.   A system of pollution permits, or tradable emission permits, established in the United States to control the emissions of sulfur dioxide from coal-fired electric utilities. The total allowable level of emissions has been declining so as to try to achieve environmental objectives and to control the problem of acid rain in eastern North America. Under the system, firms can meet their individual emission obligations by installing pollution control devices and/or purchasing emission permits from other firms. Trading of emission permits allows high-cost-of-abatement firms to reduce their costs in meeting their emission targets. See *pollution permits*.

sulfur hexafluoride (SF$_6$).   An extremely strong greenhouse gas with a global warming potential of 22 000 that is used in the electrical industry. See *global warming potential* and *greenhouse gas*.

sulfate aerosols.   Aerosols consisting of sulfur compounds that originate from greenhouse gas emissions or volcanic eruptions. These substances reduce the greenhouse effect by scattering incoming solar radiation. See *aerosol*.

sunk cost.   A forgone expenditure incurred in the past but which has no effect on present or future returns.

sunspots.   Spots on the sun that are associated with strong electro-magnetic activity. Changes in sunspot activity, which occur in regular cycles (of approximately 11 years), have been linked to changes in weather and solar radiation.

superfund.   See *CERCLA*.

superfund sites.   US hazardous waste sites defined under the Comprehensive Environmental Response, Compensation and Liability Act (CERCLA).

superphosphate.   A chemical fertilizer prepared with either a combination of sulfuric acid, calcium sulphate, and calcium acid phosphate or phosphoric acid.

supply.   The quantity available for sale of a good or service at a given point in time and at the market price.

supply curve.   See *supply function*.

supply function.   A relationship between the amount firms are willing to supply, or produce and market of a good or service, and the price of the good.

supra.   Latin term meaning "above".

supralittoral zone. The area above the littoral zone that is influenced by the spray from waves. See *littoral zone*.

surface temperature. See *global surface temperature*.

surface runoff. Rainfall that travels across the land surface into a water body. The volume of surface runoff is a principal determinant of water levels in lakes and rivers, varying with rainfall or evaporation rates.

surface water. Water found on the Earth's surface in streams, lakes, seas and other surface features.

surplus forest. Forest where the annual cut is less than the long-term yield from a forest.

surplus production model. A model of population dynamics in which the next period's biomass is calculated as the current period's biomass plus surplus production (recruitment and growth less natural mortality) less the amount harvested.

FURTHER READING
Grafton et al. (2010).

survival of the fittest. Term coined and popularized by Herbert Spencer, a British scientist of the nineteenth century. The term is often interpreted to mean that Darwinian evolution has led to higher order and superior life forms, the ultimate example being *Homo sapiens*. In reality, natural selection simply favors individuals (and their offspring) that are adapted to their particular environment and does not necessarily lead to larger or more intelligent species. See *natural selection*.

FURTHER READING
Gould (2001).

suspended solids (SS). A mixture of solids carried within a fluid.

sustainability. A concept which requires the reconciliation of economic, environmental and social interests – the three pillars of sustainability. See *weak sustainability, strong sustainability, environmental sustainability, economic sustainability*, and *sociopolitical sustainability*.

FURTHER READING
Gale and Cordray (1994).

sustainability criterion.   The principle that our actions should not make future generations any worse off than the current generation.

sustainability paradigm.   A program of resource use over time that modifies traditional cost–benefit approaches to maintain either constant capital assets (weak sustainability paradigm) or constant critical natural capital assets (strong sustainability paradigm).

FURTHER READING
Neumayer (2003).

sustainable.   The notion that an aspect of the environment or practice can be maintained into the foreseeable future.

sustainable agriculture.   Methods of farming that do not degrade the productive capacity of the land.

sustainable development.   A term popularized by the United Nations in the report *Our Common Future*, and which is defined in many different ways. It is commonly defined as a process by which current generations meet their needs and care for the environment in ways that do not compromise the needs of future generations.

FURTHER READING
Rao (2000).

sustainable development path.   See *alternative development path.*

sustainable forest management (SFM).   Forestry management that allows for the harvesting of timber, but also ensures that the forest provides needed ecological and landscape functions.

Sustainable Forestry Initiative (SFI).   A form of green or eco-labeling certification for timber products, indicating that the wood used is produced from sustainable forest management.

sustainable tourism.   Tourism that is designed to have minimal environmental impacts and which supports the local economy through actions such as staying in local hotels, eating at local restaurants, and giving back to the community in some form or another.

sustained yield.   Yield or harvest from a renewable resource that can be sustained into the future without reducing the overall size of the resource.

sweatshops. Factories or plants where the workers are employed in substandard conditions, which include but are not limited to: unfair wages, long working hours, poor health and safety standards, and child labor.

swidden. An agricultural plot cleared by slash-and-burn methods.

swidden agriculture. See *shifting cultivation.*

symbiosis. A mutually beneficial relationship between organisms. See *mutualism.*

sympatric speciation. The evolution of new species, without geographical separation, due to inbreeding or reproductive isolation of a segment of the population.

synergistic effects. When two or more substances work together to produce a greater effect than would be expected from simply adding the individual effects from each of them together.

synfuels. Shortened form of synthetic fuels and defined as liquid petroleum products that are synthesized from hydrocarbons, such as coal or tar sands.

synthetic chemicals. Chemicals synthesized artificially in laboratories or factories rather than by living organisms.

system. The interconnections between organisms and physical processes that together represent an integrated picture of the various sub-components.

systematics. The classification of flora and fauna.

system dynamics. An approach to understanding dynamic phenomena that involves modeling of the stocks, flows and feedbacks of a system and the simulation of the models under various scenarios.

system of national accounts (SNA). See *national income and product accounts.*

T

taiga. See *boreal forest*.

tailings. Crushed waste material and rock left from mining. Without some form of remediation tailings can often have an adverse effect on the environment.

takings. 1. Under the Endangered Species Act in the United States, a taking is defined as the killing, harming, or harassment of an endangered species. 2. A governmental appropriation of property rights, either in part or complete, such as the restrictions placed upon the allowable uses of land at a particular site. See *incidental taking.*

tall stacks. Large stacks up to several hundreds of meters high that are used to disperse air pollution and to minimize its impact on the local area. Tall stacks often result in greater spatial dispersion of the pollution.

tannins. Toxic substance found in plants that renders them bitter or astringent.

targeted species. A preferred species that is actively sought in the process of harvesting or capture. See *bycatch.*

tariff. A tax imposed on imports that is often based as a percentage of the import price.

tar sands. See *oil sands.*

tax. See *Pigouvian tax.*

tax credit. An allowable deduction that may be claimed in the assessment of income tax by an individual or business that relates to their expenditure on an environmentally beneficial product or services, such as household roof-top solar panels.

tax incidence. Term for the distribution of the burden of a tax such that those who initially pay the tax (such as a wholesaler with a wholesale tax) are not necessarily the persons who end up paying for the tax (consumers) after price adjustments.

taxonomy.   See *systematics*.

Taylorism.   Term describing production processes in which a distinction is drawn between management and labor, and where managers establish regimented production practices to be followed by laborers, often using assembly-line techniques. Typically associated with early mass-production processes. See *Fordism*.

TCDD.   Acronym for 2,3,7,8-tetrachlorodibenzo-*p*-dioxin that is a toxic substance used in the manufacture of herbicides. See *Seveso*.

technical change.   See *technological change*.

technical efficiency.   A measure of how close a firm is to producing the maximum amount of output from given inputs. Unlike allocative efficiency, technical efficiency is not concerned with the proportions in which inputs are used, but at a given proportion, measures the extent to which a given amount of inputs results in maximum output. See *allocative efficiency*, *economic efficiency*, and *scale efficiency*.

technical progress.   See *technological change*.

technological change.   Improvements in production processes over time which permit more outputs or services, or an improved quality of goods and services, for a given level of inputs. See *disembodied technological change*, *embodied technological change*, and *induced technological change*.

technological externality.   An externality that distorts the efficient allocation of inputs and/or outputs. See *externality* and *pecuniary externality*.

technology.   The aggregate knowledge of production processes that can be currently applied to produce goods and services.

technology-based standard.   A command and control regulatory approach in which pollution abatement is legally mandated and monitored by the government. See *command and control*.

technology forcing.   Standards that require the adoption of a particular type of technology.

technology transfer.   The transfer of technological processes or knowledge from one market or nation to another. This concept is integral to international climate policy, whereby rich countries are expected to develop and transfer mitigating technologies to poor countries so they are able to follow a sustainable development path.

temperate rainforest. Forests found in temperate regions of the world and distinguished from other temperate forests by the infrequency of fire, the predominance of evergreen species, and a high degree of structural complexity.

temperature dynamics. Changes over time in the temperature of the Earth's surface and atmosphere due to perturbations in the Earth's radiation budget.

temperature inversion. See *inversion*.

tender. An offer to contract or supply a good or service at a given level of quality, in a specified time period, and at a specific price.

tenure. A broad term that represents the property rights over land and, in particular, the duration of these rights.

TEQ. A method of measuring the toxicity of a combination of furans and dioxins and other substances relative to the toxicity of TCDD. See *TCDD*.

teratogen. Any substance that leads to physical and mental deformities in a growing fetus.

terminal time. The last period in a dynamic optimization problem.

terminal value. The value of a project or asset at terminal time.

terms of trade. Ratio of export prices to import prices. A long-term decline in the terms of trade reduces the ability of countries to pay for their imports.

terracing. The creation of level planting areas that follow the contours of elevated land and are designed to reduce soil erosion.

terrestrial radiation. The long-wave infrared radiation emitted by the Earth. See *greenhouse effect*.

Territorial User Rights in Fisheries (TURFs).   Term coined by Francis Christy to denote areas of the ocean or sea floor to which individuals or groups of individuals have property rights in terms of the harvest of marine animals. TURFs are not widely used in fisheries although they do exist for some shellfish. See *individual transferable quotas.*

Tertiary.   A geological time period which began 66.4 million years ago and lasted until about 1.6 million years before the present.

tertiary sewage treatment.   A treatment beyond primary and secondary treatment that may involve a number of processes including reverse osmosis, carbon filters and other filters to remove nitrates and chemicals from treated wastes. See *primary sewage treatment* and *secondary sewage treatment.*

tetraethyl lead (TEL).   A gasoline additive used in leaded gasoline so as to improve engine performance. Concern over lead pollution has led to a phase-out of the use of TEL in many countries.

thalidomide.   Prescribed drug widely used in the 1960s that resulted in severe physical abnormalities in infants exposed to the drug as fetuses.

theory of the second best.   Theory which states that if not all the conditions for economic efficiency are attained in an economy, it may not be optimal or desirable to ensure economic efficiency in all other conditions or sectors.

FURTHER READING
Lipsey and Lancaster (1956).

thermal expansion.   An increase in water volume due to an increase in its temperature. Warming of the Earth's oceans through climate change decreases density and increases total volume of water, thus supplementing sea level rise from glacier and ice sheet melt. See *sea level rise.*

thermal inertia.   A measure of the tendency for a substance or entity to retain its temperature in the presence of external forcing. See *inertia.*

thermal pollution.   An increase in temperature caused by anthropogenic or natural sources that can have a negative effect on an ecosystem. For example, the discharge of hot water into streams can be responsible for fish kills.

thermal infrared radiation. Commonly known as heat and often simply referred to as infrared radiation. Use of the term thermal distinguishes radiation emanating from the Earth and its atmosphere from solar infrared radiation. The interaction between solar radiation and the Earth's surface, clouds and atmosphere causes the transmission of thermal infrared radiation into the atmosphere, where it is partially absorbed by greenhouse gases or discharged into space. See *greenhouse effect.*

thermocline. A layer inside a water body where temperature changes more rapidly with depth than in surrounding layers. In oceans, thermoclines are found in between the mixed surface layer and the deep ocean.

thermodynamics. The study of different forms of energy, and especially the conversion of heat into other forms of energy, and vice versa. See *law of conservation of energy* and *second law of thermodynamics.*

thermohaline circulation (THC). Large magnitude circulation of oceanic waters arising from changes in temperature and salinity, the determinants of saltwater density. This phenomenon is driven by cooling and precipitation at high altitudes that sinks surface waters, followed by upwelling at subtropical altitudes from warming and evaporation. The resulting circulation is one of the primary means of oceanic heat transfer and, hence, affects atmospheric climate processes.

thermophile. Bacteria that live in extreme environments in terms of heat, able to exist in temperatures of up to 80° Celsius. See *hyperthermophiles.*

thinning. The removal of trees from an immature crop so as to promote growth and form among the remaining trees.

thionic flucisols. Alluvial, estaurine soils containing large concentrations of subsoil sulfuric acid when dried.

Third Assessment Report (TAR). The report of the Intergovernmental Panel on Climate Change released in 2001. See *assessment report* and *Intergovernmental Panel on Climate Change.*

third law of thermodynamics. A physical law stating that a material, substance, or system has a finite positive entropy (or level of unavailable energy). See *entropy.*

Third World. Collective term coined for developing countries.

33/50 program. A US government voluntary compliance program to reduce emissions of specified pollutants by firms by a third by 1992, and by 50 percent by 1995.

FURTHER READING
Callan and Thomas (2007).

THMs. See *trihalomethanes*.

threatened species. Species that may become endangered should current trends continue, and may be in danger of extirpation in some environments.

3°Celsius. The level of global warming scientists believe is likely to trigger tipping elements in the climate system that could initiate catastrophic climate change. See *catastrophic climate change* and *climate threshold*.

350 ppm. A long-term target for stabilization of atmospheric greenhouse gas concentration advocated by many environmental groups, climate scientists and island states. The 'ppm' stands for parts per million and is measured in terms of carbon dioxide equivalent (CO_2-e) . See *450 ppm*.

Three Mile Island. Location of a nuclear power plant in Pennsylvania, which suffered a nuclear accident and resulted in the temporary evacuation of over a hundred thousand people from the local area.

threshold. The level of a system process whereby rapid change is initiated. See *climate threshold* and *tipping point*.

threshold effect. An effect which does not arise until after a given dose or level of pollution. Many pollutants have threshold effects which will vary depending on the environment and the individuals exposed to the pollutants. See *climate threshold*.

threshold limit values (TLVs). A standard level of exposure to which most people can be exposed over time without perceived health risks. TLVs are used to help develop safe working practices for workers.

tidal power. The generation of power with electrical turbines driven by changes in sea level, and therefore water movement, due to the influence of tides. See *renewable energy*.

Tiebout hypothesis. Named after its formulator, this suggests that communities and jurisdictions can compete for residents or workers by offering them a desirable mix of public services and public goods. Jurisdictions that fail to provide a desirable mix of tax and public goods services may lose out in terms of migration of their residents to more desirable jurisdictions.

FURTHER READING
Cornes and Sandler (1996).

tight sands. Geological formation comprised of closely bound, non-porous sands.

till. Sediment deposited by glaciers.

timber concessions. Rights to harvest public timber within an area granted by the government.

timber line. The altitude or latitude beyond which trees cannot grow because of the severity of the climate.

timber utilization. The quantity and quality of timber harvested from a defined area of land. A low utilization rate often implies less sustainable forest management practices, as it often coincides with greater on-site wastes.

time preference. 1. A preference for consumption in the present rather than in the future. 2. The relative weight given to consumption, utility, or welfare in any one period relative to any other period.

time-series data. Data where observations of variables are available over time or multiple time periods.

tipping element. See *climate threshold.*

tipping fees. Charges paid to waste disposal sites or landfills for the right to dispose of specified wastes.

tipping point. The point at which a system or object begins irreversibly shifting from its present equilibrium into an alternative state. This term is commonly used in reference to tipping elements in the climate system. See *climate threshold.*

Tobit model. A method of estimation due to James Tobin that estimates a regression equation where the data is "censored" or cut off at a particular value.

FURTHER READING
Greene (2003).

tolerance.   Ability of an individual, species or ecosystem to maintain itself under a given set of conditions. For example, shade-tolerant species can survive in conditions of shade.

top-down.   A conceptual approach to a problem which begins with setting the desired outcome, followed by filling in the facilitating details afterwards. This term is often used in relation to international climate policy frameworks such as the Kyoto Protocol, whereby an overall emissions reductions target is set, national obligations are negotiated, and states then work out how to achieve their individual targets. See *bottom-up* and *climate policy*.

topsoil.   The upper and most fertile level of soils, which contains most of the organic material.

Torrey Canyon.   Name of oil tanker than ran aground off the coast of England in 1967 spilling over 100 000 tons of crude oil, and killing many thousands of marine birds and fish.

total allowable catch (TAC).   A fisheries management strategy that determines the maximum amount of fish (by individual species) that can be harvested in a fishery over a given period of time, usually defined as the fishing season.

total cost.   All costs incurred in producing a good, including both variable and fixed costs. See *total variable cost* and *total fixed costs*.

total dissolved solids (TDS).   A measure of the total quantity of solids dissolved in a solution, this is an indicator of the level of water pollution.

total economic value.   The total willingness to pay to preserve, maintain, or possess a natural resource or environmental good. The total economic value includes use and non-use values.

total factor productivity (TFP).   A broad measure of productivity defined as a ratio of output(s) to all inputs. See *productivity*.

total fixed costs. The total costs of a firm that do not vary with the level of output.

total particulate matter. A measure of the total particulates in a given volume of air, used as an indicator of the level of air pollution.

total revenue. The total receipts from fishing, or the price of fish times the quantity of fish landed.

total suspended solids (TSS). A measure of the solids suspended in a liquid, commonly used as an indicator of water pollution.

total variable cost. The total costs of a firm that vary with the level of output.

total willingness to pay. The maximum transfer, in income or wealth, that an individual would make in order to ensure a certain state of the world (e.g., the possession of a good, access to a benefit stream, the continued existence of a resource, etc.).

toxicity. The degree to which a substance can be poisonous. It includes acute toxicity (the effects due to a single exposure) and chronic toxicity (the effects due to several small-level exposures).

toxic pollutants. Materials that, upon exposure, have a detrimental effect on an organism's health.

toxic release inventory (TRI). A US database compiled out of a requirement that large manufacturers report their annual releases of defined toxic chemicals.

toxigenic E. coli. See *Escherichia coli O157:H7.*

toxins. Organically produced poisons that are usually a byproduct of bacteria.

trace gas. A gas found in trace amounts, in terms of greenhouse gas concentration, in the atmosphere. For example, the most important, carbon dioxide, currently represents about around 0.035 percent of the atmosphere.

tradable discharge permits (TDPs). See *pollution permits.*

tradable effluent permit market.   The real or theoretical market in which emissions permits or quotas can be traded freely in order to achieve a least-cost provision of a fixed level of ambient environmental quality. See *emission permit*.

FURTHER READING
Tietenberg and Lewis (2009).

tradable emission permits (TEPs).   See *emissions trading scheme* and *pollution permits*.

tradable permits.   See *emissions trading scheme* and *pollution permits*.

tradable quota system.   See *emissions trading scheme*.

tradable SO$_2$ emission allowances.   The tradable right to produce a specified quantity of airborne sulfur dioxide. See *pollution permits* and *sulfur dioxide trading*.

FURTHER READING
Tietenberg (1996).

trade-exposed, emissions-intensive industries.   Term for a country's emissions-intensive industries potentially rendered internationally uncompetitive by a domestic carbon price increasing costs of production. See *carbon leakage*.

trade-off.   A conscious decision to forgo the benefits of one resource, good, or service, in order to enjoy those of another exclusive resource, good, or service. Trade-offs also can be made between management objectives.

Trade-Related Intellectual Property Rights Agreement (TRIPS).   A World Trade Organization (WTO) agreement that provides the minimum standards of protection of intellectual property that must be undertaken by all members, such as providing copyrights for books or patents for products. A critique of TRIPS is that it fails to provide adequate protection for traditional and indigenous knowledge.

trade winds.   Prevailing winds that blow from the north-east in the northern hemisphere and the south-east in the southern hemisphere towards the equator from the mid-latitudes, and which, historically, were extensively used by sailing ships.

traditional knowledge. The body of knowledge regarding the functioning of the physical world drawn from observation and traditions passed on within different cultures. Often characterized as being distinct from information drawn from scientific inquiry.

tragedy of the commons. See *commons, tragedy of*.

transactions costs. The costs incurred in exchanging and enforcing property rights. High transactions costs can be a major impediment to trade and exchange. See *Coase Theorem*.

transboundary pollutants. Pollutants, such as sulfur dioxide, that are produced in one country but that impose environmental harm on other nations.

transects. Linear paths across the landscape used in obtaining data or samples for analysis.

transect surveys. Data or samples taken along a straight line following a particular compass bearing.

transferability. The ability of the holder of a property right to sell, exchange or bequest that property to another person or persons.

transfer earnings. The payment required to keep a factor of production in its current activity or employment.

transfer payments. Payments made by governments to individuals because of needs such as unemployment benefits, disability payments, and old-age security.

transfer pricing. The pricing of resources or materials by vertically integrated companies to minimize tax liabilities.

transformation curve. See *production possibilities frontier*.

transformative values. The properties of an environmental asset that cause preferences of an individual for that asset to change upon exposure to the asset.

FURTHER READING
Brennan (1992).

transgenic crops.   Plants produced by genetic engineering and that have genes from more than one species. See *genetically modified*.

transition economies.   A national economy in the process of transition from a centrally planned economy to a market economy. This term has historically been applied to China, Vietnam, ex-Soviet Bloc countries and many others.

transition fuels.   Non-renewable resources other than oil and gas, such as coal and uranium, that can provide an interim source of energy supply until renewable energy becomes more widely used.

transitional path.   The path taken to a steady state equilibrium. To move from one steady state sustainable yield to another requires a transitional period where harvest may be much lower (or higher depending on the direction of adjustment) than in the final resting point or steady state.

transitivity.   A requirement for logical preferences where if a bundle of goods A is preferred to bundle B, and bundle B is preferred to bundle C, it cannot be that bundle C is preferred to bundle A.

transpiration.   An important part of the water or hydrological cycle by which plants take up moisture from the soil through their root system and discharge water vapor through their leaves.

transversality conditions.   Terminal time conditions that must be satisfied at the end of a dynamic optimization problem. In the dynamic optimization problem to maximize the present value of rents from a mine, a transversality condition is that at the completion of the program (terminal time) the product of the remaining reserves and the shadow price of the resource must be equal to zero. In other words, either all the resource is extracted or it is no longer worthwhile to extract the resource.

travel-cost method.   The econometric estimation of a travel cost model in which a demand function is derived in order to estimate the welfare contribution of visits to a recreational site, or attributes of a recreational site. See *travel-cost model*.

travel-cost model.   A model of recreation demand in which the good supplied (or acquired) is the number of trips to a recreation site and the cost of acquisition is travel cost (including out-of-pocket costs and the cost of travel time).

trawl.   See *trawler.*

trawler.   A fishing vessel that employs a net (or series of nets) that is towed behind the vessel along the sea bottom or in mid-water. Trawlers commonly target groundfish species, such as cod and pollock.

treaty rights.   Rights accorded to First Nations or aboriginal people, often in return for access to land or the ceding of resources and land. Treaty rights may include access rights (the ability to continue traditional activities on ceded lands) or direct rights to land and/or resources.

tree cover.   See *crown cover*.

tree ring records.   See *dendochronology*.

Triassic.   Geological time period from 245 to 208 million years ago.

trihalomethanes (THMs).   A possible carcinogen found in some drinking water that is a byproduct of chlorination of water that has high levels of organic matter. In the United States, the maximum contaminant level of THMs is set at 0.1 milligram per liter.

trip limits.   Limits imposed on the amount of a resource that can be harvested per trip. Trip limits are imposed in some fisheries in an attempt to discourage fishers from investing in larger vessels and increasing the capacity of the fleet.

TRIPS.   See *Trade-Related Intellectual Property Rights.*

trolling.   A method of fishing whereby lures are attached to lines that are attached to a vessel and is dragged through the water. Trolling is used to catch salmonids, and also for pelagic species.

trophic level.   The level or stage in the food chain at which an organism feeds. For example, carnivores feed at a higher trophic level than do herbivores.

trophic subsidy.   The difference between the total amount of energy assimilated by green plants, and the amount of energy used by plants which is converted into organic material.

tropical.   Pertaining to a location on Earth falling between 23.5 degrees

North (tropic of Cancer) and 23.5 degrees South (tropic of Capricorn) latitude.

tropical dry forest.   Usually deciduous forests in tropical latitudes characterized by a marked dry season.

tropical hardwoods.   Deciduous tree species found in the tropics that are often commercially harvested.

tropical rain forest.   A dense evergreen forest in tropical latitudes receiving at least 2.5 meters of rain annually. Tropical rain forests are noted for their high biological diversity.

tropics.   See *tropical.*

tropism.   The direction of growth of a plant in response to an external stimulus, such as light or gravity.

tropopause.   A layer in the Earth's atmosphere ranging between 10 to 15 kilometers above sea level that represents a boundary layer between the troposphere and stratosphere.

troposphere.   Part of the atmosphere found at the Earth's surface and up to 15 kilometers (8 kilometers in polar regions) above the surface. Unlike other parts of the atmosphere, temperature declines as the distance from the Earth's surface increases.

tropospheric ozone.   Ozone located at or near the Earth's surface, and that can be harmful to plants and animals. See *ozone* and *volatile organic compounds.*

truncated normal distribution.   A normal distribution that takes on only zero or positive values, frequently used as the distribution for the technical inefficiency term in a stochastic production frontier analysis.

trypanosomiasis.   Commonly known as sleeping sickness, this is a disease endemic to many parts of Africa. The disease is contracted through the bite of the tsetse fly and is caused by a protozoa. The disease leads to various symptoms, and is frequently fatal.

t-test.   A statistical test used to test the null hypothesis that a coefficient in a regression analysis is significantly different from zero. The degrees of freedom in the test equal $n - K$, where n is the number of observations and K is the number of regressors.

FURTHER READING
Greene (2003).

Tuna–dolphin dispute. A case that involved a US law that prohibited the importation of tuna from countries where the incidental killing of dolphins exceeds a critical amount. In 1988, a lawsuit by a US environmental group argued that the United States was not upholding the law in regards to tuna imported from Mexico, where Mexican fishers used purse seine nets to catch tuna. A US federal court agreed and prohibited any further imports. The Mexican government argued that its right to sell tuna in the United States had been violated and filed an appeal under GATT. The panel that heard the appeal in 1991 agreed with Mexico, but the United States and the Mexican government reached an agreement before the GATT ruling was formally adopted.

tundra. A biome found in extreme northern latitudes north of the boreal forest. Most of the vegetation consists of grass, lichen and moss, beneath which is permafrost. See *permafrost*.

turbidity. The cloudiness in water that is usually a function of the level of suspended solids.

turbine. An engine driven by rotating blades that converts energy from steam, wind or liquid into electricity. Turbines are used in many forms of power generation, including coal and natural gas-fired, hydropower and wind.

TURFs. See *Territorial User Rights in Fisheries*.

turnpike theorems. Theorems employed in optimal growth theory that suggest that the optimal path arch towards the terminal point rather go there directly.

FURTHER READING
Takayama (1985).

turnover time. The rate at which a gas is removed from a reservoir, such as the atmosphere. Turnover time is differentiated from adjustment time, the overall decay of an increase, as a gas such as carbon dioxide can be recycled in and out of the atmosphere, whilst only losing small increments in this process by adsorption into the deep ocean. See *adjustment time* and *lifetime*.

2,4-D. 2,4-dichlorophenoxyacetic acid, a widely used broad-leaf herbicide.

2, 4, 5-T. A herbicide for controlling weeds. Its use is restricted in several countries because of concerns over the release of toxic dioxins that arise from the production and use of the herbicide.

2° Celsius. The maximum global temperature increase above pre-industrial levels that scientists predict is unlikely to initiate dangerous climate change. The 2° target has become a central focus of international climate policy, as evidenced by the text of the Copenhagen Accord. See *Copenhagen Accord.*

two-part price. Pricing structure where the per unit price paid by consumers varies according to the amount consumed. For example, consumers may initially pay high price per unit, but at higher levels of consumption the price per unit may fall if it is a decreasing block tariff or would increase if it were an increasing block tariff.

two-stage least squares (2SLS). A method of estimating simultaneous equation model, developed by H. Thiel. In the first stage, a reduced form of the model is estimated using ordinary least squares, which saves the predicted values of the endogenous variables. In the second stage, the structural equations of the model are estimated but the observed endogenous variables are replaced by their predicted values.

FURTHER READING
Greene (2003).

two-tailed test. Statistical test that uses both critical regions when testing a null hypothesis. For example, $H_0 : b_1 = 0$ is a two-tailed test if the alternative hypothesis is either $b_1 > 0$ or $b_1 < 0$. See *one-tailed test.*

two-tier price. Pricing typically adopted in regulated industries where there is a declining average costs of production such as in urban water delivery. Consumers pay a fixed connection charge and a volumetric charge that depends on the level of consumption.

Type I error. Sometimes called the alpha risk, or significance level, of a statistical test. Type I error is the probability of rejecting a null hypothesis which is true. All things being equal, the smaller the Type I error, the greater will be the Type II error in a statistical test. See *Type II error.*

Type II error.   Sometimes called the beta risk. The Type II error is the probability of failing to reject a null hypothesis which is false. See *Type I error.*

typhoid.   Bacterial disease that results in high fever and sometimes death. Typhoid is endemic in many developing countries and is contracted by eating contaminated food or drinking contaminated water.

U

ultraviolet (UV) radiation. Electromagnetic energy which has a wavelength just beyond the visible light spectrum. Much of the sun's UV radiation is blocked in the Earth's atmosphere by stratospheric ozone, but a proportion still reaches the Earth's surface. Long-term exposure to UV radiation can reduce immune responses in humans and cause skin cancers and other ailments. Ultraviolet A, B and C radiation have wavelengths of 320–400, 280–320 and 200–280 nanometers.

Umbrella Group. A group of non-EU developed nations that act as a loose coalition within UNFCCC negotiations. Members usually include: Australia, New Zealand, the United States, Japan and Canada.

uncertainty. A situation where many possible events or outcomes may occur, but it is not possible to assign probabilities to these events in advance.

underemployment. Persons who have some form of employment, part-time or seasonal, but who are willing and able to work more.

underground injection well. A term for wells that are used to store or dispose of fluids, usually polluted water or hazardous material. Environmental regulations usually require such structures to be encased in concrete and steel to prevent dispersal into groundwater.

underground source of drinking water (USDW). A term for large aquifers that provide or could provide groundwater to a public water system.

understorey. The layer of vegetation made up of shrubs and plants on the forest floor.

undervaluation. Placing too low an economic value upon an activity or resource without taking into account all the economic benefits associated with that activity or resource. For example, tropical forests yield not only timber products that tend to be underpriced for institutional reasons, but also yield a variety of non-timber products and other environmental values (such as biodiversity and carbon sequestration) that are not captured in timber prices.

unemployment rate. A measure of the number of people out of work but willing to work and looking for work. The unemployment rate is measured as a proportion of the total workforce.

UNESCO. See *United Nations Educational, Scientific and Cultural Organization.*

uneven-aged stand. A forest in which at least three stands are of significantly different ages (a difference of greater than 20 years).

ungulate. A group of herbivore mammals with hooves, such as horses and deer.

uniform distribution. A probability distribution or density function where the probability of picking any value of the random variable, defined by the distribution, is identical.

uniformitarianism. The notion that the current physical processes also existed in the geological past, and that observation of current events can provide insights about the development of the Earth.

unilateral cooperation. The strategy followed by a participant in a game who chooses to cooperate with others without any assurance that such co-operation will be reciprocated.

unilateral defection. The strategy followed by a participant in a game who chooses to deviate from any cooperative arrangement and pursue his or her own self-interest.

unimodal curves. Functional relationships that exhibit an "inverted U shape" (with a single global maximum), or a "U shape" (with a single global minimum).

unit cost measures of scarcity. Measures of the scarcity of non-renewable resources based on the costs of production or extraction.

United Nations (UN). Successor to the League of Nations established after World War II in 1945. It has 192 member nations and promotes international peace, human rights and sustainable development.

United Nations Conference on Environment and Development (UNCED). The 1992 conference that took place in Rio de Janeiro, Brazil that resulted in an environmental action plan called Agenda 21 and the United Nations

Framework Convention on Climate Change, the Convention on Biological Diversity and the Convention to Combat Desertification.

United Nations Convention on the Law of the Sea (UNCLOS).   A convention developed by a series of international meetings in the 1970s and 1980s to govern the exploitation of marine resources. The Convention established 200-nautical-mile exclusive economic zones (EEZs) from the coastline of maritime states.

United Nations Development Program (UNDP).   A United Nations (UN) agency charged with assisting other UN agencies in undertaking social and economic development in poor countries.

United Nations Educational, Scientific and Cultural Organization (UNESCO).   Formed in 1946 and charged with fostering international cooperation of scientific research. Among its many activities, UNESCO directs the Man and the Biosphere program.

United Nations Environment Program (UNEP).   Formed in 1972, the UNEP is based in Kenya and co-ordinates international efforts at monitoring and protecting the environment.

United Nations Framework Convention on Climate Change (UNFCCC). An international treaty established at the 1992 Earth Summit in Rio de Janeiro. The UNFCCC forms the basis for coordinated global action on climate change, with the stated objective to "stabilise greenhouse gas concentrations in the atmosphere at a level that would prevent dangerous anthropogenic interference with the climate system". Over 190 state signatories meet annually to formulate mitigation measures (i.e., the Kyoto Protocol), and outline broader directions, such as the Copenhagen Accord 2 degree target. Parties to the UNFCCC are distinguished between Annex I countries and non Annex I countries, with the former group initially committing to stabilizing emissions, and then later to binding reductions. See *2° Celsius, Annex I Countries, Conference of the Parties,* and *Kyoto Protocol.*

unitization.   An agreement over the use of common-pool resources, and particularly over oil fields, in which the owners of the sub-surface rights agree to let one user extract the resource and provide net benefits to all others at an agreed-to formula.

FURTHER READING
Libecap (1994).

unit root tests.   A statistical method used to test whether variables are stationary by evaluating whether the variable is following a random walk. If the null hypothesis that the variable is following a random walk cannot be rejected, then the variable is non-stationary.

FURTHER READING
Greene (2003).

universal property rights.   Situation where all property rights have been completely assigned.

unlikely.   See *likelihood.*

unstable equilibrium.   A condition in dynamic analysis when a system will not return to its original state following a perturbation. In biological systems, the minimum viable population stock is an unstable equilibrium because a positive perturbation will cause stocks to grow to the carrying capacity, while a negative perturbation leads to extinction.

upcycle.   To recycle waste into something of greater quality or higher environmental value.

uptake.   A substance's accumulation in a sink or reservoir (i.e., during carbon sequestration).

upwelling.   The ascent of cold, often nutrient-rich waters to the surface from the ocean depths. Upwelling can be caused by many factors, including wind circulation. Regions where this phenomenon occur generally support productive fisheries.

uranium.   A metal commonly found as two isotopes, uranium 238 and uranium 235. Uranium 235 is used as a fuel in nuclear fission. See *nuclear fission.*

urban heat island effect.   A phenomenon whereby heat sources, low reflectivity of constructions, vegetation removal and wind blockage in urban environments causes higher air temperatures compared to the surrounding countryside.

urbanization.   The process by which large centers of population increase in size as a proportion of the total national population. In developing countries, urbanization has occurred rapidly because of migration of persons living in rural or small communities to larger urban centers of population.

urban sprawl. The patterns under which urban areas within cities expand outside of city limits and encroach upon neighboring land in response to population increases. See *conurbation.*

Uruguay Round, The. The last of eight rounds of talks held under the auspices of GATT that focused on the removal of tariffs, the development of more liberalized trade rules, and the formation of the World Trade Organization. The round began in 1986 at Punta del Este in Uruguay and was concluded in 1994 with the Marrakesh Agreement. See *Marrakesh Agreement, The.*

US Bureau of Land Management (BLM). A part of the US Department of the Interior that manages over 100 million hectares of federal land, primarily in the western states.

US Bureau of Reclamation. A US federal agency charged with water management in the western states. It also provides hydroelectric power and is a major supplier of water.

US–Canada Air Quality Agreement. A 1991 agreement that committed the two countries to reductions in sulfur dioxide and nitrogen oxide emissions, and provided a forum for addressing transboundary air issues.

user cost. The future cost, in terms of increased extraction costs in the future, and lost market opportunities from extracting an extra unit of a resource.

FURTHER READING
Bergstrom and Randall (2010).

US National Wildlife Refuge System. A US system encompassing some 38 million hectares and 520 refuges for the protection and conservation of wildlife.

use value. From an individual's perspective, the willingness to pay for the *in situ* use of a natural resource or an environmental good.

FURTHER READING
Bergstrom and Randall (2010).

usufruct rights.   Property rights to use land or natural resources that are contingent on its use. Usufruct rights and tenure often exist with community rights where defined tracts of land may be reserved for the use of particular families or clans, but the land remains in the control of the community. Thus, upon cessation of the use of the land by a family or clan, the land may be assigned to other members of the community.

utilitarianism.   A term that operationalizes the concept of the nineteenth-century economist and philosopher, Jeremy Bentham, who argued that decisions should be made to maximize the greatest good for the greatest number. A social welfare function which maximizes aggregate utility over all individuals is called a utilitarian social welfare function. See *social welfare function*.

utility.   The well-being, or satisfaction, associated with an activity. Utility theory forms the basis of many economic models of consumer behavior.

utility function.   A relationship where utility is a function of the quantities consumed of goods and services, and other factors.

utilization factor.   See *capacity factor*.

V

vadose zone. The region between the land surface and the top of the water table.

validation. 1. The process of determining whether a model provides an accurate representation of a phenomenon for the purpose it is intended. 2. The process of evaluating the achievement of particular standards. With regards to emission inventories, validation ensures that the inventory has been constructed correctly; whereas, for offset projects, it requires an independent assessment of the project's design and calculations.

validator. A third party responsible for confirming the validity of carbon emissions and reductions.

value added. The difference between the value of a good after a production process, less the value of all the inputs in producing the good.

value added tax (VAT). A tax on the value added from one stage of the production and distribution process to another that is imposed as a percentage of the price of the good or service. Manufacturers can claim a credit on all items they purchased equal to the VAT they paid, but must pay the VAT on the goods they sell.

variable. Anything that can vary. In mathematical problems, a variable is usually denoted by a letter or symbol.

variable costs. Costs that vary with the level of output.

variable factor. A factor of production that can be varied in the short-run, such as the amount of electricity.

variable rate pricing. Pricing where the price paid per unit depends on the units used.

variance. A measure of dispersion of a variable, defined as the average of the squared deviations from the arithmetic mean.

variance–covariance matrix. A symmetric matrix where the elements along the main diagonal (a_{ij} where $i = j$) are the variances of the variables, and the covariances of variables *i* and *j* are given by the element a_{ij}, where *i* does not equal *j*.

vector.   1. A matrix consisting of one row or column. 2. Animals, such as insects, that transmit disease from one host to another.

vehicle bias.   The potential bias that may arise from the hypothetical method of payment in surveys designed to elicit the value that individuals may have for aspects of the environment. For example, a survey that elicits values by asking respondents if they would accept higher property taxes may yield different estimates of value than a survey that elicits values by proposing the use of higher user fees.

veld.   Open country with sparse shrubs or trees, found in South Africa.

Venturi effect.   The increased velocity of a fluid or gas as it passes through a restriction in its flow. In some cities, the construction of large buildings can create a Venturi effect with very high winds at particular places, such as street corners.

veratoxin.   A toxic substance produced by some types of *Escherichia coli*, including the deadly *Escherichia coli* O157:H7. See *Escherichia coli O157:H7*.

verification.   A review or an audit of a submission against a set of specific requirements. With regards to emissions, an independent *ex-post* assessment of an inventory or offset project, especially with regards to accuracy, completeness and, in the case of forestry-related activities, permanence.

Verified Emission Reduction (VER).   A carbon credit in the voluntary market that has been assessed and verified by an accredited third party. See *emissions credit*.

vertical equity.   The notion that people should be treated differently if they face different circumstances. The principle is applied in most countries that have a progressive income tax rate where individuals at higher incomes pay proportionately more in income tax than people at low incomes. See *horizontal equity*.

vertical integration.   The merger of firms at different stages of production.

vertisol.   A black soil rich in montmorillonite and that is prone to shrinking and swelling, and thus cracking.

very unlikely.   See *likelihood.*

vicarious consumption.   The enjoyment of a good or service through indirect means (e.g., by viewing a film or reading literature), or through consumption of the good by others.

Vickrey auction.   Auction named after the Nobel laureate in economics, William Vickrey, that is intended to elicit truthful bids about the value of the asset or item being sold.

Vienna Convention.   The document agreed to by 20 countries in 1985 through which the parties committed themselves to address what, at the time, was perceived to be a potential problem of stratospheric ozone depletion due to the emissions of chlorofluorocarbons (CFCs).

virgin biomass.   In fisheries, the hypothetical value of biomass before any fishing has taken place.

virtually certain.   See *likelihood.*

virtual population analysis (VPA).   A method for calculating the size of a population that is frequently applied in fisheries. It uses estimates of current harvests and natural mortality to reconstruct past levels of the resource.

FURTHER READING
Grafton et al. (2010).

virtual water.   A measure of the volume of fresh water used in the fabrication of a good or service.

virus.   A micro-organism that consists of genetic material covered in protein that reproduces inside the cells which it infects. Viruses are responsible for many diseases, including AIDS.

visceral characteristics.   Non-scientific characteristics of animal species that influence the perceived need to preserve or protect a species.

FURTHER READING
Metrick and Weitzman (1996).

viscosity.   A measure of the ability of liquids to flow, commonly measured at the speed at which a liquid flows through a tube.

volatile organic compounds (VOCs).   Reactive organic compounds that can be released from the vaporization and combustion of hydrocarbons. VOCs can act as a catalyst in the production of ground-level or tropospheric ozone, which can be harmful to plants and to animals. See *ozone*.

voluntary agreements.   Agreements reached between two parties based on mutually agreed-to responsibilites and obligations. Voluntary agreements are achieved without any legal sanctions or enforcement in the case of non-compliance.

Voluntary Emission Reduction.   See *Verified Emission Reduction*.

voluntary market.   A market for voluntary carbon offsets.

voluntary offsets.   Carbon offsets that are not part of a formal scheme such as the Clean Development Mechanism. These activities are initiated by the desire of organizations or individuals to reduce or eliminate their carbon footprint. See *carbon offset scheme*.

voluntary risk.   Risks individuals undertake on their own volition, such as crossing the road. See *involuntary risk*.

von Bertalanffy equations.   Equations that relate the length (or weight) of an individual in a population to a theoretical maximum length (or weight) defined as L_4, current age defined by t, its age at zero length (or weight) defined by t_0, and a growth parameter defined by K. Thus, the equation for growth in terms of length at time t is

$$L_t = L_4 [1 - e^{-K(t - t0)}].$$

FURTHER READING
King (1995).

von Stackelberg duopoly.   A duopoly where there exists a market leader that decides upon its production or price and, in turn, the other firm or follower bases its quantity or price decision upon this.

vulnerability.   The extent to which an environment or species is threatened or endangered by changes.

vulnerability premium.   The financial risk associated with a reliance on limited sources of supply of resources, and the possibility of disruption of those supplies.

W

wadi. A valley that remains dry except after heavy rainfall, at which time it will temporarily support vegetation.

wait-and-see scenario. An approach to environmental problems where, until firm evidence of a problem exists, no action is taken. The approach contrasts to the widely accepted precautionary principle. See *precautionary principle.*

Walkerton. A town in Ontario, Canada, the drinking water system of which was contaminated by *Escherichia coli* O157:H7 in May 2000. The contamination came from cow manure near a well that supplied water to the community and resulted in the death of seven individuals as well as making more than 2300 persons seriously ill. See *Escherichia coli* O157:H7.

Walras' law. A fundamental law of economics, named after Léon Walras, that is used in general equilibrium analysis in economics. The law states that if there are n outputs (x_i) with n prices (p_i) in an economy produced by m factors (v_j) with m factor prices (w_j) there exist only $2n + 2m - 1$ independent equations necessary to find the equilibrium. In other words, the fact that the sum of $p_i x_i$ must equal the sum of $w_j v_j$ can be used to solve for one of the unknown outputs.

FURTHER READING
Takayama (1985).

warm glow. The notion that site specific valuations of a public resource or environmental good may actually be a surrogate for values given to environmental quality in general. Such values may be a potential source of bias in contingent valuation studies.

waste-end taxes. Taxes paid on the waste discharged into the environment created by the production process. Waste end taxes may be distinguished by the volume or quality of waste discharged.

water (H_2O). One of the most abundant molecules on Earth that consists of one oxygen and two hydrogen atoms. Most of the Earth's water is found in the oceans (97 percent). The remaining water is so-called fresh water. Over three quarters of the planet's fresh water is found in snow and ice, almost a quarter is groundwater and less than 1 percent of the total fresh water is surface water found in lakes and rivers.

water balance.   See *water budget*.

water budget.   A measure of the amount of water found within an ecosystem, which takes into account how much water is coming in, how much is stored, and how much is exiting the system.

water cycle.   See *hydrological cycle*.

water entitlement.   A property right that entitles the holder to access water allocations in perpetuity. Different levels of security or reliability of property rights generate different yields or allocations.

water footprint.   See *virtual water*.

water pathogens.   Potentially harmful organisms, usually micro-organisms, that exist naturally or are introduced into bathing water (e.g., ocean, seas, lakes, etc.) and drinking water.

water quality.   A broad term for describing the chemical and biological attributes of water. Often used to express whether water is fit for human consumption or to support other organisms.

water rights market.   A market where water rights, or the rights to access and use specified volumes of water for designated periods of time, are traded.

watershed.   An area that is the drainage basin of a defined stream or river.

watershed management.   The management of the flows, discharges and withdraws into water sources in a defined watershed so as to achieve environmental and socioeconomic objectives.

water scarcity.   A situation where water demand severely exceeds supply as a result of physical and/or economic constraints. A common definition of water scarcity is less than 1000 m^3/yr of consumption per person. See *drought, economic water scarcity, physical water scarcity*, and *water stress*.

water stress.   A term applied to a community, region or country that has an annual water supply of less than 1700 m^3 per person. See *water scarcity*.

water table.   The uppermost level of groundwater, or the level below the land surface where the ground is saturated with water.

water-use efficiency. The ratio of the amount of water used by a crop to the amount of water applied.

water vapor. Water molecules in a gaseous form that constitutes one of the primary components of the atmosphere. Created by evaporation of surface water, whereupon it forms the basis of clouds, rain and snow. Water vapor plays an important role transferring heat throughout the climate system and is the most abundant greenhouse gas. In the context of global warming, this substance is associated with various feedback effects. For example, higher temperatures generate more water vapor, hence reinforcing warming. On the other hand, larger volumes of atmospheric water vapor stimulates cloud formation, potentially reflecting more solar radiation away from the Earth. See *greenhouse effect.*

watt (W). Unit of power that equals one joule per second. Power corresponds to the rate at which work is performed or energy is transmitted.

wattle jungle. In Swaziland, an overgrown and abandoned, but previously managed plantation of *Acacia mearnsii.*

wave power. A technology being developed to convert energy from ocean waves into electricity. See *renewable energy.*

weak sustainability. The notion that the environment or natural capital is, to some extent, a substitute for human-made capital. See *strong sustainability.*

FURTHER READING
Neumayer (2003).

wealth. A measure of the current value of the assets of an individual, community or society. Wealth may include tangibles, such as household goods and bank balances, as well the natural capital stock and the human capital stock.

weather. At a point in time and place, the state of the atmosphere as measured by its temperature, wind speed, precipitation, humidity, and other factors.

weighted average. An average value determined by weights. For observations $x_1, x_2, \ldots x_n$ and weights $w_1, w_2, \ldots w_n$, it is calculated as $(w_1 x_1 + w_2 x_2 + \ldots + w_n x_n)/(w_1 + w_2 + \ldots + w_n)$.

welfare aggregation.   The addition of a measure of the economic well-being of individuals to construct a measure of the overall well-being of society.

welfare change.   A measure of the change in the well-being of an individual as a result of a change in some state of the world. See *compensating surplus variation*, *equivalent variation*, and *consumer surplus*.

welfare economics.   The field of economics that investigates the interaction between the economic well-being of individuals and economic efficiency.

well-to-wheels efficiency.   A measure of the overall performance of an energy conversion chain that accounts for the unavailable or energy lost at different stages along the chain. This approach is useful for comparing the efficiency of alternative energy systems capable of providing energy services. See *energy conversion chain*.

Wellman–Lord process.   A highly effective method for removing sulfur dioxide from the emissions of coal-burning generators.

wet biomass.   The weight of living organisms at an instant in time within a given area, including any water found within the organism.

wet deposition.   The transmission of air pollutants from the atmosphere to the Earth's surface via precipitation.

wetland pollution filter.   The use of wetlands, either natural or artificial, to convert effluent and liquid waste into clean water.

wetlands.   Areas of land characterized by the presence of either standing or flowing water for at least a portion of the year, including marshes, swamps, bogs, shallow tidal areas, and ground that is waterlogged. Wetlands are important habitats for fauna, and staging areas for migrating birds.

white noise series.   A series of independent random variables that are independently and identically distributed with zero mean.

wilderness.   The 1964 United States Wilderness Act defines wilderness as "an area where the earth and its community of life are untrammelled by man, where man himself is the visitor that does not remain". More generally, wilderness is used to describe natural habitats that are largely free from human activity or development.

wilderness tourism. A subset of "ecotourism" in which destinations are primarily in remote, wilderness areas that are largely free from any development.

wildfire. A fire occurring in rural or wilderness areas, typically characterized by a large size, rapid propagation and directional changes, as well as the ability to overcome obstacles, such as roads or rivers. Although wildfires have occurred naturally throughout history, higher temperatures and lower rainfall occurring in some climate change-affected regions may increase their frequency and severity. Wildfires also contribute to global warming as they emit vast quantities of carbon dioxide into the atmosphere. See *peat fires*.

wildlife corridor. Linear areas used by wildlife to move between different habitats (e.g., river valleys used by elk to reach their summer range in the mountains and then return to their winter range in the foothills).

wildlife refuges. Locations where wildlife is protected for conservation purposes. The United States currently has 36 million hectares designated as part of its national wildlife refuge system.

willingness to accept. The compensation required to return an individual to his or her original state of economic well-being following some change (possibly hypothetical) in the world. In many cases, a willingness to accept measure is considered to be inferior to the willingness to pay measure of welfare change.

willingness to pay. The willingness of an individual to pay (i.e., give up part of one's budget) in order to secure a good or service. Willingness to pay is often used to refer to the amount that a consumer would pay for a hypothetical good, service, or change in some state of the world, particularly as described in a contingent valuation survey.

wind chill. The cooling of a body via wind which draws away latent heat. The wind-chill index measures how much body heat is lost per square meter of skin per hour and provides an equivalent (and colder) temperature that would be comparable if there were no wind.

wind power. A form of renewable electricity generated from the energy of on-shore or off-shore wind. Wind turbines range in size from small household devices to large commercial operations tens of meters high. Rapid expansion of commercial wind power is currently taking place on a global scale.

wind farm.   A grouping of wind turbines, usually providing power to a central electricity grid.

windrow.   The accumulation of slash or other vegetative debris on a forest site that has been cut for harvest. The windrow usually is intended to promote regeneration at that site.

windthrow.   See *blowdown*.

wise use movement.   A movement of various individuals, groups and communities that supports the development of natural resources. One of the goals of the movement is to ensure the primacy of private-property rights in the regulation of the environment, and to oppose large-scale government intervention.

withdrawal permits.   Permits that allow the holder to use a given volume of water for a specified period of time.

withdrawal rate.   Rate at which water is removed from an aquifer.

withdrawal right.   A property right that enables the holder to extract or harvest a resource, such as a hunting permit.

withholding period.   The period of time a food product must be withheld from sale or consumption after a treatment or exposure to a toxic substance. Many fruits have a withholding period after they have been sprayed with pesticides.

woodlot.   The forested area of privately owned land used for harvesting trees.

Working Group (WG).   A collaboration of experts or policy makers formed to address a specific topic or issue. In the context of climate change, this term may refer to the three IPCC Working Groups respectively reviewing: climate science; impacts, adaptation and vulnerability; and mitigation. This term is also applied to various Working Groups formed by the UNFCCC to address specific negotiation issues. See also

Intergovernmental Panel on Climate Change and *United Nations Framework Convention on Climate Change.*

World Bank.　See *International Bank for Reconstruction and Development.*

World Business Council on Sustainable Development (WBCSD).　A coalition of around 200 international companies representing more than 30 countries and 20 industrial sectors, who are committed to the business case for sustainable development.

World Health Organization (WHO).　A United Nations agency formed in 1948 and mandated with improving the health of the world's people.

World Trade Organization (WTO).　The successor to the GATT that was established in 1995, is mandated to expand multilateral trade. WTO has a dispute resolution mechanism to help prevent restrictive trade practices. See also *International Trade Organization (ITO)* and *General Agreement on Tariffs and Trade (GATT).*

FURTHER READING
Gallagher (2008).

X

X-efficiency. Term coined by H. Leibenstein to describe the effectiveness of managers at minimizing the costs of production for a given level of output.

xerophyte. Plant adapted to dry or desert conditions, such as a cactus.

Y

yard. Unit of length equal to 0.91 meters.

year-class. A method of defining cohorts within a population based upon the year that they were born or were spawned.

year-class strength. Term that refers to the size of a cohort in a population relative to older and younger cohorts. See *cohort*.

yellowcake. Name given to a mixture of uranium that is used as a fuel in nuclear reactors.

yellow fever. A life-threatening disease caused by a virus transmitted by mosquitoes.

younger dryas. A period of sudden cold temperatures in the northern hemisphere that began about 10 000 to 11 000 years ago and which occurred after the end of the last glacial. A hypothesis is that the sudden cooling may have arisen from the rapid melting of Greenland glaciers which, in turn, created a cold mass of water that diverted the Gulf Stream from the North Atlantic.

Z

zero discharge. No measurable emissions from a defined source.

zero emissions. See *zero discharge.*

zero population growth. An often-stated goal of some environmentalists that the human population of the world be stabilized at its current level.

zero-revenue auction. An auction designed to allocate assets or property rights among firms or users where the revenue collected from the auction is subsequently returned to owners of the newly acquired assets.

zero sum game. A mathematical construction of a set of choices where the outcomes are symmetrically opposed such that the total benefits received by one party are equivalent to the losses borne by the other. See *non-zero sum game.*

zero wealth effects. An assumption made under the Coase Theorem that, in the absence of property rights, creating and giving a property right to any one party does not affect the wealth of any of the parties, and thus does not affect the ability of the parties to reach the efficient outcome. See *Coase Theorem.*

zoning. Collective term for regulations that govern the permitted land use in defined areas. Zoning can be used to conserve green spaces within urban environments and to prevent industries from setting up their operations in residential neighborhoods.

zoogeography. The study of the spatial distribution of animals.

zoology. The study of animal life.

zooplankton. Plankton consisting of small creatures that include the juvenile stages of many larger aquatic creatures. See *plankton* and *phytoplankton.*

References

Abrams, P.A. (1995), "Monotonic or Unimodal Diversity–Productivity Gradients: What Does Competition Theory Predict?", *Ecology,* October, **76**(7), 2019–2027.

Ahrens, C.D. (2007), *Meteorology Today: An Introduction to Weather, Climate, and the Environment*, 8th edn, Belmont, California: Thomson.

Akerlof, G. (1970), "The Market for 'Lemons': Quality, Uncertainty and the Market Mechanisms", *Quarterly Journal of Economics*, **84**, 488–500.

Alberini, A. and J.R. Kahn (2006), *Handbook on Contingent Valuation*, Northampton, Massachusetts: Edward Elgar.

Aldy, J.E and R.N. Stavins (2009), *Post-Kyoto International Climate Policy*, Cambridge: Cambridge University Press.

Anderson, M.S. (1994), *Governance by Green Taxes: Making Pollution Prevention Pay*, Manchester, UK: Manchester University Press.

Archer, D. and S. Rahmstorf (2010), *The Climate Crisis: An Introductory Guide to Climate Change*, Cambridge: Cambridge University Press.

Arnold, J.E.M. and J.G Campbell (1986), "Collective Management of Hill Forests in Nepal: The Community Forest Development Project", in *Proceedings on the Conference on Common Property Management*, National Academy Press, pp. 425–454.

Arrow, K. and A. Fisher (1974), "Environmental Preservation, Uncertainty, and Irreversibility", *Quarterly Journal of Economics*, **88**(1), 312–319.

Barbier, E.B. (ed.) (1993), *Economics and Ecology*, London: Chapman and Hall.

Barro, R.J. and X. Sala-i-Martin (2004), *Economic Growth*, 2nd edn, Cambridge, Massachusetts: MIT Press.

Barry, R.G. and R.J. Chorley (2003), *Atmosphere, Weather and Climate*, 8th edn, London: Routledge.

Baumol, W.J. and W.E. Oates (1988), *The Theory of Environmental Policy*, 2nd edn, Cambridge: Cambridge University Press.

Bergstrom, E. and A. Randall (2010), *Resource Economics: An Economic Approach to Natural Resource and Environmental Policy,* 3rd edn, Northampton, Massachusetts: Edward Elgar.

Bierman, H.S. and L. Fernandez (1993), *Game Theory with Economic Applications*, Reading, Massachusetts: Addison-Wesley Publishing.

Binmore, K. (2007), *Playing for Real: A Text on Game Theory*, Oxford: Oxford University Press.

Birmingham, E., C.W. Dick and C. Moritz (eds) (2005), *Tropical*

Rainforests: Past, Present and Future, Chicago: Chicago University Press.

Bloom, A.L. (2004), *Geomorphology: A Systematic Analysis of Late Cenozoic Landforms*, 3rd edn, Long Grove, Illinois: Waveland Press.

Boserup, E. (2005), *The Conditions of Agricultural Growth*, London: Allen & Unwin.

Brennan, A. (1992), "Moral Pluralism and the Environment", *Environmental Values*, **1**, 15–33.

Brown, G. and R. Mendelsohn (1984), "The Hedonic Travel Cost Method", *Review of Economics and Statistics*, **66**, 427–433.

Brown, J.H., D.W. Mehlman and G.C. Stevens (1995), "Spatial Variation in Abundance", *Ecology*, October, **76**(7), 2028–2043.

Callan, S.J. and J.M. Thomas (2007), *Environmental Economics and Management: Theory, Policy, and Applications,* 4th edn, Mason, Ohio: Thomson.

Campbell, N.A., J.B. Reece, L.A. Urry, M.L. Cain, S.A. Wasserman, P.V. Minorsky and R.B. Jackson (2010), *Biology*, 9th edn, San Francisco: Benjamin Cummings.

Carson, R. (2002), *Silent Spring*, New York: Houghton Mifflin.

Chiang, A.C. (2000), *Elements of Dynamic Optimization*, Long Grove, Illinois: Waveland Press.

Chow, G. (1960), 'Tests of Equality between Sets of Coefficients in Two Linear Regressions', *Econometrica*, July, **28**(3), 591–605.

Clark, C. (2010), *Mathematical Bioeconomics*, 3rd edn, New York: John Wiley & Sons.

.Coase, R.H. (1960), "The Problem of Social Cost," *Journal of Law and Economics*, **3**, 1–44.

Cohen, J. and I. Stewart (2000), *The Collapse of Chaos*, New York: Penguin Books.

Colborn, T., D. Dumanoski and J.P. Myers (1997), *Our Stolen Future*, New York: Plume/Penguin.

Conrad, J. (1995), "Bioeconomic Models of the Fishery", in *The Handbook of Environmental Economics*, Cambridge, Massachusetts: Basil Blackwell, Chapter 18.

Cornes, R. and T. Sandler (1996), *The Theory of Externalities, Public Goods and Club Goods*, 2nd edn, Cambridge: Cambridge University Press.

Cropper, M. and F.G. Sussman (1990), "Valuing Future Risks To Life", *Journal of Environmental Economics and Management*, **20**(2), 160–174.

Daly, H.E. (1973), "The Steady-State Economy: Toward a Political Economy of Biophysical Equilibrium and Moral Growth", in H.E.

Daly (ed.), *Toward a Steady-State Economy*, San Francisco: W.H. Freeman and Company.

Daly, H.E. (1996), "Consumption: Value Added, Physical Transformation, and Welfare", in R. Costanza, O. Segura and J. Martinez-Alier (eds), *Getting Down to Earth*, Washington DC: Island Press, 49–59.

Darwin, C. (1859), *On the Origin of Species by Means of Natural Selection or Preservation of Favoured Races in the Struggle for Life*, 1st edn, London: Murray.

Davis, P. (2000), *The Fifth Miracle: The Search for the Origin and Meaning of Life*, New York: Simon Schuster.

Dawkins, R. (2006), *The Selfish Gene*, Oxford: Oxford University Press.

de Bruyn, S.M. (2000), *Economic Growth and the Environment: An Empirical Analysis*, Dordrecht: Kluwer Academic Publishers.

Devall, B. and G. Sessions (1985), *Deep Ecology*, Salt Lake City, Utah: Peregrine Smith.

Devlin, R.A. and R.Q. Grafton (1998), *Economic Rights and Environmental Wrongs: Property Rights for the Common Good*, Cheltenham, UK: Edward Elgar.

Eggertsson, T. (1990), *Economic Behavior and Institutions*, Cambridge: Cambridge University Press.

Ekins, P. (2000), *Economic Growth and Environmental Sustainability: The Prospects for Green Growth,* London: Routledge.

Fetter, C.W. (2001), *Applied Hydrogeology*, Upper Saddle River, New Jersey: Prentice Hall.

Field, B.C. (2008), *Natural Resource Economics: An Introduction*, 2nd edn, Long Grove, Illinois: Waveland Press.

Fisher, A. and R. Raucher (1984), "Instrinsic Benefits of Improved Water Quality: Conceptual and Empirical Perspectives", in V.K. Smith and A.D. Witte (eds), *Advances in Applied Microeconomics*, Greenwich, Connecticut: JAI Press.

Food and Agriculture Organization of the United Nations (2009), *State of the World's Forests*, Rome: Food and Agriculture Organization of the United Nations.

Ford, A. (2009), *Modeling the Environment: An Introduction to System Dynamics Modeling of Environmental Systems*, 2nd edn, Washington DC: Island Press.

Frank-Kamenetskii, M.D. (1997), *Unraveling DNA: The Most Important Molecule of Life*, Reading, Massachusetts: Addison-Wesley.

Frankel, J. (2009), 'Addressing the Leakage/Competitiveness Issues in Climate Change Policy Proposals', in L. Brainard and I. Sorkin (eds), *Climate Change, Trade and Competitiveness: Is a Collision Inevitable*, Baltimore, Maryland: Brookings Institution Press.

Freeman, S. and J.C. Herron (2007), *Evolutionary Analysis*, 4th edn, Upper Saddle River, New Jersey: Prentice Hall.

Gale, R.P. and S.M. Cordray (1994), "Making Sense of Sustainability: Nine Answers to 'What Should Be Sustained'", *Rural Sociology*, **59**(3), 311–332.

Gallagher, K.P. (ed.) (2008), *Handbook on Trade and the Environment*, Northampton, Massachusetts: Edward Elgar.

Garnaut, R., S. Howes, F. Jotzo and P. Sheehan (2008), "Emissions in the Platinum Age: The Implications of Rapid Development for Climate-change Mitigation", *Oxford Review of Economic Policy*, **24**(2), 377–401.

Garnaut, R. and A.C. Ross (1975), "The Neutrality of the Resource Rent Tax", *The Economic Record*, **55**, 193–201.

Garrod, G. and K.G. Willis (1999), *Economic Valuation of the Environment: Methods and Case Studies*, Northampton, Massachusetts: Edward Elgar.

Georgescu-Roegen, N. (1973), "The Entropy Law and the Economic Problem", in H.E. Daly (ed.), *Toward a Steady-State Economy*, San Francisco: W.H. Freeman and Company.

Gleick, J. (2008), *Chaos: Making a New Science*, London: Penguin.

Gould, S.J. (2001), 'Reconstructing (and Deconstructing) the Past', in S.J. Gould (ed.), *The Book of Life*, Toronto: Viking.

Grafton, R.Q. (1996), "Individual Transferable Quotas: Theory and Practice", *Reviews in Fish Biology and Fisheries*, **6**, 5–20.

Grafton, R.Q. (2000), "Governance of the Commons: A Role for the State?", *Land Economics*, November, **76**(4), 504–517.

Grafton, R.Q., W. Adamowicz, D. Dupont, H. Nelson, R.J. Hill and S. Renzetti (2004), *The Economics of the Environment and Natural Resources*, Malden, Massachusetts: Blackwell Publishing.

Grafton, R.Q. and R.A. Devlin (1996), "Paying for Pollution: Permits and Charges", *Scandinavian Journal of Economics*, **98**, 275–288.

Grafton, R.Q., R. Hilborn, D. Squires, M. Tait and M. Williams (2010), *Handbook of Marine Fisheries Conservation and Management*, New York: Oxford University Press.

Grafton, R.Q., T. Kompas and R. Hilborn (2007), "The Economics of Overexploitation Revisited", *Science*, **318**, 1601.

Grafton, R.Q. and H.W. Nelson (1997), "Fishers' Individual Salmon Harvesting Rights: An Option for Canada's Pacific Fisheries", *Canadian Journal of Fisheries and Aquatic Sciences*, **54**, 474–482.

Grafton, R.Q., L.K. Sandal and S.I. Steinshamn (2000), "How to Improve the Management of Renewable Resources: The Case of Canada's Northern Cod Fishery", *American Journal of Agricultural Economics*, **82**, 570–580.

Grafton, R.Q. and J. Silva-Echenique (1997), "How to Manage Nature? Strategies, Predator–Prey Models, and Chaos", *Marine Resource Economics*, **12**, 127–143.

Grafton, R.Q., D. Squires and K.J. Fox (2000), "Private Property and Economic Efficiency: A Study of a Common-Pool Resource", *The Journal of Law and Economics*, **43**(2), 679–713.

Granger, C.W.J. (1969), "Investigating Causal Relations by Econometric Models and Cross-Spectral Models", *Econometrica*, **37**, 424–438.

Greene, W.H. (2003), *Econometric Analysis,* 5th edn, Upper Saddle River, New Jersey: Prentice-Hall.

Griffiths, A.J.F., S.R. Wessler, R.C. Lewontin, W.D. Fixsen and S.B. Carroll (2008), *Introduction to Genetic Analysis*, 9th edn, New York: W.H. Freeman.

Grossman, M. (1972), "On the Concept of Health Capital and the Demand for Health", *Journal of Political Economy*, **80**(2), 223–255.

Guttorp, P. (1995), *Stochastic Modelling of Scientific Data*, London: Chapman and Hall.

Hanley, N. and E.B. Barbier (2009), *Pricing Nature: Cost–Benefit Analysis and Environmental Policy*, Cheltenham, UK: Edward Elgar.

Hardin, G. (1968), "The Tragedy of the Commons", *Science*, **162**, 1143–1248.

Hartman, R. (1976), "The Harvesting Decision When a Standing Forest Has Value", *Economic Inquiry*, **14**, 52–58.

Hartwick, J.M. (1977), "Intergenerational Equity and the Investing of Rents from Exhaustible Resources", *American Economic Review*, **67**, 972–974.

Hartwick, J.M. and N. Olewiler (1998), *The Economics of Natural Resource Use*, Reading, Massachusetts: Addison-Wesley.

Hazilla, M. and R.J. Kopp (1990), "The Social Cost of Environmental Quality Regulations: A General Equilibrium Analysis", *Journal of Political Economy*, **98**(4), 853–873.

Hill, C., W. Griffiths and G. Judge (2001), *Undergraduate Econometrics*, New York: John Wiley & Sons.

Holling, C.S. (1973), "Resilience and Stability of Ecosystems", *Annual Review of Ecology and Systematics*, **4**, 1–23.

Holmberg, J., K. Robert and K. Erikson (1996), "Socio-ecological Principles for a Sustainable Society", in R. Costanza, O. Segura and J. Martinez-Alier (eds), *Getting Down to Earth*, Washington DC: Island Press, pp. 17–48.

Katz, M.L. and H.S. Rosen (1998), *Microeconomics*, 3rd edn, Boston, Massachusetts: Irwin McGraw-Hill.

Kaufman, L. and K. Mallory (1993), *The Last Extinction*, Cambridge, Massachusetts: MIT Press.

Kennedy, P. (2003), *A Guide to Econometrics*, 5th edn, Cambridge, Massachusetts: MIT Press.

Keynes, J.M. (1936), *The General Theory of Employment, Interest, and Money*, London: Macmillan.

Kimmins, J.P. (2004), *Forest Ecology: A Foundation for Sustainable Management*, 3rd edn, Upper Saddle River, New Jersey: Prentice Hall.

King, M. (2007), *Fisheries Biology, Assessment and Management*, Malden, Massachusetts: Blackwell.

Krebs, C.M. (2009), *Ecology*, San Francisco: Benjamin Cummings.

Krugman, P., M. Obstfeld and M. Melitz (2010), *International Economics: Theory and Policy*, 9th edn, Upper Saddle River, New Jersey: Prentice Hall.

Lenton, T.M., H. Held, E. Kreigler, J.W. Hall, W. Lucht, S. Rahmstorf and H.J. Schelnhuber (2008), "Tipping Elements in the Earth's Climate System", *Proceedings of the National Academy of Sciences*, **105**, 1786–1793.

Libecap, G.D. (1994), *Contracting for Property Rights*, New York: Cambridge University Press.

Lipsey, R.G. and K. Lancaster (1956), *The General Theory of the Second Best*, Cheltenham, UK: Edward Elgar.

Lovelock, J.E. (1990), "Hands Up for the Gaia Hypothesis", *Nature*, March, **344**, 100–102.

Lovelock, J.E. (2000), *Gaia: A New Look at Life on Earth*, 2nd edn, Oxford: Oxford University Press.

Lutgens, F.K., E.J. Tarbuck and D. Tasa (2011), *Essentials of Geology*, 11th edn, Upper Saddle River, New Jersey: Prentice Hall.

Mankiw, N.G. (2009), *Principles of Economics*, 5th edn, Mason, Ohio: Cengage Learning.

McConnell, K.E. (1990), "Models for Referendum Data: The Structure of Discrete Choice Models for Contingent Valuation", *Journal of Environmental Economics and Management*, **18**(1), 19–34.

Meadows, D.H., D.L. Meadows, J. Randers and W.W. Behrens III (1974), *The Limits to Growth*, 2nd edn, New York: Signet.

Metrick, A. and M.L. Weitzman (1996), "Endangered Species Preservation", *Land Economics*, February, **72**(1), 1–16.

National Geographic (March 1999), "Ten Years After *Exxon Valdez*", **195**(3), 98–117.

National Research Council (1999), *Sharing the Fish: Toward a National Policy on Individual Fishing Quotas*, Washington DC: National Academy Press.

Neumayer, E. (2003), *Weak Versus Strong Sustainability: Exploring the Limits of Two Opposing Paradigms,* 2nd edn, Cheltenham, UK: Edward Elgar.

New, M.G., D.M. Liverman, R.A. Betts, K.L. Anderson and C.C. West (2011), "Four Degrees and Beyond: The Potential for a Global Temperature Increase of Four Degrees and its Implications", *Philosophical Transactions of the Royal Society A*, **369**(1934), 6–19.

Nordhaus, W.D. and E.C. Kokkelenberg (1999), *Nature's Numbers: Expanding the National Economic Accounts to Include the Environment*, Washington DC: National Academy Press.

Organization for Economic Co-operation and Development (1999), *Trade Measures in Multilateral Environmental Agreements*, Paris: OECD Publications.

Ostrom, E. (1990), *Governing the Commons: The Evolution of Institutions for Collective Action*, Cambridge, Massachusetts: Cambridge University Press.

Ostrom, E., R. Gardner and J. Walker (1994), *Rules, Games and Common-Pool Resources*, Ann Arbor, Michigan: The University of Michigan Press.

Perkins, D.H., S.C. Radelet and D.L. Lindauer (2006), *Economics of Development*, 6th edn, New York: W.W. Norton.

Perman, R., Y. Ma and J. McGilvray (2003), *Natural Resource and Environmental Economics*, 3rd edn, London: Pearson.

Perrings, C. (1991), "Reserved Rationality and the Precautionary Principle: Technological Change, Time and Uncertainty in Environmental Decision Making", in R. Costanza (ed.), *Ecological Economics: the Science and Management of Sustainability*, New York: Columbia University Press, pp. 153–166.

Phillips, A.W. (1958), "The Relationship between Unemployment and the Rate of Change of Money Wages in the United Kingdom, 1861–1957", *Economica*, **25**, 283–299.

Pielou, E.C. (1998), *Fresh Water*, Chicago: University of Chicago Press.

Ramanathan, R. (1992), *Introductory Econometrics*, 2nd edn, New York: Dryden Press.

Rao, P.K. (2000), *Sustainable Development: Economics and Policy*, Malden, Massachusetts: Basil Blackwell.

Rawls, J. (1971), *A Theory of Social Justice*, Cambridge, Massachusetts: Harvard University Press.

Romer, D. (2006), *Advanced Macroeconomics*, 3rd edn, New York: McGraw-Hill.

Rosen, S. (1974), "Hedonic Prices and Implicit Markets: Product Differentiation in Pure Competition", *Journal of Political Economy*, **82**, 34–55.

Ruddiman, W.F. (2007), *Earth's Climate: Past and Future*, 2nd edn, New York: W.H. Freeman.

Schneider, S. (2008), "Geoengineering: Could we or Should we Make it Work?", *Philosophical Transactions of the Royal Society A*, **366**(1882), 3843–3862.

Schumacher, E.F. (2010), *Small is Beautiful: Economics as if People Mattered*, New York: Harper and Row.

Sheffrin, S.M. (1996), *Rational Expectations*, 2nd edn, Cambridge: Cambridge University Press.

Shone, R. (2002), *Economic Dynamics,* 2nd edn, Cambridge, Massachusetts: Cambridge University Press.

Simon, C.P. and L. Blume (1994), *Mathematics for Economists*, New York: W.W. Norton.

Sinclair, A.R.E., J.M. Fryxell and G. Caughley (2006), *Wildlife Ecology, Conservation and Management*, Malden, Massachusetts: Blackwell.

Smith, D.M., B.C. Larson, M.J. Kelty, P. Mark and S. Ashton (1997), *The Practice of Silviculture: Applied Forest Ecology*, New York: Wiley.

Smith, R.L. and T.M. Smith (2008), *Elements of Ecology*, 7th edn, San Francisco: Benjamin Cummings.

Stern, N. (2008), "The Economics of Climate Change", *American Economic Review*, **98**(2), 1–37.

Studenmund, A.H. (2010), *Using Econometrics: Practical Guide,* 6th edn, Upper Saddle River, New Jersey: Prentice Hall.

Taiz, L. and E. Zeigler (2010), *Plant Physiology*, 5th edn, Sunderland, Massachusetts: Sinauer Associates.

Takayama, A. (1985), *Mathematical Economics*, 2nd edn, Cambridge, Massachusetts: Cambridge University Press.

Thornton, J. (2000), *Pandora's Box: Chlorine, Health, and a New Environmental Strategy*, Cambridge, Massachusetts: MIT Press.

Tietenberg, T.H. (1996), *Environmental and Natural Resource Economics*, 4th edn, New York: Harper Collins.

Tietenberg, T.H. and L. Lewis (2009), *Environmental Economics and Policy*, 6th edn, Upper Saddle River, New Jersey: Prentice Hall.

Tomar, M. (1999), *Quality Assessment of Water and Wastewater*, New York: Lewis Publishers.

Van Kooten, G.C. (1993), *Land Resource Economics and Sustainable Development: Economic Policies and the Common Good*, Vancouver, BC: University of British Columbia Press.

Van Kooten, G.C. and E.H. Bulte (2000), *The Economics of Nature*, Oxford: Blackwell Publishers.

Vogler, J. (2000), *The Global Commons: Environmental and Technological Governance*, 2nd edn, Chichester, UK: John Wiley & Sons Ltd.

Walters, C.J. and R. Hilborn (1976), "Adaptive Control of Fishing

Systems", *Journal of The Fisheries Research Board of Canada,* **33**, 145–159.

Weitzman, M. (2009), 'On Modeling and Interpreting the Economics of Catastrophic Climate Change', *Review of Economics and Statistics,* **91**(1), 1–19.

Wiesbrod, B. (1964), "Collective Consumption Services of Individual Consumption Goods", *QJE,* **77**(3), 71–77.

Williams, S.L. (1995), "Surfgrass (*Phyllospadix torreyi*) Reproduction: Reproductive Phrenology, Resource Allocation and Male Rarity", *Ecology,* **76**(6), 1953–1970.

World Bank (2007), "Biofuels: The Promises and the Risks", in World Development Report 2008, pp. 70–71.

World Commission on Environment and Development (1987), *Our Common Future*, Oxford: Oxford University Press.

Appendix 1: Greek Alphabet

LETTER	NAME
A α	Alpha
B β	Beta
Γ γ	Gamma
Δ δ	Delta
E ε	Epsilon
Z ζ	Zeta
H η	Eta
Θ θ	Theta
I ι	Iota
K κ	Kappa
Λ λ	Lambda
M μ	Mu
N ν	Nu
Ξ ξ	Xi
O o	Omicron
Π π	Pi
P ρ	Rho
Σ σ	Sigma
T τ	Tau
Υ υ	Upsilon
Φ φ	Phi
X χ	Chi
Ψ ψ	Psi
Ω ω	Omega

Appendix 2:   Roman Numerals

I	1
II	2
III	3
IV	4
V	5
VI	6
VII	7
VIII	8
IX	9
X	10
XX	20
XXX	30
XL	40
L	50
LX	60
LXX	70
LXXX	80
XC	90
C	100
CC	200
CCC	300
CD	400
D	500
DC	600
DCC	700
DCCC	800
CM	900
M	1000

Appendix 3: Système Internationale (SI) Units and Other Measurement Conversions

SI Units	Conversions
LENGTH	
Meter (m)	1 m = 3.2808 ft 1 ft = 0.3048 m
Centimeter (cm)	1 cm = 0.3937 in 1 in = 2.54 cm
Kilometer (km)	1 km = 0.6214 mile 1 mile = 1.6093 km
Micrometer (μm)	1 μm = 10^{-6} m
	1 angstrom = 10^{-10} m
	One degree of latitude = 111.1 km
	= 69.1 miles
	= 60 nautical miles
	One nautical mile = 1.15 statute miles
	= 1.85 km
MASS	
Kilogram (kg)	1 kg = 2.2046 lb 1 lb = 0.4536 kg
Gram (g)	1 g = 0.0353 oz 1 oz = 28.3495 g
Metric tonne (t)	1 tonne = 10^3 kg = 1.10 short ton
	= 2204.622 lbs
	1 gigatonne = 1 billion kg
	= 10^9 kg
TEMPERATURE	
Kelvin (K)	°C + 273.15 = K
	(°F + 459.67) × (5/9) = K
	°C = (°F − 32) × (5/9)
	°F = (°C × (9/5)) + 32
	Change (Δ)1 K = Δ1°C = Δ1.8°F
	Δ1°F = Δ0.56°C = Δ0.56 K
AREA	
Square meter (m²)	1 m² = 10.764 sq ft
	1 sq ft = 0.0929 m²
Square centimeter (cm²)	1 cm² = 0.155 sq in
	1 sq in = 6.4516 cm²
Square kilometer (km²)	1 km² = 0.3861 sq miles
	1 sq mile = 2.5899 km²
Hectare (ha)	1 ha = 10 000 m²
	= 2.471 acres
	1 acre = 43 560 sq ft = 0.4047 ha

SI Units	Conversion
VOLUME Cubic meter (m³)	$1\,m^3 = 35.3\,cu\,ft$ $1\,cu\,ft = 0.028\,m^3$ $1\,cm^3 = 0.061\,cu\,in$
Cubic centimeter (cm³)	$1\,cu\,in = 16.39\,cm^3$ $1\,liter\,(l) = 10^{-3}\,m^3$ $= 0.264\,gal\,(US)$ $= 0.22\,gal\,(Imperial)$ $1\,barrel = 158.983\,liters$
WORK AND ENERGY Joule (J)	$1\,calorie\,(cal) = 4.186\,J$ $1\,ft\,lb\,force\,per\,sec = 1.36\,J\,per\,sec$ $1\,J = 0.738\,ft\,lb\,force\,per\,sec$ $1\,BTU = 1055\,J$
POWER Watt (W)	$1\,W = 1\,J\,per\,sec$ $1\,horse\,power\,(hp) = 33\,000\,ft\,lb\,per\,min$ $= 746\,W$

TOPIC SPECIFIC UNITS AND CONVERSIONS

CLIMATE CHANGE Gigatonne carbon dioxide (GtCO₂)	$1\,gigatonne\,carbon\,(GtC) = 3.67\,GtCO_2$
Atmospheric greenhouse gas concentration (ppm)	1 ppm of atmospheric CO_2 by volume ≈ 2.13 atmospheric GtC

Global Surface Temperature
  Increase (Change in °C)
	°C above pre-industrial levels $\approx$ °C above 1990
	  levels $+ 0.5$ °C

WATER Gigaliter (GL)	$1\,GL = 1000$ megaliters (ML) $= 1$ billion (or 10^9) liters $= 264\,172\,052\,gal\,(US)$ $= 810.71$ acre foot $= 0.001\,km^3$
Cubic meter per second (m³/sec)	$1\,m^3/sec = 86.41\,ML\,per\,day$ $= 31.536\,GL\,per\,year$ $= 15\,850\,gal\,(US)\,per\,min\,(gpm)$ $= 35.314\,cu\,ft\,per\,sec\,(cfs)$
Acre-inch	$1\,acre\text{-}inch = 27154\,gal\,(US)$ $= 102.79\,m^3$
Acre-foot	$1\,acre\text{-}foot = 1233.482\,m^3 = 1.233\,ML$

SI Units	Conversion
ELECTRICITY	
Rate of Consumption or Generation (Power):	
Kilowatt (kW)	1 kW = 1000 W = 1000 J per sec
Megawatt (MW)	1 MW = 1000kW = 1 million W
Gigawatt (GW)	1 GW = 1000 MW = 1 million kW
Terawatt (TW)	1 TW = 1000 GW = 1 million MW
Aggregate Consumption or Generation (Energy)	
watt-hour (Wh)	1 Wh = 3600 J
	= 0.5 W consumed for 2 hours
Kilowatt-hour (kWh)	1 kWh = 3 600 000 J
	= 10 W consumed for 100 hours

Appendix 4: Prefixes of the Système Internationale (SI)

Prefix	Abbreviation	Multiplication factor
exa	E	10^{18}
peta	P	10^{15}
tera	T	10^{12}
giga	G	10^{9}
mega	M	10^{6}
kilo	k	10^{3}
hecto	h	10^{2}
deca	da	10^{1}
deci	d	10^{-1}
centi	c	10^{-2}
milli	m	10^{-3}
micro	μ	10^{-6}
nano	n	10^{-9}
pico	p	10^{-12}
femto	f	10^{-15}
atto	a	10^{-18}

Appendix 5: Common Abbreviations

A	ampere
Bq	becquerel
cm	centimeter
°C	degree Celsius
d	day
dB	decibel
g	gram
GWh	gigawatt hour
ha	hectare
h	hour
Hz	hertz
J	joule
kg	kilogram
km	kilometer
km²	square kilometer
km/h	kilometers per hour
kPa	kilopascal
kt	kilotonne
kV	kilovolt
kWh	kilowatt hour
l	litre
m	meter
m²	square meter
m³	cubic meter
mm³	million cubic meters
mm	millimeter
mg	milligram
mol	mole
mSv	millisievert
µg	microgram
ng	nanogram
ppm	parts per million
ppb	parts per billion
ppt	parts per trillion
s	Second
SIC	Standard Industrial Classification
t	tonne
TJ	terajoule
W	watt

Appendix 6:   Geological Time

ERA PERIOD	MILLION YEARS BEFORE PRESENT	SIGNIFICANT EVENTS
Cenozoic		
Quaternary	0–1.6	
Tertiary	1.6–66.4	
Mesozoic		
Cretaceous	66.4–144	
Jurassic	144–208	First birds
Triassic	208–245	First dinosaurs and mammals
Late Paleozoic		
Permian	245–286	
Carboniferous	286–360	First reptiles
Devonian	360–408	First insects
Early Paleozoic		
Silurian	408–438	First land plants
Ordovician	438–505	First fishes
Cambrian	505–570	First shellfish and corals
Precambrian Time		
Proterozoic Era	544–2500	
Archean Era	2500–3800	
Hadean Time	3800–4550	

Note:
The Carboniferous period is sometimes split into the Pennsylvanian Period (286–320 million years before present) and the Mississippian Period (320–360 million years before present).

SUBDIVISIONS OF THE CENOZOIC ERA

PERIOD EPOCH	THOUSAND YEARS BEFORE PRESENT	SIGNIFICANT EVENTS
Quaternary		
Holocene	0–10	
Pleistocene	11–1600	First modern humans
Tertiary		
Pliocene	1600–5300	
Miocene	5300–23 700	
Oligocene	23 700–36 600	
Eocene	36 600–57 800	
Paleocene	57 800–66 400	

Improved dating techniques and further study of stratigraphic sequences and fossils will require the adjustment of the periods and epochs given in Appendix 6.

For further information, consult the University of California (Berkeley) Museum of Paleontology's *Web Geological Time Machine* at http://www.ucmp.berkeley.edu/help/timeform.html. For a review of the concept of geologic time and how it is measured, see the USGS publication *Geologic Time* at http://pubs.usgs.gov/gip/geotime/.

Appendix 7: Selected Environmental Treaties and Conventions

ANTARCTIC

Name: *Agreed Measures for the Conservation of Antarctic Fauna and Flora*
Initiated: 2 June 1964 (Brussels, Belgium)
Ratified: 1 November 1982
Description: An agreement to implement the principles and purposes of the Antarctic Treaty.
FURTHER INFORMATION: iea.uoregon.edu
                        www.ecolex.org

Name: *The Antarctic Treaty*
Initiated: 1 December 1959 (Washington, USA)
Ratified: 23 June 1961
Description: A treaty first signed in 1959 that encourages scientific research in the Antarctic in the absence of territorial and sovereignty claims by nations.
FURTHER INFORMATION: Vogler (2000)
                        iea.uoregon.edu
                        www.ecolex.org

Name: *Convention on the Conservation of Antarctic Marine Living Resources*
Initiated: 20 May 1980 (Canberra, Australia)
Ratified: 7 April 1982
Description: An agreement designed to protect the marine resources in the Antarctic Convergence and, in particular, to offer a sanctuary to whales in Antarctic waters.
FURTHER INFORMATION: Vogler (2000)
                        iea.uoregon.edu
                        www.ecolex.org

Name: *Convention on the Conservation of Antarctic Seals*
Initiated: 1 June 1972 (London, UK)
Ratified: 11 March 1978
Description: An agreement designed to conserve and ensure the sustainable use of seals in the Antarctic ecosystems.

FURTHER INFORMATION: Vogler (2000)
iea.uoregon.edu

Name: *Convention on the Regulation of Antarctic Mineral Resource Activities*
Initiated: 2 June 1988 (Wellington, New Zealand)
Ratified: Not ratified
Description: An agreement designed to regulate mining within the Antarctic.
FURTHER INFORMATION: Vogler (2000)
iea.uoregon.edu

Name: *Protocol to the Antarctic Treaty on Environmental Protection (PREP)*
Initiated: 4 October 1991 (Madrid, Spain)
Ratified: 14 January 1998
Description: An agreement designed to bring together the various Antarctic agreements on natural resource use and ensure comprehensive conservation of the Antarctic environment.
FURTHER INFORMATION: Vogler (2000)
iea.uoregon.edu

ATMOSPHERE

Name: *Adjustments and Amendment to the Montreal Protocol on Substances that Deplete the Ozone Layer (London Amendment)*
Initiated: 23 June 1990 (London, UK)
Ratified: 1 January 1992
Description: A 1990 amendment to the 1987 Montreal Protocol which accelerated the phase-out reductions on the production of chlorofluorocarbons (CFCs) that destroy stratospheric ozone.
FURTHER INFORMATION: Vogler (2000)
iea.uoregon.edu

Name: *Amendment to the Montreal Protocol on Substances that Deplete the Ozone Layer (Copenhagen Amendment)*
Initiated: 25 November 1992 (Copenhagen, Denmark)
Ratified: 1 January 1994
Description: A 1992 amendment to the 1987 Montreal Protocol that accelerated the phase out of production of chlorofluorocarbons (CFCs) which breakdown stratospheric ozone.

FURTHER INFORMATION: Vogler (2000)
                       iea.uoregon.edu
                       www.ecolex.org

Name: *United Nations Framework Convention on Climate Change (UNFCCC)*
Initiated: 9 May 1992 (New York, USA)
Ratified: 21 March 1994
Description: A convention signed by over 160 countries at the 1992 Earth Summit in Rio de Janeiro, and which came into force in March 1994 following its ratification by over 50 nations. Annex I countries to the convention committed themselves to return their greenhouse gas emissions to 1990 levels by the year 2000. This commitment was modified in the 1997 Kyoto Protocol. The ultimate objective of the UNFCCC is to stabilize greenhouse gas concentrations at a level that would prevent dangerous anthropogenic interference with the climate system.
FURTHER INFORMATION: unfccc.int
                       iea.uoregon.edu
                       www.ecolex.org

Name: *Kyoto Protocol*
Initiated: 11 December 1997 (Kyoto, Japan)
Ratified: 16 February 2005
Description: The 1997 agreement by parties to the United Nations Framework Convention on Climate Change for Annex I (or developed) countries to collectively commit to a 5 percent reduction in greenhouse gas emissions by 2008–12, relative to 1990 levels. This reduction was to be achieved through binding national emissions constraints, in conjunction with mechanisms for reducing mitigation costs, such as emissions trading and projects in developing, or non Annex I, countries. Although the Protocol represented an important first step towards coordinated global action, global emissions rose significantly over this period. A major drawback was the non-participation of the United States, who refused ratification because large developing country emitters, such as China and India, were not subjected to binding commitments.
FURTHER INFORMATION: unfccc.int
                       www.ecolex.org
                       www.cop4.org/kp/kp.html

Name: *Copenhagen Accord*
Initiated: 18 December 2009 (Copenhagen, Denmark)
Ratified: Not ratified

Description: The official UNFCCC pronouncement emanating from the COP-15. Despite great public expectation prior to this meeting, a binding international agreement on climate change mitigation was not formulated. However, the Copenhagen Accord provided a formal target for limiting global warming to 2°C, and acknowledgement that large global cuts in emissions were required. Although accompanying national emissions reductions pledges at the time were not commensurate with the overall goal, the Copenhagen Accord paved the way for further advancement of international negotiations.

FURTHER INFORMATION: unfccc.int

Name: *The Montreal Protocol (on Substances that Deplete the Ozone Layer)*
Initiated: 16 September 1987 (Montreal, Quebec)
Ratified: 1 January 1989
Description: An agreement first signed in 1987 in Montreal, Canada, under which signatories agreed to reduce by half their production of ozone-damaging chlorofluorocarbons (CFCs) by the year 2000, and that continued the process begun in 1985 with the Vienna Convention for the Protection of the Ozone Layer. The Protocol was subsequently strengthened in 1990, and again in 1992, with the London and Copenhagen Amendments under which all CFC production would stop by 2000.

FURTHER INFORMATION: Organisation for Economic
                        Co-operation
                        and Development (1999)
                        Vogler (2000)
                        iea.uoregon.edu
                        www.ecolex.org
                        ozone.unep.org/

Name: *Treaty Banning Nuclear Weapon Tests in the Atmosphere, in Outer Space, and Under Water*
Initiated: 5 August 1963 (Washington, USA)
Ratified: 10 October 1963
Description: An agreement to end the testing of nuclear weapons, except in underground tests, that was originally signed by the United States, the United Kingdom and the Soviet Union.

FURTHER INFORMATION: iea.uoregon.edu
                        www.ecolex.org

Name: *US–Canada Air Quality Agreement (Agreement between the Government of Canada and the Government of the United States of America on Air Quality)*
Initiated: 13 March 1991 (Ottawa, Ontario)
Ratified: Not applicable
Description: A 1991 agreement that committed the two countries to reductions in sulfur dioxide and nitrogen oxides emissions, and provided a forum for addressing transboundary air issues.
FURTHER INFORMATION: iea.uoregon.edu

Name: *Vienna Convention for the Protection of the Ozone Layer*
Initiated: 22 March 1985 (Vienna, Austria)
Ratified: 22 September 1988
Description: An agreement between 20 countries in 1985 that committed them to address what, at the time, was perceived to be a potential problem of stratospheric ozone depletion due to the emissions of chlorofluorocarbons (CFCs).
FURTHER INFORMATION: Vogler (2000)
                       iea.uoregon.edu
                       www.ecolex.org

BIODIVERSITY

Name: *Convention on Biological Diversity*
Initiated: November 1988
Ratified: September 1993
Description: Came into force on 29 December 1993 after 30 ratifications. Initially assembled by UNEP and officially coined as the Convention on Biological Diversity in May 1992 in Nairobi. Convention was opened for signatures in Rio at the UN conference on Environment and Development. Its three objectives are :

- The conservation of biological diversity
- The sustainable use of the components of biological diversity
- The fair and equitable sharing of the benefits arising out of the utilization of genetic resources

FURTHER INFORMATION: www.cbd.int/convention/about.shtml

Name: *Cartagena Protocol on Biosafety*
Initiated: 29 January 2000
Ratified: 11 September 2003
Description: International agreement on the safe handling of, transport and use of living modified organisms (LMOs).
FURTHER INFORMATION: bch.cbd.int/protocol/

Name: *Nagoya Protocol on ABS*
Initiated: 29 October 2010
Ratified: Not ratified
Description: Adopted by the Conference of the Parties to the Convention on Biological Diversity at its tenth meeting on 29 October 2010 in Nagoya, Japan. International agreement aimed at fair and equitable sharing of the benefits arising out of the utilization of genetic resources.
FURTHER INFORMATION: http://www.cbd.int/abs/

Name: *Aichi Biodiversity Targets*
Initiated: Meeting from 18–29 of October 2011
Ratified: Not applicable
Description: A 10-year plan by all countries to save biodiversity. It includes twenty targets around conserving biodiversity. Created during the 10th meeting of the Conference of the Parties (within the Convention on Biological Diversity).
FURTHER INFORMATION: www.cbd.int/doc/strategic-
                                      plan/2011-2020/Aichi-Targets-EN.pdf

CULTURE

Name: *Convention for the Protection of the World Cultural and Natural Heritage*
Initiated: 16 November 1972 (Paris, France)
Ratified: 17 December 1975
Description: An agreement designed to protect cultural and natural heritage sites of world significance.
FURTHER INFORMATION: iea.uoregon.edu
                                      www.ecolex.org

Name: *Declaration of the United Nations Conference on the Human Environment (UNCHE)*
Initiated: 5 June 1972 (Stockholm, Sweden)

Ratified: Not applicable
Description: A declaration with 26 principles that sets an agenda for sustainable development and the environment.
FURTHER INFORMATION: Vogler (2000)
www.biblebelievers.org.au/gc1972.htm

ENVIRONMENT AND TRADE

Name: *The Marrakesh Agreement Establishing the World Trade Organization*
Initiated: April 1994
Ratified: April 1994
Description: One of the agreements reached at the conclusion of the round of talks that started in Uruguay in 1986 under the auspices of GATT. The agreement established the framework for the WTO which embodied the series of rules reached under the previous GATT agreements and also contained a preamble recognizing the need to ensure sustainable development and environmental protection. A statement released by attendees at Marrakesh, called The Decision on Trade and Environment, states "There should not be, nor need be, any policy contradiction between upholding and safeguarding an open, non-discriminatory and equitable multilateral trading system on the one hand, and acting for the protection of the environment, and the promotion of sustainable development on the other."
FURTHER INFORMATION: www.wto.org/english/docs_e/legal_e/04-wto_e.htm

Name: *The North American Agreement on Environmental Co-operation (NAAEC) (side agreement to NAFTA)*
Initiated: 13 September 1993
Ratified: 1 January 1994
Description: An agreement that creates a framework to better conserve, protect and enhance the North American environment through cooperation and effective enforcement of environmental laws.
FURTHER INFORMATION: iea.uoregon.edu
www.ecolex.org
www.naaec.gc.ca/

Name: *North American Free Trade Agreement (NAFTA)*
Initiated: 17 December 1992 (Ottawa, Ontario)
Ratified: 1 January 1994

Description: An agreement signed between the United States, Canada and Mexico in 1992 to promote free trade among the three countries. The agreement includes various environmental clauses and commits the signatories to promote sustainable development.
FURTHER INFORMATION: iea.uoregon.edu
www.ecolex.org
www.nafta-sec-alena.org/

FORESTRY

Name: *The Helsinki Process*
Initiated: Not applicable
Ratified: Not applicable
Description: A meeting between European countries, called the Second Ministerial Conference on the Protection of Forests in Europe, held in Helsinki in 1993, which identified the general guidelines for sustainable forestry management in Europe, and which led to the subsequent development of a series of criteria and indicators in later meetings.
FURTHER INFORMATION: www.iisd.ca/linkages/forestry/hel.html

Name: *International Tropical Timber Agreement*
Initiated: 18 November 1983 (Geneva, Switzerland)
Ratified: 1 April 1985
Description: An agreement to ensure consultation and cooperation in the production, harvesting and consumption of tropical timber and products and to ensure the sustainable use of tropical forests.
FURTHER INFORMATION: iea.uoregon.edu
www.ecolex.org

Name: *International Tropical Timber Agreement, 1994*
Initiated: 26 January 1994 (Geneva, Switzerland)
Ratified: 1 January 1997
Description: An agreement by producing and consuming nations of tropical timber to ensure that, by the year 2000, exports of tropical timber only originate from sustainably managed forests.
FURTHER INFORMATION: iea.uoregon.edu
www.ecolex.org

Name: *The Montreal Process*
Initiated: Not applicable
Ratified: Not applicable

Description: An initiative to develop a series of criteria and indicators for sustainable forest management for temperate forests outside of Europe.

FURTHER INFORMATION: www.iisd.ca/linkages/forestry/mont. html

MARINE AND WATER RESOURCES

Name: *78 Agreement Between The United States and Canada on Great Lakes Water Quality (As Amended through 16 October 1983)*
Initiated: 22 November 1978 (Ottawa, Ontario)
Ratified: Not applicable
Description: A 1978 accord that updated a 1972 agreement between the United States and Canada to reduce the pollution entering the Great Lakes (Superior, Michigan, Huron, Erie and Ontario) that are shared between the two countries.
FURTHER INFORMATION: iea.uoregon.edu
www.ijc.org

Name: *Agreement Concerning Interim Arrangements Relating to Polymetallic Nodules of the Deep Sea Bed*
Initiated: 2 September 1982 (Washington, USA)
Ratified: 2 September 1982
Description: An agreement designed to avoid conflicts and overlaps of mining activities on the deep sea floor.
FURTHER INFORMATION: iea.uoregon.edu
www.ecolex.org

Name: *Agreement for the Implementation of the Provisions of the United Nations Convention on the Law of the Sea Relating to the Conservation and Management of Straddling Fish Stocks and Highly Migratory Fish Stocks*
Initiated: 4 December 1995 (New York, USA)
Ratified: 11 December 2001
Description: An agreement designed to secure enhanced management of high seas resources through provisions of monitoring under regional organizations, and the ability to adjudicate disputes under a variety of international arenas.
FURTHER INFORMATION: iea.uoregon.edu
www.ecolex.org

Name: *Agreement on Port State Measures to Prevent, Deter and Eliminate Illegal, Unreported and Unregulated Fishing*
Initiated: 22 November 2009
Ratified: Not ratified
Description: Agreement aimed at preventing the illegal use of marine life and to ensure sustainable practices and long-term conservation of these fish.
FURTHER INFORMATION:　www.ecolex.org/server2.php/libcat/
　　　　　　　　　　　　　docs/TRE/Multilateral/En/tre154601.
　　　　　　　　　　　　　pdf

Name: *Agreement Relating to the Implementation of Part XI of the United Nations Convention on the Law of the Sea of 10 December 1982*
Initiated: 28 July 1994 (New York, USA)
Ratified: 16 November 1994
Description: See United Nations Convention of the Law of the Sea.
FURTHER INFORMATION:　sedac.ciesin.org/pidb
　　　　　　　　　　　　　www.ecolex.org

Name: *The Control of Eutrophication of Waters (OECD)*
Initiated: 14 November 1974
Ratified: Not applicable
Description: An agreement to ensure the protection of the quality of water resources in member countries of the Organisation for Economic Co-operation and Development (OECD).
FURTHER INFORMATION:　iea.uoregon.edu

Name: *Convention on the Conservation and Management of Fishery Resources in the South East Atlantic Ocean (SEAFO)*
Initiated: 20 April 2001
Ratified: 13 April 2003
Description: Aimed at the conservation and sustainable use of living marine resources in the South East Atlantic.
FURTHER INFORMATION:　www.ecolex.org/server2.php/libcat/
　　　　　　　　　　　　　docs/TRE/Multilateral/En/TRE001384.
　　　　　　　　　　　　　txt

Name: *Convention concerning Fishing in the Black Sea*
Initiated: 7 July 1959 (Varna, Bulgaria)
Ratified: 21 March 1960
Description: An agreement between Bulgaria, Romania and the Soviet Union to rationally utilize the fishery resources of the Black Sea.

FURTHER INFORMATION: iea.uoregon.edu
www.ecolex.org

Name: *Convention for the Establishment of an Inter American Tropical Tuna Commission*
Initiated: 31 May 1949 (Washington, USA)
Ratified: 3 March 1950
Description: An agreement to establish a permanent commission to help ensure the sustainability of yellow fin and skipjack tuna. The Commission members include the United States, Canada, Costa Rica, Ecuador, France, Japan, Mexico, Nicaragua, Panama and Vanuatu.
FURTHER INFORMATION: iea.uoregon.edu
www.ecolex.org

Name: *Convention for the Prohibition of Fishing with Long Driftnets in the South Pacific*
Initiated: 24 November 1989 (Wellington, New Zealand)
Ratified: 17 May 1991
Description: An agreement to prevent the use of drift nets in the South Pacific.
FURTHER INFORMATION: iea.uoregon.edu
www.ecolex.org

Name: *Convention for the Protection and Development of the Marine Environment of the Wider Caribbean Region*
Initiated: 24 March 1983 (Cartagena de Indias, Colombia)
Ratified: 30 March 1986
Description: An agreement to ensure the protection and sustainable management of the marine environment in the Caribbean region.
FURTHER INFORMATION: iea.uoregon.edu
www.ecolex.org
www.cep.unep.org/

Name: *Convention on Fishing and Conservation of the Living Resources of the High Seas*
Initiated: 29 April 1958 (Geneva, Switzerland)
Ratified: 20 March 1966
Description: An agreement to help ensure the sustainable use of marine resources.
FURTHER INFORMATION: iea.uoregon.edu
www.ecolex.org

Name: *Convention on the High Seas*
Initiated: 29 April 1958 (Geneva, Switzerland)
Ratified: 30 September 1962
Description: An agreement to codify rules on international law on the high seas.
FURTHER INFORMATION: iea.uoregon.edu
www.ecolex.org

Name: *Convention on Wetlands of International Importance Especially as Waterfowl Habitat (The Ramsar Convention on Wetlands)*
Initiated: 2 February 1971 (Ramsar, Iran)
Ratified: 21 December 1975
Description: An agreement by 122 countries to facilitate the conservation and wise use of wetlands.
FURTHER INFORMATION: iea.uoregon.edu
www.ecolex.org
www.ramsar.org

Name: *International Convention for the Regulation of Whaling (International Whaling Convention)*
Initiated: 2 December 1946 (Washington, USA)
Ratified: 10 November 1948
Description: A convention signed in 1946 that established the International Whaling Commission, mandated to preserve whale stocks. The Commission has been highly controversial and failed to prevent the near extinction of several whale species. More recently, a moratorium on commercial whaling has helped some species to increase in number but controversy still remains, as some members of the commission would like to ban all commercial whaling in perpetuity while other countries would like to resume commercial whaling.
FURTHER INFORMATION: Vogler (2000)
iea.uoregon.edu
www.ecolex.org

Name: *United Nations Convention on the Law of the Sea*
Initiated: 10 December 1982 (Montego Bay, Jamaica)
Ratified: 16 November 1994
Description: A convention developed by a series of international meetings in the 1970s and 1980s to govern the exploitation of marine resources. The Convention established, among other things, 200 nautical mile exclusive economic zones (EEZs) from the coastline of maritime states.

FURTHER INFORMATION: Vogler (2000)
iea.uoregon.edu
www.ecolex.org

POLLUTION

Name: *Agreement Between the Government of the United States of America and the Government of Canada Concerning the Transboundary Movement of Hazardous Wastes*
Initiated: 28 October 1986 (Ottawa, Ontario)
Ratified: 8 November 1986
Description: An agreement between the United States and Canada on the appropriate treatment and transport of hazardous wastes.
FURTHER INFORMATION: iea.uoregon.edu

Name: *Agreement Concerning the Protection of the Waters of the Mediterranean Shores*
Initiated: 10 May 1976 (Monaco)
Ratified: 1 January 1981
Description: An agreement between France, Monaco and Italy to ensure cooperation and co-ordination of policies to improve the coastal waters of the three nations.
FURTHER INFORMATION: iea.uoregon.edu
www.ecolex.org

Name: *Bamako Convention on the Ban of the Import into Africa and the Control of Transboundary Movement and Management of Hazardous Wastes Within Africa*
Initiated: 30 January 1991 (Bamako, Mali)
Ratified: 22 April 1998
Description: An agreement to prevent the dumping or improper disposal of hazardous wastes in Africa.
FURTHER INFORMATION: iea.uoregon.edu
www.ecolex.org

Name: *Basel Convention on the Control of Transboundary Movements of Hazardous Wastes and their Disposal*
Initiated: 22 March 1989 (Basel, Switzerland)
Ratified: 05 May 1992
Description: A 1992 convention, commonly referred to as the Basel Convention, which ultimately led to an agreement to phase out the

dumping and inappropriate disposal of hazardous wastes in developing countries.
FURTHER INFORMATION: Organisation   for   Economic   Co-operation and Development (1999)
iea.uoregon.edu
www.ecolex.org
www.unep.ch/basel/index.html

Name: *Convention Concerning Safety in the Use of Asbestos*
Initiated: 24 June 1986 (Geneva, Switzerland)
Ratified: 16 June 1986
Description: An agreement to prevent the potential health hazards associated with the use of asbestos.
FURTHER INFORMATION: iea.uoregon.edu
www.ecolex.org

Name: *Convention for the Protection of the Mediterranean Sea against Pollution*
Initiated: 16 February 1976 (Barcelona, Spain)
Ratified: 12 February 1978
Description: An agreement designed to help protect the Mediterranean marine environment.
FURTHER INFORMATION: iea.uoregon.edu
www.ecolex.org
www.unepmap.gr/

Name: *Convention for the Protection of the Rhine against Chemical Pollution*
Initiated: 3 December 1976 (Bonn, Germany)
Ratified: 1 March 1983
Description: An agreement between countries bordering on the Rhine River to help prevent chemical spills and pollution.
FURTHER INFORMATION: iea.uoregon.edu

Name: *Convention on Environmental Impact Assessment in a Transboundary Context*
Initiated: 25 February 1991 (Espoo, Finland)
Ratified: 10 September 1997
Description: An agreement to use Environmental Impact Assessments so as to prevent transboundary deterioration of the environment.
FURTHER INFORMATION: iea.uoregon.edu
www.ecolex.org

Name: *Convention on Long-Range Transboundary Air Pollution*
Initiated: 13 November 1979 (Geneva, Switzerland)
Ratified: 16 March 1983
Description: An agreement to reduce long-range transboundary air pollution.
FURTHER INFORMATION: iea.uoregon.edu
                        www.ecolex.org

Name: *Convention on the Prevention of Marine Pollution by Dumping Wastes and Other Matter (London Dumping Convention)*
Initiated: 29 December 1972 (London, UK)
Ratified: 30 August 1975
Description: First signed in 1972, and ratified by 75 countries, the Convention prohibits the dumping of specified wastes at sea, including radioactive wastes.
FURTHER INFORMATION: Vogler (2000)
                        iea.uoregon.edu
                        www.ecolex.org

Name: *Convention on the Prevention of Marine Pollution from Land-Based Sources*
Initiated: 4 June 1974 (Paris, France)
Ratified: 6 May 1978
Description: An agreement designed to prevent land-based pollution of the marine environment.
FURTHER INFORMATION: iea.uoregon.edu

Name: *Convention on the Prohibition of the Development, Production and Stockpiling of Bacteriological (Biological) and Toxic Weapons and on their Destruction*
Initiated: 10 April 1972
Ratified: 26 March 1975
Description: An agreement designed to prevent the production and storage of chemical weapons.
FURTHER INFORMATION: iea.uoregon.edu
                        www.ecolex.org

Name: *European Agreement on the Restriction of the Use of Certain Detergents in Washing and Cleaning Products*
Initiated: 16 September 1968 (Strasbourg, France)
Ratified: 16 February 1971
Description: An agreement designed to protect freshwater environments,

particularly for recreational and wildlife purposes, by controlling the use of detergents.
FURTHER INFORMATION: iea.uoregon.edu
www.ecolex.org

Name: *International Convention on Civil Liability for Oil Pollution Damage*
Initiated: 29 November 1969 (Brussels, Belgium)
Ratified: 19 June 1975
Description: An agreement designed to standardize the rules of compensation for oil spills.
FURTHER INFORMATION: iea.uoregon.edu
www.ecolex.org

Name: *Hong Kong International Convention for the Safe and Environmentally Sound Recycling of Ships, 2009*
Initiated: 11 May 2009 (Hong Kong, China)
Ratified: Not ratified
Description: Focus on ensuring that ships, when being recycled, do not cause unnecessary harm to human health and safety and the environment.
FURTHER INFORMATION: www.ecolex.org

Name: *International Convention on the Establishment of an International Fund for Compensation for Oil Pollution Damage*
Initiated: 18 December 1971 (Brussels, Belgium)
Ratified: 16 October 1978
Description: An amendment to a previous agreement that helps ensure that the owners of oil cargoes bear a responsibility for paying compensation for oil spills.
FURTHER INFORMATION: iea.uoregon.edu

Name: *International Convention for the Prevention of Pollution from Ships, as modified by the Protocol of 1978 relating thereto (MARPOL 73/78)*
Initiated: 2 November 1973 (London, UK)
Ratified: 2 October 1983
Description: An agreement that regulates the discharge of petroleum products and other pollutants at sea.
FURTHER INFORMATION: Vogler (2000)
iea.uoregon.edu
www.ecolex.org

Name: *Protocol Concerning Marine Pollution Resulting from Exploration and Exploitation of the Continental Shelf*
Initiated: 29 March 1989 (Kuwait)
Ratified: 17 February 1990
Description: An agreement to help prevent marine pollution from exploration and exploitation of the sea-bed.
FURTHER INFORMATION: iea.uoregon.edu
www.ecolex.org

Name: *Protocol for the Prevention of Pollution of the Mediterranean Sea by Dumping from Ships and Aircraft*
Initiated: 16 February 1976 (Barcelona, Spain)
Ratified: 12 February 1978
Description: An agreement to help prevent the disposal of wastes in the Mediterranean sea.
FURTHER INFORMATION: iea.uoregon.edu
www.ecolex.org

Name: *Protocol for the Protection of the Mediterranean Sea against Pollution from Land-Based Sources*
Initiated: 17 May 1980 (Athens, Greece)
Ratified: 17 June 1983
Description: An agreement to prevent pollution in the Mediterranean from various land-based sources.
FURTHER INFORMATION: iea.uoregon.edu
www.ecolex.org

Name: *Protocol Relating to Intervention on the High Seas in Cases of Pollution by Substances other than Oil*
Initiated: 2 November 1973 (London, UK)
Ratified: 30 March 1983
Description: An agreement that allows coastal states to act on the high seas to prevent pollution, other than oil.
FURTHER INFORMATION: iea.uoregon.edu
www.ecolex.org

Name: *Protocol to the 1979 Convention on Long-Range Transboundary Air Pollution Concerning the Control of Emissions of Volatile Organic Compounds or Their Transboundary Fluxes*
Initiated: 18 November 1991 (Geneva, Switzerland)
Ratified: 29 September 1997

Description: An agreement to reduce the emissions of volatile organic compounds.
FURTHER INFORMATION: iea.uoregon.edu
www.ecolex.org

Name: *Protocol to the 1979 Convention on Long-range Transboundary Air Pollution on Further Reduction of Sulphur Emissions*
Initiated: 14 June 1994 (Oslo, Norway)
Ratified: 5 August 1998
Description: An agreement to reduce the emissions of sulfur dioxide and their long-range transportation.
FURTHER INFORMATION: www.ecolex.org
www.unece.org/env/lrtap/

Name: *Protocol to the 1979 Convention on Long-range Transboundary Air Pollution on Long-term Financing of Co-operative Program for Monitoring and Evaluation of the Long-term Transmission of Air Pollutants in Europe*
Initiated: 28 February 1984 (Geneva, Switzerland)
Ratified: 28 January 1988
Description: An agreement on the monitoring of long-range pollutants in Europe.
FURTHER INFORMATION: iea.uoregon.edu
www.ecolex.org
www.unece.org/env/lrtap/

Name: *Protocol to the 1979 Convention on Long-range Transboundary Air Pollution on Persistent Organic Pollutants*
Initiated: 24 June 1998 (Aarhus, Denmark)
Ratified: 23 October 2003
Description: An agreement to reduce the emissions of persistent organic pollutants and their long-range transportation.
FURTHER INFORMATION: www.ecolex.org
www.unece.org/env/lrtap/

Name: *Protocol to the 1979 Convention on Long-range Transboundary Air Pollution on the Reduction of Sulphur Emissions or their Transboundary Fluxes by at Least 30 Percent*
Initiated: 8 July 1985 (Helsinki, Finland)
Ratified: 2 September 1987
Description: An agreement by some European countries to reduce sulfur dioxide emissions by 30 percent by 1993.
FURTHER INFORMATION: iea.uoregon.edu
www.ecolex.org

Name: *Protocol to the International Convention on the Establishment of an International Fund of Compensation for Oil Pollution Damage*
Initiated: 19 November 1976 (London, UK)
Ratified: 22 November 1994
Description: An agreement to establish international compensation for the effects of oil pollution damages.
FURTHER INFORMATION: iea.uoregon.edu
www.ecolex.org

Name: *Rotterdam Convention*
Initiated: 10 September 1998 (Rotterdam, the Netherlands)
Ratified: 24 February 2004
Description: The Rotterdam Convention lists a set of procedures for providing adequate information to importing countries about potentially hazardous internationally traded materials.
FURTHER INFORMATION: www.ecolex.org

SUSTAINABLE DEVELOPMENT

Name: *Convention on Biological Diversity*
Initiated: 22 May 1992 (Rio de Janeiro, Brazil)
Ratified: 29 December 1993
Description: The 1992 Convention on biological diversity which allows for the free trade of genetic resources while providing ways for rich countries to finance biodiversity conservation.
FURTHER INFORMATION: iea.uoregon.edu
www.biodiv.org

Name: *International Convention to Combat Desertification in those Countries Experiencing Serious Drought and/or Desertification, Particularly in Africa*
Initiated: 14 October 1994 (Paris, France)
Ratified: 26 December 1996
Description: An agreement to help countries facilitate plans to address the problems of desertification within their countries, as well as to encourage technology transfer to help countries overcome the deleterious effects of desertification.
FURTHER INFORMATION: iea.uoregon.edu
www.iisd.ca/linkages

Name: *The Rio Declaration on Environment and Development*
Initiated: 13 June 1992 (Rio de Janeiro, Brazil)
Ratified: 13 June 1992
Description: A declaration of 27 principles issued at the 1992 United Nations Conference on Environment and Development, including the principle of sustainable development.
FURTHER INFORMATION: Vogler (2000)
iea.uoregon.edu

Name: *Treaty for Amazonian Cooperation*
Initiated: 3 July 1978 (Brasilia, Brazil)
Ratified: 2 August 1980
Description: An agreement by Bolivia, Brazil, Colombia, Ecuador, Guyana, Peru, Suriname and Venezuela to ensure the sustainable development of the Amazon region and an equitable division of its benefits.
FURTHER INFORMATION: iea.uoregon.edu
www.ecolex.org

WILDLIFE

Name: *African Convention on the Conservation of Nature and Natural Resources*
Initiated: 15 September 1968 (Algiers, Algeria)
Ratified: 16 June 1969
Description: An agreement to promote the conservation, utilization and development of the soil, water, plant and wildlife resources of the African continent.
FURTHER INFORMATION: iea.uoregon.edu
www.ecolex.org

Name: *Agreement on Conservation of Polar Bears*
Initiated: 15 November 1973 (Oslo, Norway)
Ratified: 26 May 1976
Description: An agreement to protect the polar bear through appropriate management measures.
FURTHER INFORMATION: iea.uoregon.edu

Name: *Agreement for the Establishment of a General Fisheries Council for the Mediterranean*
Initiated: 24 September 1949 (Rome, Italy)
Ratified: 3 December 1963

Description: An agreement to help achieve international cooperation in the use of the fisheries in the Mediterranean.

FURTHER INFORMATION: iea.uoregon.edu

www.ecolex.org

Name: *ASEAN Agreement on the Conservation of Nature and Natural Resources*

Initiated: 9 July 1985 (Kuala Lumpur, Malaysia)

Ratified: Not applicable

Description: An agreement among members of the Association of Southeast Asian nations to jointly and separately help conserve their natural capital.

FURTHER INFORMATION: iea.uoregon.edu

Name: *Benelux Convention on Nature Conservation and Landscape Protection*

Initiated: 8 June 1982 (Brussels, Belgium)

Ratified: 1 October 1983

Description: An agreement between the Netherlands, Luxembourg and Belgium to cooperate in the management of their natural heritage.

FURTHER INFORMATION: iea.uoregon.edu

www.ecolex.org

Name: *Convention on Nature Protection and Wildlife Preservation in the Western Hemisphere*

Initiated: 12 October 1940 (Washington, USA)

Ratified: 1 May 1942

Description: A Pan-American agreement to help conserve the flora and fauna in the western hemisphere.

FURTHER INFORMATION: iea.uoregon.edu

www.ecolex.org

Name: *Convention on the African Migratory Locust*

Initiated: 25 May 1962 (Kano, Nigeria)

Ratified: 25 July 1968

Description: An extension to a 1952 agreement signed by 22 African countries to include the control of all other migratory species locusts in Africa.

FURTHER INFORMATION: iea.uoregon.edu

www.ecolex.org

Name: *Convention on the Conservation of European Wildlife and Natural Habitats*
Initiated: 19 September 1979 (Berne, Switzerland)
Ratified: 1 June 1982
Description: An agreement by 21 European countries and Burkino Faso and Senegal to co-operate to conserve European wildlife.
FURTHER INFORMATION: iea.uoregon.edu
www.ecolex.org

Name: *Convention on the Conservation of Migratory Species of Wild Animals, as amended 1985*
Initiated: 23 June 1979 (Bonn, Germany)
Ratified: 1 November 1983
Description: An agreement by 39 countries to protect migratory wildlife species.
FURTHER INFORMATION: iea.uoregon.edu
www.ecolex.org

Name: *Convention on the International Trade in Endangered Species of Wild Flora and Fauna (CITES)*
Initiated: 3 March 1973 (Washington, USA)
Ratified: 1 July 1975
Description: A convention initiated in 1973 that seeks to protect endangered plant and animal species from extinction. CITES has been signed by over 100 countries and has been an important mechanism in preventing (or at least restricting) the trade in animal parts, such as ivory.
FURTHER INFORMATION: Organisation for Economic Co-operation and Development (1999)
iea.uoregon.edu
www.ecolex.org
www.cites.org/

Name: *Convention Relative to the Preservation of Fauna and Flora in their Natural State*
Initiated: 8 November 1933 (London, UK)
Ratified: 14 January 1936
Description: An agreement by 11 countries and dominions to protect flora and fauna, particularly in Africa, through the control of the trade of trophies, restrictions of certain types of hunting and protection of listed species.

FURTHER INFORMATION: iea.uoregon.edu
www.ecolex.org

Name: *European Convention for the Protection of Animals During International Transport*
Initiated: 13 December 1968 (Paris, France)
Ratified: 20 February 1971
Description: An agreement by 22 European countries to prevent the suffering of animals during transportation across their borders.
FURTHER INFORMATION: iea.uoregon.edu
www.ecolex.org

Name: *International Convention for the Protection of Birds*
Initiated: 18 October 1950 (Paris, France)
Ratified: 17 January 1963
Description: An agreement to protect birds, especially migratory birds and those species in danger of extinction. The convention includes articles about when birds should not be hunted.
FURTHER INFORMATION: iea.uoregon.edu
www.ecolex.org

Name: *International Plant Protection Convention*
Initiated: 6 December 1951 (Rome, Italy)
Ratified: 3 April 1952
Description: An agreement to prevent the spread of plant diseases and to co-ordinate measures for preventing pests moving between the signatory countries.
FURTHER INFORMATION: iea.uoregon.edu
www.ecolex.org

Name: *Plant Protection Agreement for the Asia and Pacific Region*
Initiated: 27 February 1956 (Rome, Italy)
Ratified: 2 July 1956
Description: An agreement between 26 countries to prevent the spread of plant diseases in the Asian and Pacific regions by certification of the trade of plants, and special measures to prevent the introduction of South American leaf blight to rubber trees.
FURTHER INFORMATION: iea.uoregon.edu
www.ecolex.org

Name: *Statutes of the International Union for Conservation of Nature and Natural Resources (as amended)*

Initiated: 5 October 1948 (Fontainebleau, France)
Ratified: 5 October 1948
Description: An agreement that formed the International Union for the Conservation of Nature (IUCN) and that now includes 139 countries represented by both governments and non-governmental organizations. The mandate of the IUCN is to conserve the integrity and diversity of nature.
FURTHER INFORMATION: www.ecolex.org
                                                www.iucn.org

Name: *UN Commission on Sustainable Development*
Initiated: December 1992
Ratified: Not applicable
Description: Promotes sustainable development, more specifically focusing on the environmental, social and economic concerns around sustainable development. The commission's goals are to implement Agenda 21 (www.un.org/esa/dsd/agenda21/), The Johannesburg Plan of Implementation (www.un.org/esa/sustdev/documents/WSSD_POI_PD/English/POIToc.htm) and the Barbados Programme of Action for Sustainable Development of Small Island Developing States (www.un.org/documents/ga/conf167/aconf167-9.htm).
FURTHER INFORMATION: www.un.org/esa/dsd/dsd/dsd_index.
                                                shtml

Name: *United Nations Global Compact*
Initiated: 26 July 2000
Ratified: Not applicable
Description: Over 8700 signatories in over 130 countries, the Global Compact is an initiative that brings together companies that want to align their business practices with the 10 principles of the UN Global Compact. These Principles focus on four main areas of human rights, labour, environment and anti-corruption.
FURTHER INFORMATION: www.unglobalcompact.org/
                                                AboutTheGC/index.html

Name: *United Nations Principles for Responsible Investment (UNPRI)*
Initiated: 2005
Ratified: Not applicable
Description: Efforts by investment organizations to identify principles of responsible investment.
FURTHER INFORMATION: www.unpri.org/

Name: *World Charter for Nature (1982)*
Description: A declaration to ensure nations follow principles of conservation and ensure the functioning of natural systems.
FURTHER INFORMATION: iea.uoregon.edu